TELESCOPE R*x*

Telescope R*x*

is dedicated to
all of my five children

who turned out so well
in spite of me.

TELESCOPE R*x*

The **BIG** Book
on
Equipping, Maintaining and Using a Telescope

Part I – Tips for Optimizing Your Telescope
and
Part II – Scientific Observing Projects

by

P. Clay Sherrod

Printed by LuLu – Publishing and eBook Company
Copyright 2017, P. Clay Sherrod
Printed in the United States of America
All rights reserved, Published 2017

Arkansas Sky Observatories Publications

ISBN: 978-1-365-74579-9

"There are even some stars so remote that their light will reach the Earth only when Earth itself is a dead planet, as they themselves are dead, so that the living Earth will never be visited by that forlorn ray of light, without a living source, without a living destination. Often on fine nights when the park of this establishment is vacant, I amuse myself with this marvelous instrument [the telescope], I go upstairs, walk across the grass, sit on a bench in the Avenue of Oaks – and there, in my solitude, I enjoy the pleasure of weighing the rays of dead stars……"

- Villiers de L'Isle-Adam, *Tomorrow's Eve*

CONTENTS – TELESCOPE Rx

PART ONE – Tips for Optimizing Your Telescope

1. Page 11 - Introduction: Your House Gorilla - Page
2. Page 17 - Opening Windows to Your Night Sky
3. Page 27 - The Human Eye
4. Page 35 - Daylight Observing
5. Page 41 - Telescope Magnification
6. Page 45 - Telescope Resolution
7. Page 49 - Seeing Conditions and Transparency
8. Page 55 - Star Testing Your Telescope
9. Page 59 - Practical Guide to Eyepiece Selection
10. Page 69 - A Guide for the Wise Selection of Telescope Accessories
11. Page 93 - Telescope Limiting Magnitude
12. Page 97 - Summer and Winter Temperature Precautions
13. Page 103 - Controlling Heat Currents – Air Convection
14. Page 113 - Telescope Power Supplies and Voltage Considerations
15. Page 125 - Lightning and Surge Protection + Pier Construction
16. Page 133 - Pollen Alert: Protecting you Telescope
17. Page 137 - ASO Fine Optics Cleaning Methods – Parts I and II
18. Page 151 - Polar Alignment - Clay's Kochab Clock
19. Page 167 - Polar Alignment Without Polaris
20. Page 169 - Polar Alignment – Drift Method
21. Page 171 - Polar Alignment – Iterative Computerized Method
22. Page 173 - Telescope Setting Circles – How to Use Them
23. Page 183 - Balancing an Equatorial Telescope

PART TWO – Scientific Observing Projects for Your Telescope

24. Page 187 - Techniques for Observing Meteors –Lists of Annual Showers
25. Page 209 - Scientific Studies of the Sun
26. Page 219 - Our Romance with the Moon
27. Page 231 - Observations of the Planet Venus
28. Page 241 - Observations of the Planet Mars
29. Page 265 - Observations of the Planet Jupiter
30. Page 279 - Observations of the Planet Saturn
31. Page 285 - Studies of Comets and Meteors
32. Page 299 - Comet Magnitude Determinations via CCD
33. Page 303 - Observing the Companion to the Bright Star Sirius
34. Page 309 - Observations of Algol – The Demon Star
35. Page 313 - Observations of the Globular Clusters – a Directory
36. Page 321 - Guide to Observing the Variable Stars
37. Page 333 - Your Search for Novae and Supernovae
38. APPENDIX – An Introduction into Piggyback Astrophotography

Abstract and Summary

There are two truths about amateur astronomy: there is a vast repository of bad information out there about how to buy, equip, and maintain a telescope, and then there is a huge void of information as to actually how to use them properly. The Internet today is full of experts who freely offer to give advice from their vast storehouse of experience when, in reality, we find that many of these people do not even own a telescope.

During the past six decades I have spent all of my hours either using telescopes are advising people how to properly use them. Indeed, I am quite well known as "Dr. Clay", the guy who knows everything about telescopes. In those years, I have never purposely given out bad information in the event I did not know a correct answer. In those cases, I either admitted I did not have the answer, or I would find the answer within a short time.

When the average citizen starts out in their pursuits of studying the beautiful night sky, the excitement is high and the thrill of discovery and sight drives them.

Then, after a mixture of failure at the telescope, inability to find the objects they pursue, or equipment which does not operate as it should, too many of these folks throw in the towel and are done with their pursuit of amateur astronomy....OR they turn to the Internet and get advice on how to spend their money. That will make it all better.

The investment in our love of skygazing is high, unless we choose to study the sky with the tools that God gave us (even that is covered in this book). All too often, we are led to believe that we did not have enough equipment, or have the wrong equipment or we are not doing things right.

Telescope Rx is a book intended to level the playing field: to provide solid and practical advice on everything from setting up a telescope, equipping it with the proper power and accessories, then turning that telescope into a nightly research tool with rewards for every night you wish to pursue.

It is a "good book" packed with all of the information that has been sought from me by thousands of confused and misled stargazers.

Use this book as your directory to properly outfit your telescope without spending lots of money; how to understand the way nature works and sometimes prevents us from seeing all that our telescopes are capable of; pitfalls to avoid in purchasing, and ultimately your guide to pursue some serious scientific studies with your telescope after you have had your long look around.

The sky is out there for all of us to study and enjoy. Through your proper understanding of how to do those studies, you mind, spirit and enthusiasm will grow.

                                        Much success and the clearest nights await......Doc Clay

Inside the observatory at H45 Petit Jean Mountain South Arkansas Sky Observatories

Chapter 1

YOUR HOUSE GORILLA

It is nothing short of amazing how easily and quickly amateur astronomers can enter the realm of scientific and discovery astronomy. But to reach that step, several things must happen:

1) you must develop a routine that will allow your casual interest to transition into a meaningful and focused study of the celestial objects that you are most interested in studying;

2) your skills with the equipment that you have – no matter how modest or extensive – must evolve so that you almost "forget" that your telescope and ancillary equipment are even there and your attention is maintained on your dedicated pursuit.

3) the telescope and associated equipment must work according to design and you must know the tricks and skills to maintain that equipment.

There is nothing more frustrating that having mechanical trouble at night.....a first night of failure will become the ultimate excuse to avoid setting up for a night of discovery the second night.

Telescopes serves as a guide of practical tips and techniques to get the most out of not only your telescope and associated equipment, but also YOU as the observer. Telescopes have limitations – so does the observer. Recognizing the necessary shortcuts and small tidbits of knowledge to make you a better observer – make your telescope a better instrument – will improve your skill set and motivate you for each successive night with the telescope.

Telescopes will show activities and give educational steps into the use of your astronomical equipment for true research in the science of astronomy, including where to submit your observations, how to report discoveries, how to streamline your telescope and equipment exactly to fit your scientific pursuits. Every amateur - or "non-professional" - astronomer reaches the proverbial **brick wall** with his or her interest in stargazing. Casual viewing leads to the desire to capture a distant galaxy in a photograph, astronomical imaging leads to more and more sophistication in equipment and ultimately there is THAT DAY where each of us realizes, that someone out there has already "taken that picture" and likely there are many superior to what you have just captured.

It is that brick wall that ultimately leads to unused equipment with excuses every night to not spend the time taking it out and setting it up for viewing or imaging. We have had our look around.....the splendors are all still there, so now what do I do?

Look around....are you stepping around the Gorilla in the house to get from one room to the other? You know.....that monstrous thing that you KNOW is there, but simply refuse to come to terms that it exists. We have all had those from time to time, those gorillas, and they stick around because each of us never wants to be the one who screams "....Look! the Emperor has no clothes!"

For those who were never blessed with the child's story, a practical-joking tailor convinced the highest emperor of the lands that his new set of clothes were so remarkably beautify that only a fool or idiot would not see them for the beauty that they held.....ever the Emperor himself pretended to "see" the magnificent dress, lest he be ridiculed a fool by his own people. His parade to premier the "clothing" was preceded by his knights going through the streets announcing that only fools would not be able to see the Emperor's new clothes, so all that watched as the old man passed by, of course, were held in wonderment of what they saw. It was not until an innocent youth of four years pronounced the truth: "Look! The Emperor has no clothes!"......then all joined in the merriment and laughter of a naked Emperor parading the streets of his kingdom.

He was the gorilla in the room. He was there, they knew he was naked, but no one wanted to admit it for that would be to admit that each was a fool.

How about that 'Gorilla' in your house? Are you going to admit that it is foolish to keep it there and continue to ignore it night after night?

Come on, now....admit it. I have many times. What was early-on your pride and joy, the most huge telescope in all the neighborhood, so large that many garbage cans can fit inside. So large in fact that the only place to store it is - well, in the house, back in a corner and surely it will make a wonderful conversation piece among guests who visit. As the dust collects, and insects have constructed tiny condominiums throughout the hidden recesses, what once was a magnificent photo-gobbling machine is now the white elephant of your living room.

Night after night goes by and there it sits: "Darn, it might be cloudy....."; "....they may be calling for dew tonight....."; "pollen sure is thick right now..."; "I bet the neighbor will wait and turn on his light just as soon as I get this all set up in the back yard."

So, instead of exploring the heavens with this masterpiece of monoliths, you tune into *'Naked and Afraid'* instead, all the while ignoring the Gorilla in the Room.

Thus it is nearly all of the time when we "macho-purchase" a telescope instead of thinking this hobby through from the onset; let's face it, Astronomy is a very serious, and a very expensive, hobby. But, then again, we belong to a local astronomy club and once a year we all set up telescopes in a parking lot out on the edge of town and have a star party....the guy with the biggest telescope is always the favorite. Even the length of the stepladder is impressive to the school children who line up for precarious, body-bending and perilous views of celestial objects.

"Wow....uh-huh....yeah, I see it."

And then the youngster is off the ladder never to return and the next one launches skyward in search of an aerial eyepiece.
The parents all want to know how much it cost.....how much does it weigh.....how long does it take to set it up.

And: What in the world do you do with it when you get it home?
"Uh.....well it sits in the corner with the rest of my family."
At that point you have gone full circle with this difficult realization:

"I have a Gorilla in my House."

In 50 years of consulting with astronomy folks of all levels and interests I have seen the *"House Gorilla Affliction"* more than any other single reason be the one factor that ultimately forces a sky-lover to abandon his or her lifelong interest in astronomy.
To prevent this, I have always had some very, very simple words of advice for those wanting to get their "first telescope" and these will be the most valuable and money-saving words that you will ever read:

1) Buy a telescope for the age that will be using it; simply put, do not buy a computerized, narrow field and heavy telescope for a youngster just starting out. Little kids need a grab-and-go, wide field, easy to point to any bright object and let them believe that they are mastering what will very quickly become a complicated hobby. Do not be intimidated with the modern GO-TO concept of computerized telescopes....you will have more reward in one night than I used to in one month. That builds confidence and encourages more use.

2) Think before you buy: what do you want to achieve out of this hobby? Do you want it to grow as your knowledge and interests grow? Will I be content with seeing things with my eye, or will I want to capture them digitally? Do you want to contribute to the science of astronomy, or entertain yourself and others with the wonders of space? And, most importantly once you have targeted a potential telescope and equipment to purchase: "How often will I use this if I get it?"

3) What conditions do you have to use a telescope? Do you need something portable so that you might escape the city lights? Will humidity, wind, cold, mosquitoes, all interfere with the amount of time that I would otherwise use this equipment?

4) Without hesitation, never buy your telescope at a local department store and always purchase from a reputable dealer who will stand with you as you learn and grow with this hobby.

5) How LONG will this equipment last me before I want to move up to something else?

6) NEVER under buy once you have honed in the parts of astronomy that interest you the most.....always think about the 10-year-rule". What will you want to do in ten years?

7) And, finally, if I get a sophisticated (or large) telescope capable of years of sights and discovery, will I spend the time necessary to climb the steep learning curve as I advance?

Believe it or not, a really good starting point to find out which telescopes never make the House Gorilla status is to search through the used telescope databases. You will be surprised how many of the same types appear over and over again, and typically end up selling for half of what the owner originally paid for it. For starters (and this is always a very dangerous observation to make, as there are many who CAN see the Emperor's clothes!) you will always see an abundance of:

* high end APO refractors costing in the thousands of dollars ("I waited for five years for delivery of this telescope and it is absolutely perfect in every way...never been out of the box!" HUH?)
* enormous Dobsonian telescopes, complete with folding ladders and many times utility trailers for transport (honestly, how many times are you willing to do this?)
* very sophisticated-looking telescopes with lots of chrome gears and cables and shiny tubes on suspiciously familiar mounts (all made in the same Chinese factories and sold under different labels)
* and even some incredibly high-end large research grade telescopes and mounts, all of which most of the time have intimidated and out-muscled their owners.

Now, I have absolutely nothing against any of these types of telescopes....all have their merits and values among astronomy enthusiasts. But they appear on the market for a reason, and that reason is that they were NOT a good fit for the original owner.

For example, the quality APO refractors are the most efficient for spectacular pin-point images of wide field, Milky Way, comets and similar astrophotography. But perhaps the owner forgot to add in the cost of ownership of a guiding system, and adequate high quality mount, a high resolution, large chip CCD camera, not to mention that they live in downtown Chicago where the stars don't shine.

Or the Dobsonian ("Dobs")....great for crowd pleasing, these enormous cement-mixer-appearing giants appear as the photon buckets to end all telescopes. But, even when a tracking system is incorporated, and a GO TO computerized access is installed on the Dobs, they quickly reveal themselves as far too shaky for photography, research and in many cases even high power viewing.

And then there are telescopes that are being sold for the wrong reasons: the classic example is the recent "fad" with "GEMs" - German Equatorial Mountings. Many thousands of fork-mounted excellent telescopes (did you hear me say 'best bang for the buck'?) have provided the stepping stones and in many cases the final workhorse telescopes for astronomers worldwide for nearly 50 years. These are now being shunned because of the Internet, where the self-proclaimed experts have deemed them unsuitable for "real astronomy". Poppycock.....although I use a high end GEM on one of my observatory telescopes, there is nothing better and more efficient than a properly tuned fork mount. Even the manufacturers have fallen victim to the fickleness of the buying public, replacing their once quality fork mounts with German Equatorial models that talk to you while you use them, set themselves up and suggest that the observer needs no input in the operation of this telescope other than to set it outside.

Bottom line to this is simple: never, ever trust the advice of someone who owns a telescope. He is wearing the Emperor's clothes; he is not about to admit that his telescope has shortcomings, for fear of appearing foolish. At any sky function, or star party....even trade show....the guys you are talking to are going to recommend what THEY own and/or make. It is the truth and it is simple.

Always trust a reliable dealer, because they have no dog in the fight....they have all the brands, all the types. Get to know someone and be completely honest about what YOU want now and in the future

*   *   *

In all honesty, pretty much everyone in astronomy is indeed fickle: we love to buy, trade and sell equipment, just as avid photographers do. However, in nearly every situation, a telescope owner will express some exact reason of disappointment in selling a telescope (but of course, never to the potential buyer of that telescope).

In no other hobby - I always cringe when I use that word,, 'hobby', because in astronomy nearly all of us are educators, researchers, inventors and motivationalists - has the technology advanced as rapidly as astronomy; in order to keep up with the growing innovations, one simply must buy new equipment and sell old equipment. It is the nature of the beast.....rather a bit of 21st Century hunting and gathering. Demands for larger telescopes, wider fields, larger CCD chips, advanced computer guiding and plate solving, and - finally - the exponential surge of non-professional astronomers into the fields of research/science astronomy.

There is NO hobby on this planet outside of astronomy where amateurs (we will hereby call ourselves 'non-professional' ["NP"] astronomers since the only difference in us and "the big guys" is that we do not get paid to do what we love, and we have to spend our own time and our own money doing so.

Eventually you will learn that there are dozens of high-impact areas of research and true science in astronomy where YOU can make a difference and actually contribute to the repository of science throughout your life. You can take your telescope - no matter how large or small and no matter what type - and do real science from your back yard, or your small remote observatory under dark skies.

So now you have had your "look around" the neighborhood of the Heavens....you have taken pretty pictures of much of it.....attended about as many star parties and your feet can stand.....

Let's show you how to turn your telescope into an astronomical laboratory, researching the thing that you love the most and adding value to the body of science.

A few simple steps along with guidance and your perseverance can turn your House Gorilla into a Screaming Sky Machine.

Hopefully this Two-Part **Telescope Rx** will not only allow you to optimize your telescope to the highest performance, thus providing you with a trouble free instrument that yields pleasure and reward rather than frustration – but also in Part II give you the guidance to turn your telescope from casual interest to genuine pursuits into the science and discoveries of this wonderful world of Astronomy.

*   *   *

Chapter Two

Opening the Windows to the Nighttime Skies

"The cosmos is all that is or ever was or ever will be. Our feeblest contemplations of the Cosmos stir us - there is a tingling in the spine, a catch in the voice, a faint sensation, as if a distant memory, or falling from a height. We know we are approaching the greatest of mysteries....."
— Carl Sagan, first paragraph of <u>Cosmos</u>, 1980

As we stand amidst the technology of the 21st Century, the yearning by many human explorers goes unsatisfied and the adrenalin of discovery remains untapped, un=flowing through the pathways of life. All of the lands have been explored....there are Wal-Mart plastic sacks being picked up by Sherpa from the heights of Everest. The oceans have been sailed to lands now known and the beaches no longer await discovery but serve the labor-weary during vacations.

Those of us who are fortunate to have grown up during the great Space Race of the 1950's and 1960's remember the thrill and rush as one after another challenge was met and frontier broken in the great abyss above us, yet today very little awaits discovery in our immediate celestial neighborhood.

In fact, our technology - through social media and modern sci-fi - has pretty much numbed us to the "tingling in the spine" that was very much once a part of each and every discovery that a common man would make in his life. We are no longer amazed, nor thrilled, nor even interested perhaps, in some things we are exposed to in the media. Nothing is a miracle, it is expected.

A bit depressing you think? Not so fast: YOU may not even be reading this if it were not for the fact that you are the exception to the rule in modern-day explorations. You, unlike your neighbors and friends, are yearning to enrich your life through a study of the sky; you are awaiting the day when your efforts result in that first discovery and contribution to the archives of science. In fact, it is this common bond that draws all of us into *astronomy* at the very beginning. At some point, you have said to yourself, *"I have always been interested in looking at the sky...."* and now you are ready for your ship to depart to those other worlds through the very technology that may keep us in our comfortable chairs at night.

Never in the history of science has there been such a vast wealth of equipment, knowledge, and leadership available for the average guy to make a true contribution to science. And among all the sciences, there is not a greater opportunity than in astronomy for the contributions of the Non-professional ("NP") astronomer. The nighttime sky is packed with horizons never seen, changes being made that have hitherto been seemingly impossible in our world of physics....discoveries. Real discoveries.

This chapter of *Telescopes* may seem to you at first to be a long narrative of common sense, nothing more. However, there is a wonderfully special mind-set of the non-professional astronomer that must be put in synchronization with the rest of his/her lifestyles. In order for you to be successful in contributing to the science of astronomy you must establish parameters and a discipline perhaps unmatched in all of your other endeavors.

YOUR PERSONAL MOTIVATION

In coming installments, I am going to reveal many of the exciting areas of real astronomical science in which YOU can make contributions and feel the personal reward of contributing part of your life to science. In these guides, you will learn what you need, how to go about taking scientific data, processing of the information and all of the avenues and links that you will need to submit your observations and measurements to the appropriate professional and academic repositories for each of these studies.

But there is a bit of an issue with all of us making these explorations and awaiting discoveries. First of all, we have "real lives"....jobs, families, life commitments which must be met. One thing that I have noted in my fifty years of working with astronomy folks of all levels is that we are conscious and dedicated, not just to the sky but to our families and outside lives as well. That is what makes us such good practitioners of discovery.

This means that there are conflicts in time (we all must observe at night for the most part) in which we would otherwise love to spend with our families. There is also a huge conflict of your body's personal and physiological source of ENERGY to drive you to do what you love the best. On the best of days there are times when it is all too easy to decide to forego your time at the telescope for a great football game on television, or a good night's rest. On the worst of days, it takes oh so little to keep us indoors and out of the frosty night air.

What motivates YOU and what incentives can you create for yourself to carefully balance your love for astronomy, the adrenalin rush of discovery and the grounding reality of your "real life?"

Most of us have a telescope and some quality ancillary equipment to go with it that can get us started toward identifying and dedication to an astronomical project. In fact, most of you probably have everything that you need right now to get started in some specialized field of astronomical research. In this series I can show you *how* to develop your research protocol based on your equipment, but the one thing that I cannot do is to *motivate you* to stay focused and put in the time necessary to be a leader in the "NP" astronomer ranks.

No, it does not require every nighttime hour, nor does it require every night at the telescope; but it does take discipline and a very strict protocol....in addition to the learning curve for each and every endeavor. Getting involved in serious astronomical science DOES require your dedication and willingness to not roll over in the warm bed when the alarm goes off at 2 a.m.

IDEAS TO MAKE THE ROUTINE EASIER

Believe it or not, there ARE ways to self-motivate in this avocation that really do work. I know because I have used them for over five decades now and have been hugely successful mentoring hundreds of "NP" astronomers to high success in the ranks of astronomical research contributed. In a normal "working job" your incentive is normally monetary reward.....in your astronomical endeavors, you are actually paying to do what you love the most. A bit ironic, right?

Consider each of the following aspects to simplify and focus your efforts in astronomy. Every aspect covered following is equally important for your success. Be realistic in terms of both your time and your equipment, and establish your projects based on what you know will be positive input into the scientific astronomical equipment. All things are possible, but not all are possible with each and every telescope and observer. Remember: the worse scenario is that in which you accomplish nothing and quickly burn out of your desire to continue. Set goals so that you can experience SUCCESS within a short time.

1. <u>Evaluate your equipment and its potential as well as its limitations</u>
You have, in all likelihood, a nice telescope with accessories in place. It might be a large wide field Dobsonian or a sophisticated Catadioptic model with a longer focal length. Can both be used for the same projects? Well....the simple answer is "possibly". But not likely. As you have used your equipment, you very likely have noted that sometimes some objects appear better than others with each telescope. In fact, I have had experiences where large comets actually appear brighter and with more detail in a finderscope than in the main telescope.

Long focal length SCT, catadioptic and refractors of f/8 and longer will always be more suitable for observing programs of the sun, planets and moon; on the other hand telescopes with either a native focal ratio of f/8 and faster - or those equipped with a quality focal reducer to provide a faster overall focal ratio will provide wider and brighter fields, with exposure times being shorter, making acquisition of data many times much easier and faster.

No doubt once you settle on a program of study and scientific contribution, you are going to need to add some ancillary equipment to your existing main telescope and its armament. Since all aspects of modern research involve digital data acquisition and reduction, there is no doubt that your camera that you match to your telescope system will be as equally important as the main optical system. Your dealer can greatly assist in matching the perfect camera for your project of choice. There is simply no getting around it: if you are going to do science with your equipment, you are going to have to invest in astrophotography.

As mentioned, longer focal length systems can be excellent for studies of planets, using "webcams" for rapid (1/60th second is not uncommon) acquisition of hundreds and thousands of blurry images of a planetary body, all merged via a computer program into one final incredibly sharp high definition result. Nearly ALL of modern astronomy's planetary imaging archives are being filled by "NP" astronomers.

Deep sky studies with larger aperture and faster systems can be rewarding. Such studies (discussed in a later installment) might include asteroid/near earth object astrometry (position and orbit), photometry (brightness changes as the large minor planet rotates), supernovae searches of other galaxies, exoplanet discovery and confirmation, variable star studies, and physical morphology studies of comets all the way to 20th magnitude. Literally the sky is the limit. We will help you decide on what is suited best for your equipment and interests in this science series.

2. Examine your observing environment and conditions

Just as all telescope systems are not suitable for all projects and goals, neither are the local conditions from one geographic area to another. Locations throughout any area will vary with the terrain in terms of air steadiness (necessary for stable high resolution imaging of planets, for example), dark sky index (rural areas of course rule in this regard), artificial lighting (which can be sometimes controlled with calm neighborly appeals), and moisture content. Arid, dry and dark locations of course are far more desirable than urban lighted areas with pollution and moisture in the air. But we most times must deal with the hand dealt and learn to adapt our projects to our environment. Only through such acceptance and adaptation can you find success in scientific pursuits.

It is reasonable to perhaps plan for a remotely operated observatory in the future....this is the new wave of "NP" astronomers with thousands of such facilities already operating worldwide. It is unreasonable on the other hand to expect to be able to transport all of your equipment to a dark site on a regular basis. Remember: family and responsibilities first.....space conquest next.

Interestingly, stable air for planetary imaging and study is typically found beneath strong inversion layers of polluted urban atmosphere, so there is hope for city dwellers. For your supernovae searches, comet and asteroid measuring, the darker the skies, the fainter the objects you can monitor.

3. Explore your possible projects based on your equipment and your observing conditions

Now you have a hint that you CAN come up with projects no matter what type of equipment and what conditions you must operated that equipment in. There are true science endeavors awaiting for all locations and applications. Many times adding just a few critical accessories - a filter, reducer, change in camera - can make all the difference in terms of utilizing your existing equipment within the environment that you must observe. More on this later.

4. Set realistic goals for your science

You must set goals for your astronomy science and stick with them, strive to reach them and assess along the way. The goals must be realistic and not fantasy; never base what you should expect from your efforts and your equipment through "forums" in which many self-appointed experts assail their unrealistic successes. When you set out - as an example - to discover supernovae in distant galaxies, never set a goal for "one supernovae per month". It simply is not going to happen. Instead, you goal should be to assign an arbitrary number of galaxies for examination on a regular basis, perhaps 25, within that month, changing your targets as the seasons change. At the end of each month, write yourself up a report (if you have a web or Facebook outlet, put the report out there) stating which galaxies were studied and that NO supernovae were found within that period. That, in itself, is not failure....it is success in knowing that those galaxies did not have supernovae activity!

Similar goals apply to all branches of astronomical research; those goals will be realistically outlined in future installments.

5. <u>Examine ways to simplify setting up</u>

This is a key and incredibly important step in success. The more difficult and lengthy your set-up time, the less motivation (as well as time) there is to observe. This typically does not manifest itself immediately; trust me on this one. It is something that grows with time, and suddenly you find those "excuses" growing each and every night to skip your study of the night sky in lieu of comfort and relaxation.

If everything must be unboxed, moved outdoors, set up, aligned, calibrated, focused and tested, you are in for a long night. In some cases however, this is the only option available to thousands of devoted "NP" astronomers who are currently contributing. They must essentially set up from scratch each and every night; so how do they do it? Normally, they plan (see #6, below) and plan carefully, choosing exact nights to which they will devote their scientific pursuits.

There are many ways for you to simplify setting up, some more realistic than others, but all should be considered. Briefly, some of these are:

a) a designated observing area in your yard, shielded from external lighting, wind and lights from your own house;

b) a pad, or cleared and level area in which you can set up your telescope quickly;

c) a dedicated all-weather power cord from house outlet to your observing area, ready when you are, in a water-protected box;

d) consider if you are using a portable pier or tripod marking the ground or concrete pad with "leg marks" for quick setup;

e) if possible, consider installing an inexpensive concrete block (or Sonotube), or pipe permanent pier at your observing station, this resulting in only the telescope being placed onto the base with near instant polar alignment!

f) if you are in an open and windy location, consider having some down-and-dirty plywood windscreens set off to the sides which can quickly be set in place, but only if they can be set up firmly....a sheet of plywood in the blowing wind can quickly end your astronomical pursuits;

g) if you know that your next few days are going to be relatively weather-free, invest in a telescope cover (NEVER cover with plastic bags or tarps) that is made for this application; you can cover your telescope and bring your delicate laptop, camera and other equipment indoors;

h) have all of your ancillary equipment is a "ready case" that has everything in one location, easily accessible and ready to place on the telescope; some observers have constructed handy "cabinets" which contain all their accessories, laptop, power source and even a small working desk with dim red lamp.....makes all the difference in the world.

i) and, of course the ultimate is to have your own small observatory set up, telescope and equipment awaiting whatever time you can spend there. This may not seem realistic to you now, but more and more "NP" astronomers are going this route and tens of thousands of private observatories are now operating across the globe by folks just like you.....most of whom never thought they would have such a dream setup!

6. <u>Make a realistic schedule considering: Family, job, moonlight, targets</u>

Setting a schedule that you can stick with, without stressing yourself out, is paramount. After all, we are all into astronomy because we love to explore the night sky. We are not paid astronomers and we are not "working for anybody." We are doing this for self-satisfaction and self-reward. So, never overschedule yourself and expect more out of your time than you can deliver. If you can only devote one weekend per month at a dark sky site then set that as your schedule. If you want to target only the week around the new moon, then make the most out of that week.

In scheduling however you must consider your target objects: for example, the planets are not all in the sky at the same time, and will be in optimal observing altitude only certain times and dates during the years, even that changing year-to-year. The galaxies that you want to examine for supernovae are likely going to dominate in the spring and summer months in terms of early evening, and be scarce during winter months. If embarking on asteroid and comet studies, there is never a lack of those, no matter what time of night, date or year.

As seasoned "NP" astronomers get even more seasoned over the years, we come to realize that enjoyment of doing all this is ONLY accomplished if kept in a proper proportion to family life and balanced with your worldly duties; those come first. When you are satisfied that the family is happy and content and that your employer is happy with what you are providing to him or her, only then can you truly enjoy your astronomical endeavors.

7. <u>Establish a self-assessment method for you to see and experience your results</u>

I already touched on setting GOALS for your astronomy studies. Also noted was that not every night at the telescope will lead to a discovery. Nonetheless, your efforts are valuable beyond description. Let us imagine for the moment that you have selected 25 fairly faint NGC galaxies to monitor over the course of a three month period. After those three months you will start on a new set of 25. Not getting ahead of myself in this discussion, but you will take a suitable CCD image of each galaxy to serve as your "Base Catalog"......those will be your comparison images. If you spend just one night a week for three months, taking a similar photograph of each galaxy, you have a good chance of discovering a "new star" supernova in one of those galaxies....many have been found by "NP" astronomers in Messier 51 (The Whirlpool Galaxy) in just the past two decades.

However, chances are that you will NOT discover that new star in your first three months out. The odds are against you. So, does that mean you have done "nothing?" Wasted your time? Heavens no! There are two ways of looking at what you have done for the past three months, even though it did not lead to discovery:

- First, you have confirmed for the scientific community that NO supernovae within reach of your equipment have occurred in those selected 25 galaxies in a three month span;
- Second, consider "what if".....what if you had not been watching the galaxies, assuming there were not going to be any outbursts. But indeed there was, in one galaxy! And your apathy prevented YOU from discovering that new supernova! That, my friend, is the worst case scenario and nothing can plant your heart deeper in your tummy than that very situation.

Self-assessment of your study is one of the most important aspects of what you will be doing. You must *record* your observations, positive or negative, for the archives and annals of science. One of exciting aspects of this series is that you will be provided with the tools and links through which your observations can routinely be placed among others, whether those others be "NP" efforts or those in the professional ranks. The direct access to archives is there for you - planetary high resolution images, solar activity, cataclysmic variable star changes, supernovae searchers, exoplanet discoveries and confirmations, asteroid and comet contributions. Just posting all of this on Facebook or Flicker may seem fine and a quick way to draw attention to what you are doing, but it has zero impact on the science in which you are contribution.

*   *   *

But absolutely none of this matters if your equipment and your observing skill set is not honed to its optimal level. Understanding what steps must be taken to get your equipment into top shape, the methods for setup and alignment, the preparations and knowledge that you must have as an observer must be first perfected.

Telescopes will get these small things that "go bump in the night" out of the way for you and soon you will "forget" your equipment and your focus will truly be on the stars.

*   *   *

Telescope R*x*

PART ONE

Optimizing

Getting the most out of your Telescope

Chapter Three

The Ultimate Astronomical Instrument: The Human Eye

During the past years I have written much on how your astronomical telescope is governed by certain "functions" and have gone into great detail about *limiting magnitude* of stellar objects which you can expect to see, *sky transparency*, the *steadiness* of the air, and observing "tricks" for discerning even the faintest of detail on planets and deep sky objects.

From these I hope you have gleaned some appreciation for the "work" that your telescope can provide you.....but, on the other hand, we have up until now MISSED discussion the MOST IMPORTANT component of your telescope system: *YOUR EYES*.

EYESIGHT AND TELESCOPE PERFORMANCE

Your telescope can provide the finest of astronomical images, many times approaching and surpassing its theoretical limits on light grasp, resolution and magnification. Yet, consider this. A lot of what you see through your telescope: the moon, Venus, Mars, Jupiter, Saturn, the Andromeda galaxy and Orion nebula, are ALL visible to the eye WITHOUT your telescope!

Keep in mind that this is not just a "visual observing" testament to the human eye; without good eyesight, your image processing, research and evaluations of all that you do would be far more difficult

The telescope simply makes celestial objects easier to see. Indeed, the most sophisticated telescope, when coupled with visual observing, is nothing more than a fancy pair of eyeglasses. An extension of the eye itself.

Look at the two charts in the overview of *Seeing* and *Transparency* published in this book: there are two charts of star clusters, the Hyades (for naked eye and binocular acuity testing) and the Pleiades (for telescopic acuity). ACUITY is merely your eye's (coupled with the telescope in the case of the latter) ability to see the faintest star possible.

Imagine this: if all of our eyes were the same, we WOULD NOT HAVE TO TEST to see what our acuity really is! Y ours would be the same as mine (Heaven forbid), and mine the same as everyone else's. But they are all **considerably different** for many reasons including:

1) Natural eye sensitivity - you are born with this, some people having much more acute eyesight than others; age and color of the eyes actually plays into this;
2) Age - the older you become, the less acute your vision can be;
3) Eye Fatigue - just like when reading a book or using a computer, your eyes can become fatigued when using a telescope for high intensity or prolonged viewing;

4) <u>Position of the eye</u> - in relation to what it is being viewed (*averted* vs. direct viewing);
5) <u>Dark adaptation</u> - the eye becomes considerably more sensitive to subtle light variations when totally dilated and equalized to its dark surroundings; and,
6) <u>Eye defect and disease</u> - AND DON'T WRITE THIS ONE OFF! We all have at least some degree of defect in our eyes as you will come to agree!

THE NATURE OF OUR "FINEST EYEPIECE"

Your eye is an eyepiece that you do not need to carry in the accessory case. You never need to clean it, but it gets dirty. It does not have an accessory case, but it is very well protected. It is subject to the extreme cold and heat, but never fogs over. It is not multi-coated "nine times" but provides the ultimate in light transmission and a minimum of light scatter.

The only drawback to this marvelous telescope accessory, this human eye? You cannot trade how it works. Do not do this at home, or your eye's warranty will likely be voided.

In the accompanying figure you will note that the eye gathers light (just like the telescope - see the red lines), focuses the image through a LENS and transfers it to an awaiting screen, the RETINA. The retina is much like a piece of photographic film which receives the light of the image (upside down) and passes the data through the OPTICAL NERVE into the brain where it is turned upright (at least in our minds) and fine tuned, just like a computer imaging program does for CCD recording of deep sky objects.

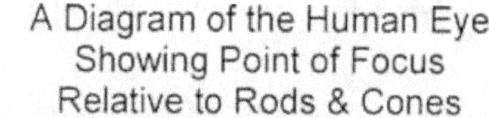
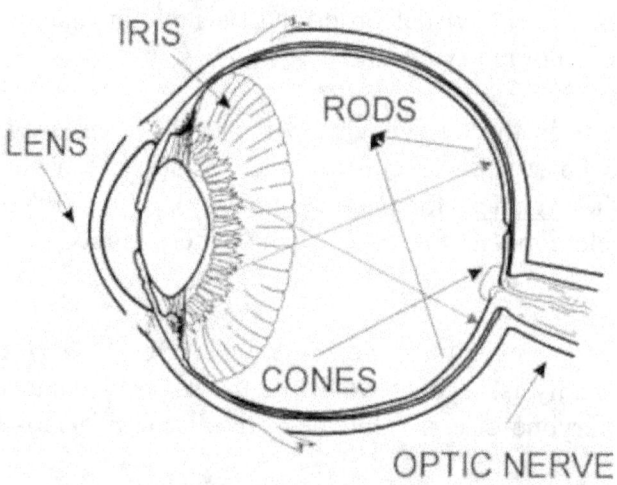

Illustrations by P. Clay Sherrod

In some of my reports covering viewing fine planetary detail or very faint stars near the threshold of your telescope's capability, I have mentioned many times a technique known as "averted vision," whereby an observer can not look DIRECTLY at an object (which focuses the light on the eye's blind spot) but rather slightly to one side of the object, yet still be perceptible of the image itself. Such "averted vision" allows one to use the eye's more sensitive receptors.

The blind spot in your right eye, if that is the eye you observe with, is located to the right hand side of direct center, so concentrating your vision slightly to the left will increase your ability to see finer detail and color. It is the opposite for the left eye.

The very center of your eye is known as the YELLOW SPOT which contains only slightly sensitive CONES. Surrounding them are the more sensitive rods in combinations with more cone receptors.

The light falls upon RODS (about 120 million of them in each eye) and CONES (only 7 million); Rods are more than 10,000 times more sensitive to light than are the cones, which are color sensitive in daylight and only in very good light.

So, literally speaking, it is best to look "out of the corner of your eye" when observing planets, faint deep sky objects and very faint stars; attempt to keep the light focused away from the "blind spot" immediately centered in your eye!

The idea of eating carrots for Vitamin A is not taken lightly; vitamin A allows you to maintain the sensitivity of both rods and cones as your age increases. There are many easy maintenance steps that you can do on a regular basis to MAXIMIZE your eyes' ability to provide the very best image possible, even if you are not born with exceptionally good vision. Such steps are discussed in the section "EYE MAINTENANCE" following.

1) NATURAL EYE SENSITIVITY
Some people are born with more sensitive eyes than others. Indeed, although the jury is still out on this one, there appears to be evidence to suggest that people with "lighter color" eyes (i.e., blue and green) are more sensitive to subtle color and detail than those with darker eyes. Eye color, as we all know, is NOT an effect of the actual focusing nor image forming (the retina area), but rather due to the pigmentation of the eye's IRIS.

Hence, it may be difficult to accept that a lighter colored IRIS (which serves primarily to "stop down" the eye for bright light and "open it up" for darkened conditions) could have any direct bearing on retinal activity.

Nonetheless, some people are far more visually acute than others; indeed, if using a six-inch telescope, some people can see stars of magnitude 13.5, while others of average vision may be doing good to see a 12.5 magnitude star. The same discrepancy will occur in planetary viewing of fine detail as well.

Many times such lack of visual perception is due only to eye defect or disease which is discussed in Section 5.

2) AGE OF THE OBSERVER

It is unfortunately true that our eyes become less sensitive as we age, primarily from loss of receptors (physically deteriorating) and from loss of sensitivity of those that still work! Many times sensitivity can be DECREASED by any person doing repetitive conditions such as:

a) working or playing in bright sunny conditions without eye protection;
b) untreated internal eye disease such as Glaucoma;
c) exposure to prolonged eye focus, such as computer work or assembly line work;
d) excessive reading and/or drawing;
e) very close activity such as sewing, model-building, electronics, jewelry making and more.

One fact of the loss of eye sensitivity due to excessive rod and cone loss is plain and indisputable: once they are gone....they are gone. Eyesight is something that cannot be "revitalized" with a pill or a drop if there is slow degradation of rod or cone reception.

One final thought on aging eyes: the eye tends to age more rapidly beginning at the very center, in the aforementioned "yellow spot," and moves gradually outward. Consequently, amateur astronomers with less-than-perfect eyesight might attempt to sharpen their "averted vision" skills even more and greatly improve their observing prowess.

3) and 4) EYE FATIGUE - EYE POSITION

Probably the most plaguing of all eye issues for amateur astronomers is the EASIEST TO REMEDY: "eye fatigue." Not that I have to tell most of you, but sure signs of eye fatigue - and LOSS of acuity for astronomical viewing (yes, the two go hand-in-hand more than any other factor) include:

a) dry itching feeling behind the eyelids;
b) a seeming inability to keep the eyelids "propped open"
c) excessive tearing or - oddly - excessive dryness;
d) matting of this mucous in the corners of the eye;
e) headaches "behind" the eye sockets; and,
f) difficulty focusing for long periods and/or difficulty concentrating.

I have always told my astronomy students to get a good rest before an observing project, particularly if it is a visual one. This is not to rest the body so much as it is to rest the mind and the eyes - remember, THEY are an essential part of your telescope!

Try to practice this rule, particularly if you have had a hard day and physically feel tired (yet you STILL want to observe): we know that our telescope MUST equalize to the outside air in both summer and winter, so we plan ahead and put them out about one- to two-hour PRIOR to viewing, depending on the size of the primary lens or mirror. You may as well NOT waste your time if you do not allow your telescope to "cool down;" so, knowing that you HAVE to do that....why not set up the telescope outdoors, come back in and relax.

DO NOT READ A BOOK! Don't read the paper, and minimize the evening news; simply let your eyes REST. You'll be surprised just how this technique not only rests your eyes for a night's viewing....it rests YOUR MIND and attitude as well. Try it...you'll be surprised at the results.

Eye fatigue also sets in after a hour or so of strenuous viewing through the telescope, or particularly "guiding" a long exposure photograph; your eyes are really taxed to the limit and so is your concentration after about one hour, so be sure to give yourself some breaks.

Remember that we use our eyes constantly, every waking minute. And like other organs of the body they require rest over and above merely sleeping at night.

There are two big problems that are paramount in telescope viewing ability and pleasure that you experience when looking through your telescope. Both become exaggerated over prolonged periods of using your eyes.

a) **DRY EYES** - some people, myself included, have a condition in which the eye does not produce enough natural lubricant (or "tears"); in some cases, no tears whatsoever can be produced. In other people with "normal" tearing, the production of tears can slow or even cease after prolonged stress on the eye. This condition is known as "dry eye," and is more serious for you observing than may first appear.

Tears are a necessary part of the eye's ability to FOCUS; the LENS of the eye of course does most of the focusing, but it is designed to depend on tears to do a small fraction of the focus; if there are no tears to coat the eye, then your ultimate focus will most definitely be "off."

There are two methods to replenish tears for any eye:
1) "artificial tears", drops that are available at any pharmacy; use as much as necessary and as often as needed; and,
2) "rolling your eyes," a technique that will also be used in the following discussion concerning "*eye floaters;*" merely close your eyelids and roll the eyes gently one way, and then the other.....repeat and leave eyes closed for a brief moment. Then return to viewing.

You will surprised what a difference having a lubricating film across the eye will do for visual acuity and your ability to simply focus like the pros!

b) EYE "FLOATERS" - Just like in a crime scene at a lake, a "floater" with the eye is a dead body; only the "body" is a detached cone or rod (see discussion above) that has come loose from the retina and floats amidst the thick *AQUEOUS HUMOR*, the liquid filling the eyeball.

As gross as it sounds, we ALL have floaters but some have more - and larger - than others.

Typically, unless two or more floaters become entangled which they very often do permanently (I've had two sets since I was 17 years old that are like old friends to me now), you will not notice them. However, large floater masses are clearly visible when
1) looking at bright sky or bright objects on a sunlit day;
2) when reading a book on your lap under a bright light; and,
3) when tee-ing off a golf ball on a bright day.

Bulletins of the Arkansas Sky Observatory
TELESCOPIC OBSERVING
and
"Eye Floaters"

A - The eye in normal viewing conditions

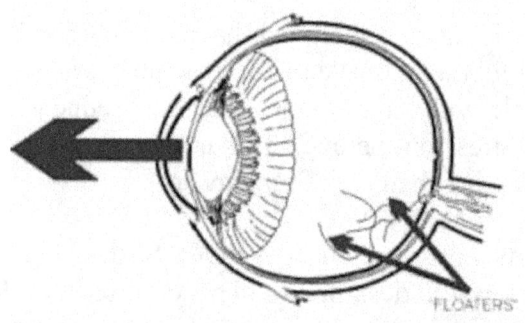

B - The eye when viewing through an astronomical telescope

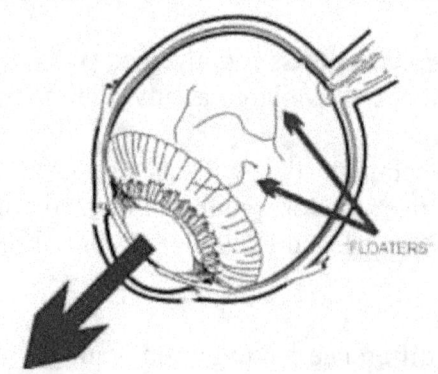

Notice I did not include looking at a computer screen; seems a lot like reading the book, doesn't it? Not with floaters as can be clearly seen in the attached diagram.

Floaters are an absolute nuisance for visual astronomy; you will be looking for a tiny festoon on Jupiter and here comes this horrible looking, amoebae-shaped creature floating across the field of view! And then it stops....dead in its tracks right in front of your image of Jupiter!

You can clearly see why this happens in the diagram: floaters are suspended in the aqueous humor and move only according to gravity and rapid motions of the eyeball. When you are looking at the computer screen, the television, or at a ball game, the floaters respond to gravity and settle like pond sediment at the bottom of the eye cavity.
On the other hand, looking at the lower drawing, you can see that if your are reading a book on your lap, looking down to drive a golf ball, or - LOOKING IN YOUR TELESCOPE, the floaters respond again to gravity, only this time right between your eye lens and the retina! You SEE them!

To get rid of pesky floaters, move back away from the telescope and rotate your eyes in their sockets rapidly back and forth while holding your eyes skyward (of course, leave the eyes in your head for safety). This will shuffle them up and they will simply move out of the way.

They WILL, of course return in time. Merely do the same procedure again. Rest assured....they WILL NOT go away. Like I said....in time they will become easily recognized "old friends."

5) OPTICAL DEFECT AND DISEASE
No eye is perfect and we all suffer from some degree of optical defect and/or disease. Indeed, we ALL have floaters....that is "optical defect." Another optical defect that will interfere with your focusing and acuity is a corneal scratch or scar.

People who work outdoors, such as construction workers, lifeguards and lumberjacks frequently suffer from another optical defect in which the "yellow spot" immediately at the center of the retina has become burned and scarred from exposure to sunlight over very long periods of time. This is particularly common for persons living on or near bright beaches and those who enjoy boating as a hobby.

Eye disease unfortunately can take many forms and affect very difficult viewing, such as that necessary for astronomy. Eyesight correction requirement is a form of eye disease, requiring only the fitting of proper glasses.

IF YOU WEAR GLASSES, take them off to observe; do not attempt to focus with them on for three very important reasons:
1) you cannot enjoy the full field of view provided by your telescope/eyepiece combination;
2) your visual acuity is much better without the glasses, simply focus for your eyes using the telescope focus to compensate for your eyeglasses; and,
3) your optics of your multi-hundred/thousand-dollar telescope are NOT BETTER than the cheap lenses of you glasses!

REGARDING DISEASES OF THE EYE which can affect your telescopic viewing: the many minor/major infections on the outside (cornea) of the eye will greatly hamper your ability to see the faintest star or the finest detail; drops help greatly, but in many cases, physicians recommend complete REST of the eye, and observing with your telescope is NO EXCEPTION to this "prescription."

Perhaps the most serious and detrimental to astronomical use of you eyes is the widespread internal eye disease of *GLAUCOMA*, resulting in considerably increased internal pressure of the eye; this has an end result of focusing difficulty, poor retinal perception, loss of peripheral vision, and many other adverse conditions. However, ophthalmologists can prescribe eye drops that, if taken properly, can maintain a near-normal eye performance both in astronomy use and in everyday life.

I hope this helps you to better understand how important the eye is in astronomy; there is not a day that goes by that I don't look in my accessory case and remember WHY I have been given the privilege to even look there in the first place! I have the greatest accessory available - my eyesight!

Treat YOUR precious accessory with the best of care and there will never be any need to "upgrade" or trade in on the latest version....they just don't get any better!

* * *

Chapter 4

DAYLIGHT OBSERVING

Introduction

Most of you have in your possession a telescope that is truly a "wonder of the world" in that it can determine locations of objects, tell you where they are in the sky....and ZERO IN on them....without you ever looking through the telescope. The modern computerized "GO TO" telescopes can take you on a tour of the majority of Messier objects in one night's observing....zip through hundreds of NGC-listed galaxies and tour the stars by name at your leisure.

But consider the following scenario:

1) you work the night shift and the sky is "all gone" when you finally get home, and the telescope sits in its corner awaiting your first slew to knock the cobwebs from the gearwork;

2) you've had a bad bout of weather and it seems like every night is cloudy, but the days are clear...what gives?

3) Venus and Mercury are both showing extremely favorable and large crescent shapes during their inferior conjunctions but they are "too close" to the sun to see in the morning or evening sky;

4) you have just heard about a very wonderful lunar occultation of the bright star Antares by the moon....but the darned thing happens just after 3 p.m. in the afternoon, and the moon is but a thin crescent and you can't see it with the naked eye!

5) or, you're just bored to death and cannot wait until nightfall to "look at" something.

If this sounds like you....then **DAYTIME OBSERVING** may be just for you!

'GO TO' TELESCOPES AND DAYLIGHT OBSERVING –

On extremely clear - "deep blue" - days there are a remarkable array of celestial objects that can actually be observed during daylight. The trouble has always been knowing WHERE to look for them.

In the "old days" we had to rely on setting circles to find the objects, and had no way of semi-accurately aligning to north for precision pointing, even using those mechanical circles. Also, the only reference point from which to "set" the circles would be the sun itself.

Nonetheless, that archaic method of finding daytime objects was satisfactory....until computerized telescopes could find them for us with a lot less hassle...and a lot more accurately.

Among the objects that can be observed in daylight (very deep clear skies needed....low humidity, no haze, no pollution) are:

- Venus - Mercury - Mars at Perihelion - Saturn (when the sun is very low) - the Moon - the brightest Stars - "sungrazing" comets - and, of course...the Sun.

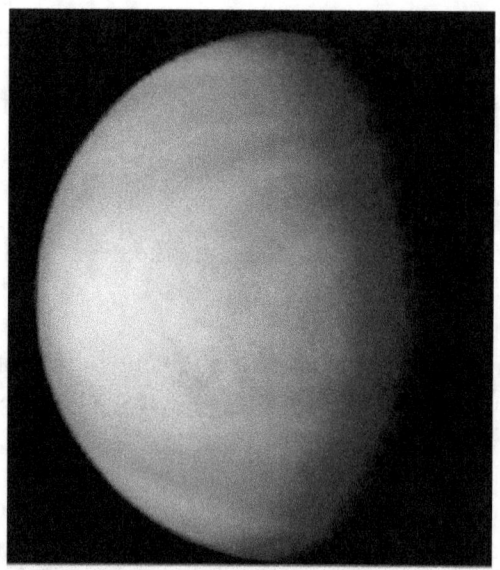

Venus in visible light
Photo Courtesy NASA

On the bright inferior planets Venus and Mercury more detail amidst the scintillating brightness of the planet can be seen during daylight than at night. The photo above (courtesy NASA) shows Venus in visible light (as your eyes would see it) and demonstrates some of the cloud patterns that can actually be seen in amateur telescopes during daylight observing.

In fact, the best time to observe the fascinating crescent phase of Venus and when both it and Mercury are LARGEST as we see from Earth is when it is at INFERIOR CONJUNCTION, very much between the Earth and sun and consequently always see during brightened skies.

Likewise, occultations of the brighter stars by the moon DO take place frequently in daylight hours; and there are daylight times that these bright stars can be OCCULTED by asteroids, thus causing them to seemingly "blink out" of visibility....all this CAN be observed without much difficulty with your GO TO telescope.

Following is a brief description of the two most important aspects of daylight observing:

1) how to acquire (find) the objects with a GO TO scope; and,
2) how to properly observe these objects for maximum definition and least sky/sun glare.

SETTING UP YOUR TELESCOPE FOR DAYLIGHT OBSERVING –

Finding objects in daylight is easy with your computerized telescope, and not at all unlike using the telescope at night. Patience at the eyepiece is the key factor as your eyes will play tricks on you as you attempt to actually "see" the object superimposed in a bright blue sky.

NOTE: other than Mercury, Venus, Jupiter and the Moon, objects will NOT be visible in either the 8 x 21 nor the 8 x 50 finder equipped on most telescopes. You will be required to use the lowest magnification possible for your main telescope and attempt to acquire the object in its field of view!

Follow these easy steps for your next daytime astronomy outing:

1) Achieve focus the night before with your lowest power, widest field eyepiece. The most difficult of daylight observing is attempting to determine FOCUS. If you are not focused just right, NOTHING will be visible, even if it is dead-center! An out-of-focus image blends much too quickly into the bright blue sky that you will see in your eyepiece! Another alternative to prior-night focusing (it will be right at the last place you observed a nighttime object...there is no difference in daylight and nighttime celestial focusing of course...just make sure you use the SAME eyepiece as the night before, as no eyepieces are truly parfocal) will be to use the SUN to focus on prior to setting up; this should ONLY be done when a proper solar filter of optical glass is in place over the front of the telescope!

2) Set up your telescope per the instructions, just as you would at night: level the tripod, place the telescope in "Home Position", initialize, and then do the "Setup/Align/Easy...." two star method;

3) When AutoStar or similar controller takes you to each of the two alignment stars, merely press "enter" when you hear the beep; do not attempt to even locate the stars....if your setup in home position and leveling was accurate, you will be a lot closer than you might imagine!

4) Unless you really messed up, AutoStar/NexStar, etc. will now come back and tell you "Alignment Successful" and your clock drives will engage and you are ready to GO TO something during the day!

5) Just like at night (let's say you are looking for Venus), merely press "Select/Object/Solar System/Venus.." enter and the telescope will take you remarkably

eyepiece; give the telescope time to complete its GO TO slew and do not do anything until you hear the "beep" and the object's data is listed on the AutoStar/controller screen;

6) If you do not find it immediately, allow your eyes time to adjust to the brilliant sky and lack of contrast...the object may, indeed, be in the field of view! If you still do not see it and are reasonably sure it is NOT in the field of view, hold down the "ENTER" key for about 3 seconds and do the "Spiral Search" which will initiate the telescope to begin moving in ever-larger squares outward until you stop it by pressing "Mode."

WAYS TO INCREASE YOUR CONTRAST AND OBJECT VISIBILITY –

Once you have found your object you may have difficulty getting a good image for three reasons:

1) during daylight, the radiant heat from the ground typically interferes with steady seeing; likewise the telescope tube assembly is absorbing heat and those internal currents will greatly degrade the image on some occasions;

2) sunlight in your eyes interferes with your eyesight and your ability to concentrate on the image;

3) there is considerable glare and light flares that are entering your optical tube as a result of direct sunlight.

Unfortunately there is nothing that can control #1) above; we are at the mercy of our atmosphere and daily climate; however, observing early in the morning hours is preferred prior to ground warming to steady air currents; likewise, you can attempt to observe with your telescope IN SHADE (under a porch or a tree) and aimed into the open sky toward your object, to reduce telescope heating.

As for sunlight in your eyes, use the same trick that "old timer" photographers had to use....put a dark cloth over your head and shield your eyes from direct sunlight. This is even a good option for nighttime observing as well. It is amazing the difference you will see in both image contrast and your ability to focus/concentrate on what is being shown!

Figure 1 shows the effect of #3) above, the serious problem of extraneous sunlight reflecting off of your front telescope element as well as entering the tube itself; both the reflection and the intrusion of light will totally break down your image sharpness.

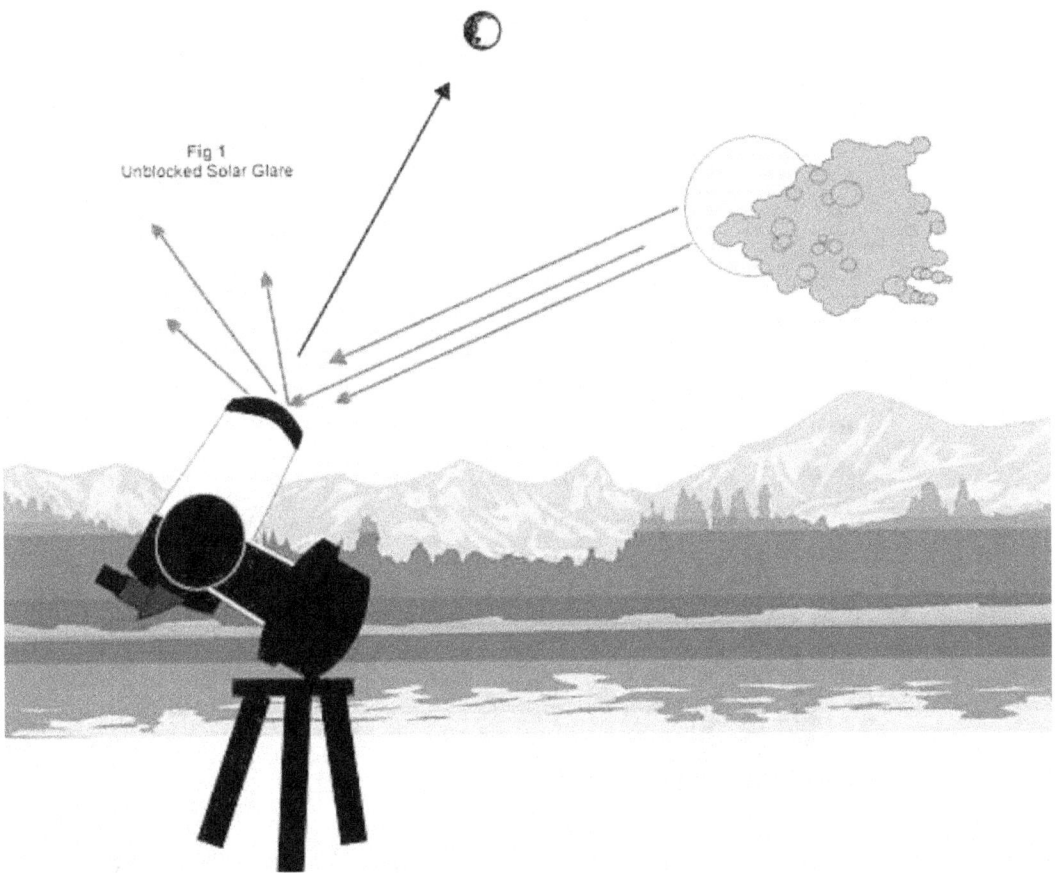

Fig 1
Unblocked Solar Glare

As you can see, the nature of the design of the all Schmidts and Maksutovs is such that the front glass element is exposed well into the outer opening of the tube and thus very easily in the path of direct sunlight. The Newtonian series does not have such as big a problem as the main optical component (the mirror) is located well toward the bottom and shielded from the light!

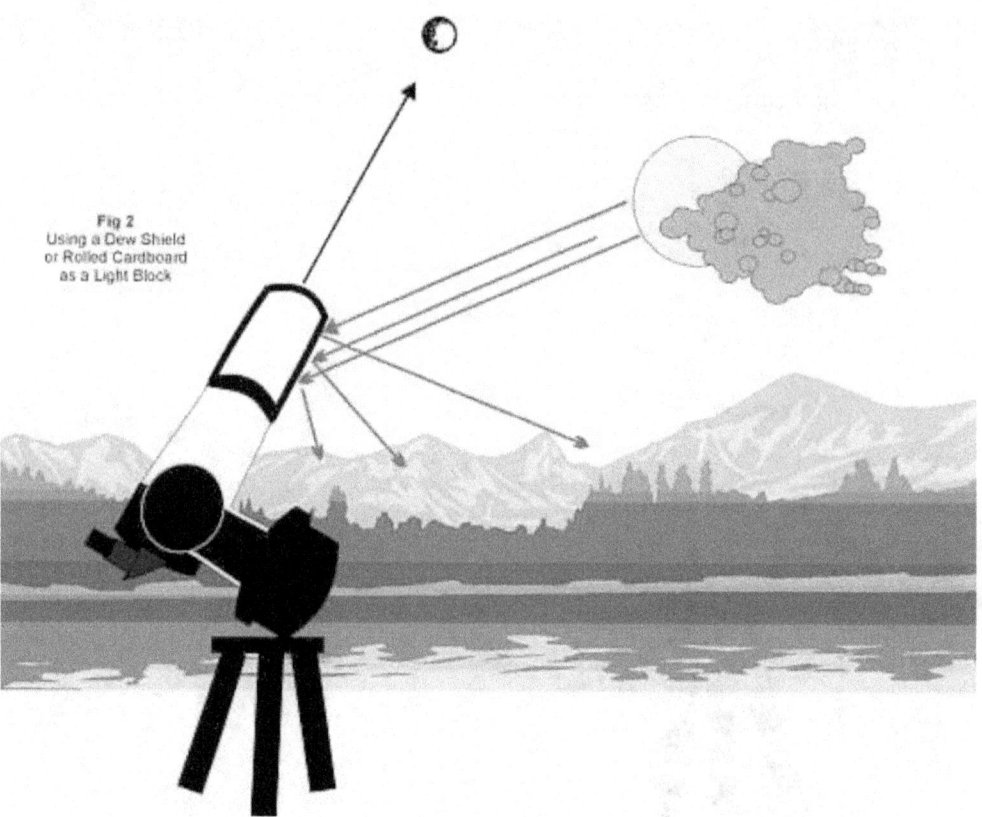

Fig 2
Using a Dew Shield
or Rolled Cardboard
as a Light Block

The very simple solution to greatly enhance your contrast and image quality on the all telescopes is to simply provide a light shield that blocks direct sunlight from striking the front optical element as shown in Figure 2. This can be accomplished by a simple "dew shield" that is commercially available in all colors, fabrics and feels, or simply by rolling up a piece of "poster board" into a tube which slips down over the front of the optical tube assembly of your telescope!

To adequately block extraneous solar rays, the rule of thumb is that the shield must be TWO TIMES longer than the lens/corrector plate is in diameter....thus if you have a 3.5" telescope, then make your cardboard tube about 7" in total length extended beyond your front surface.

You will be amazed at the difference in clarity and contrast so simple a device makes. As a matter of fact this is an EXCELLENT tool to also use when observing celestial objects that are VERY CLOSE TO THE BRIGHT MOON at nighttime as well. Anytime the moon is quarter or more illuminated in the sky and you wish to observe even a bright objects such as a planet, a light shield will greatly improve your observing.

So, not only do our GO TO telescopes open up the frontier of distant galaxies and intriguing nebulous filaments of deep space in the darkened recesses of night....they also allow us the freedom of exercising a little astronomy in daylight hours as well.

Chapter 5

TELESCOPE - MAGNIFICATION

How Far Can I See? The Myths and Moderations of Magnification

In this book to help you understand better the major functions of a telescope, I have stressed three categories that are important responsibilities of your telescope:

1) **LIGHT GRASP** - the telescope's ability to enlarge the capability of your own eye, collecting more light so that you can see fainter objects more clearly, this increasing your "Limiting Magnitude," or how faint you can see as well as increasing your....
2) **RESOLUTION** - the ability of your scope to discern very close and delicate markings, also a result of the size of the telescope's lens or mirror, but also limited by such things as the steadiness of the air and your own visual acuity;
3) **MAGNIFICATION** - the last and least important (YES, I said "...the least important!") of ALL functions of any astronomical telescope.

This is NOT to say that magnification isn't important to you....certainly it is. It is the gift we have been given to magnify Saturn to be able to see the rings clearly, to make out the phases of Venus, to resolve a fuzzy glow in Hercules into a ball of hundreds of thousands of stars.

Of course "magnification" is important....*just not THAT important.*

Let's say, for the sake of starting, that the human eye has a magnification of "1"; that means that if your friend standing next to you is 6-feet tall, you "see" him as six feet tall, a ratio of 1:1.

By contrast we will equate that to common 35mm cameras. A *"single lens reflex"* camera (the kind you look through the lens, not a viewfinder) typically comes equipped with a 50 millimeter (**50mm**) lens which provides an image scale on the film of **1:1** - remember that ratio? It is the same ratio as the human eye, so the standard camera lens provides a real-life portrayal of image scale on the film.

Are you with me so far? If so, then let's add a **telephoto** lens to the same camera of, say **150mm**, or 3-times (*50 divided into 150 = 3x*) the magnification over the standard 1:1 ratio of the regular lens.

So you're closer to the object you are photographing, or so it seems, with the telephoto lens...that's great! Yep, but there are some drawbacks, just like there are in your telescope:

1) when you increase magnification, the BRIGHTNESS of the object you are observing diminishes in direct proportion to the increase;

2) magnification DECREASES CONTRAST after a point and thus provides a more "washed-out" appearance of your object, and consequently begins to lose the saturation of pure color out of the object;
3) not only does "magnification" increase the size of your object (if you observing, say Jupiter at 150 power, you are in essence bringing the image to where it APPEARS 150 closer to you than with the naked eye), but you are magnifying EVERYTHING ELSE as well. This means that you are increasing the effect of tiny vibrations by 150x, you are increasing unsteady air moving between you and what you're looking at by 150x and you are increasing your eye fatigue.

But....all that being said....magnification is **not** a "bad" thing. It's like many other vices in life. It's okay - and even beneficial when used in moderation. It's ***when you overdo it***, that the effect can be detrimental. And most amateur astronomers new to the wonderful hobby do, indeed, try to overdo it.

It's not your fault! We see advertisements for little bitty telescopes that are "454 Power!" To prove that it is "454 Power" there are images all over the boxes and ads for these scopes (taken by large telescopes by a guy with a long beard and government subsidies) showing EXACTLY what "454 Power" looks like. Oh, come on. The modern consumer is too smart to fall for that anymore.

So....how much magnification is *TOO MUCH* magnification?

Well, we hear a lot of people - reputable telescope manufacturers, educators, other amateur astronomers - who all suggest that (and I can almost quote): "...*how much power you can use depends on how big your telescope is...*"

Well part right, like being half-crazy. There's a LOT more to magnification than simply how BIG the telescope is. If we were on the moon, for example, free of moisture, dust, pollution, haze and AIR, observing with a tiny Questar could reveal as much as the largest telescopes could on this Earth with its heavy blanket of air. That is but one limitation among many:

1) the atmosphere (the steadier the night, the higher magnification you can use) - See my discussion concerning Seeing and Transparency on this web site.
2) the telescope size;
3) the telescope quality;
4) the type of eyepiece used;
5) the object you are looking at.

First, remember the "bad seeing rule." If the stars twinkle overhead, then restrict your observing to either low power or deep sky objects at a maximum of around 100 power REGARDLESS of what telescope you use; indeed, smaller telescopes CAN USE MORE POWER PER INCH than large ones in times of bad seeing without noticing the detrimental effects of "bad seeing."

When the air is very, very steady (like stagnant polluted air in summer), you can use 50 x to even 80x per inch aperture (175x on a 90mm scope or 250x on a 125mm scope). The transparency ("clear-ness") of the sky has NOTHING to do with your ability to use magnification, nor does moonlight or streetlights or barking dogs. Only your telescope (aperture and quality) and the atmosphere steadiness m, matters.

Here is the **golden rule of magnification** (maximum values given PER INCH APERTURE):

1) No bright stars twinkle rapidly (naked eye) near the horizons nor anywhere across the sky - Great conditions: up to 75x to 90x per inch aperture;
2) Bright stars twinkle on horizon, but no higher in the sky - Good conditions, 50x to 75x per inch aperture;
3) Bright stars twinkle from the horizon to nearly half-way to the zenith (directly overhead) - Average night, 25x to 50x tops per inch aperture;
4) Bright stars twinkle to the naked eye all across the sky (usually when it is deep clear) - Very poor conditions, limit your observing to 100x TOTAL power or less; a good night for deep sky observing.

CONCERNING EYEPIECES - There are many "super-duper wide field" high power eyepieces on the market today which are very expensive, and indeed very good. But I have yet to find one as good as a medium power eyepiece coupled with a top-quality 2x Barlow or telenegative lens. High power observation is just that - looking at something CLOSE. Wide fields are nice, but not necessary. It has been my experience that the wide field short focal length eyepieces (for high power) suffer from light loss and detailed resolution. Using a lower power and more simple (i.e., I love Plossls and even the older Orthoscopics) eyepiece has the advantage of reducing eye fatigue and giving you some extension AWAY from pressing your face against the telescope!

Nonetheless, if you ever have the opportunity to view through one of the modern 65-degree or larger apparent field of view eyepieces – perhaps at a star party or viewing with friends – you will never be happy with low quality eyepieces again. They are truly spectacular, nothing less than like looking through a porthole into deep space.

My suggestion on eyepieces is very simple: you need a low power, medium power and a high power eyepiece of good quality. All three are important. Add to that collection a very good Barlow amplifying lens and you are set. NO need for dozens of specialty eyepieces.

In terms of 1-1/4-inch or 2-inch, this is even more simple: you need ONE two-inch diameter eyepiece (providing that your telescope has provision to accept the larger eyepieces, many do not!). Your lowest power eyepiece, if possible, should be a very high quality wide field two-inch eyepiece. If you want to make a substantial investment in eyepieces, this is the one to purchase.

OBJECT YOU ARE LOOKING AT - This is an important, and often overlooked, aspect of selection of magnification. Deep sky objects are "extended diffuse" objects, meaning that their brightness is scattered over an (sometimes great) area. Using very high magnification results in losing contrast, and thereby some of the finer detail of faint extended objects; in addition many of the deep sky objects are larger than can be accommodated in short focus eyepieces. ALSO...remember I mentioned that the field of view becomes DARKER when one increases magnification? Very true with deep sky objects. Up to about 100x or 150x, regardless of telescope used, the increased power actually INCREASES the object contrast by slightly darkening the sky around it; however, after that is the point of diminishing return.

Of course, the planets and moon are also "extended objects," but they suffer from no lack of brightness, so the 75x to 90x per inch on a steady night is great for such observing.

In short, never over-rate "power." It is the least important thing that your telescope does for you. Indeed, I pity the poor youngster, so excited about astronomy and his or her first telescope. They pull out the 4mm eyepiece and the triple power barlow (454x) and aim the darned thing at the sky, simply awaiting their first glimpse of the beauty of God's creation.

What do they see? *Nothing*.

But try convincing the same novice to DROP the magnification down to 25x or 40x.....and the thrill is gone.

I mean, if you cannot MAGNIFY SOMETHING, what the heck do I even want a TELESCOPE for?

Good luck, and keep it down (the power, that is)....please.

* * *

Chapter 6

TELESCOPE PERFORMANCE - RESOLUTION
Splitting Hairs - The Resolving Power of Your Telescope

You will always hear – no matter what the size - how your telescope is limited in "*light gathering*" simply by the size of the total surface area of the glass or mirror used as the objective (or primary optical element) of the scope.

Indeed - and somewhat unfortunate - "big IS usually better" in astronomical telescopes, but this does not mean that smaller instruments cannot reveal striking images, resolved minute double stars, peer at galaxies millions of light years distant, see tiny craterlets within larger craters and, yes, even discern the SHAPE of a rapidly orbiting satellite as it streaks by.

To me, the "satisfaction" that you feel with your telescope, and hence the enjoyment of astronomy is three-fold:

1) Being able to "see" enough variety of bewildering objects in the cosmos to keep your appetite whetted for yet another night. Even with the SMALLEST telescope there are MORE objects visible than you can possibly see throughout the sky in your entire lifetime (yes, even with GO TO computer capability!);

2) Knowing the LIMITS of your optical instrument, from the eye...to binoculars....to your telescope; you cannot be satisfied if you THINK you should be seeing more than your telescope is capable of revealing! In large part, many of the advertisements for popular telescope and the packaging that we see emblazoned in full color suggests that we can see a whole lot more than we actually can to the eye. Many times, the result of this is disappointment.

3) Now that you know you can see many, many objects, and you have acclimated yourself to the fact that those objects will likely appear different in your 3-inch telescope than they do in Hubble Space Telescope photos, the third - and in my opinion the GREATEST - satisfaction is being able to SHARE your enthusiasm. If YOU are not excited about that great view of Saturn through your telescope, how in the heck are you going to get Ma and Pa outside on a cold night to "share" in your luke-warm grandeur?

Aside from talking about telescope limitations, let's talk for just a brief minute about *YOUR LIMITATIONS*. No, I am not talking about social, economic, appearance nor educational limitations....I am talking about your ability to actually FIND the stuff in your telescope you should be seeing!

Those who get a nice scope and give up early on the hobby quickly are those who can find the moon, maybe Saturn and Jupiter....Venus is easy....perhaps accidentally stumble across the Orion Nebula. Maybe they got that telescope just because they read in the paper that the "comet of the decade," *Comet Whatshisname*, is about to appear in the evening sky.

Once the comet is gone.....what are they going to look at? Indeed, if you want a bargain in a telescope, trying looking through the classifieds right after a popular comet has disappeared from view!

RESOLVING POWER AS A FUNCTION OF WHAT YOU 'CAN SEE'

If you haven't already, check out my discussion on **limiting magnitude** inn the GUIDES on this web site. "Limiting magnitude" is how FAINT you can see with any optical instrument: your eye, binoculars, telescope. The larger the telescope typically, the fainter object you can see. Telescopes are like rain buckets; you're simply going to get more water in one that has a bigger opening than the smaller one next to it.

The light-gathering power is also a determining factor in the *RESOLUTION* of the telescope; all things equal, such as optical quality and telescope design, a larger telescope will "resolve" finer detail. What does this mean?

Good resolution means that you can clearly see Cassini's division in the rings of Saturn, while a smaller telescope cannot make out the black thin gap. It means that you might glimpse the gossamer "festoons" in the equatorial belt of Jupiter and be able to "split" a double star to see BOTH stars as individual points of light instead of an elongated oval of two connected images. If there are hundreds of very tiny craters inside of Plato on the moon, a larger scope usually has a chance of seeing more of them than the smaller telescope if the craters drop BELOW THE RESOLVING LIMIT of the smaller scope.

FACTORS THAT DETERMINE TELESCOPE RESOLUTION

The size (diameter of the objective) of a telescope is not the only factor in resolution; others play into the equation as well:

1) <u>Aperture</u> - the larger the telescope the smaller an object can be seen and the closer the spaces between objects (I.e., double stars) that can be distinguished;
2) <u>Optical quality</u> - like in any item, a quality optical system will resolve finer detail than one of mediocre or poor quality;
3) <u>The Atmosphere</u> - read carefully my article concerning "seeing and transparency" so that you can determine the quality of the air around you. The largest telescope at a star party cannot resolve to its limit (discussed below) if the air is unsteady;
4) <u>The Observer</u> - YOU. A keen eye is necessary to reach your resolving limit; things that affect your eyesight are: 1) normal eyesight; 2) eye fatigue or disease; 3) extraneous light such as streetlights; 4) actually "knowing" what to look for; 5) use of alcohol and/or tobacco; 6) knowing how to use "averted vision." (discussed following); and, 7) AGE, with younger eyes being much more keen than older ones!

Let's say that all of the above factors are "good" and you are at a star party in which there are many telescopes - all of the SAME quality but in many different sizes. For this example, and for you to better understand how "resolving power" comes into play for all things astronomical, we will only use the example of DOUBLE STARS.

[**NOTE**: There are many excellent books which list challenging double stars for tests with any telescope size, from 2.4" to 24" aperture. One is Robert Burnham's *CELESTIAL HANDBOOK,* a three-part alphabetical reference to every constellation and interesting objects within them. Another excellent choice is *The Constellations: Sky Tours for Computerized Telescopes*, by P. Clay Sherrod.

The acid test for a telescope is the resolving power, but no one should be slighted who owns a small telescope....again, just know and understand your limitations are reserve your efforts within YOUR telescope's capabilities!

With all the telescopes at the star party, here is the list of apertures and here are the double stars (hypothetical under excellent steadiness) that they are looking at, clearly resolved, or "separated with a dark space between each star." The separation is given in *SECONDS OF ARC*.

As a reference, the MOON is one-half **degree**, or 30 MINUTES of arc; just like on the clock, there are 60" (the symbol ["] means arc seconds, and the symbol ['] means arc minutes) in ONE ARC **MINUTE**, meaning that the expanse of the moon as we see it is 30 (minutes arc) x 60 (seconds arc in each arc minute) or: 1800 arc seconds across. Sounds big, huh? It IS when you consider what even the smallest telescope is capable of resolving.

Your EYE by itself is capable of resolving something as small (or as close as) TWO MINUTES (2') ARC in the sky. That in itself, knowing that the Moon is only 30 arc minutes across is a pretty small space!

So here is a report on HOW CLOSE a double each of our participants can just separate into two stars:

Ramon's 2.4" refractor -   **1.9"** arc (wow....that's close if the moon is 1800" arc!)
Susan's 3.5" Maksutov -   **1.3"** arc (we're really closing in!)
Dilbert's 4" refractor -   **1.14 "** arc (see how much closer the stars are getting?)
Bruce's 4.5" Newtonian -   **0.10"** (now we're cookin' below one second arc!) Clay's 5" Maksutov - 0.91" (this is getting exciting!)
Laura's 6" Newtonian -   **0.76"** (getting into some serious doubles here)
Michael's 8" Schmidt -   **0.57"** (these stars must be getting faint!)
Bocephus' 10" Schmidt -   **0.45"** (does it even matter after they're so close?)
Halley's 12" Newtonian -   **0.38"** (she beat everybody, but her scope is a mess)

In actuality, there are few nights when those limits can be reached. The closest resolution (or finest detail) that your telescope can make out is limited by an astronomically-accepted "law" known as "***DAWE'S RULE.***"

If there were no atmosphere to look through and we lived in a perfect world, Dawes' limit tell us to simply DIVIDE the number "**4.56**" by the aperture of your telescope IN INCHES to get the values for the star party above. Likely, they were having their party on the moon with no air, since the steadiness of the air is the greatest limiting factor to reach "Dawes' Limit."

I know you are wondering, because it seems to be so important: "What <u>MAGNIFICATION</u> is necessary to reach the maximum resolution of my scope?" Actually, in this case, magnification DOES help to a degree and ONLY up to a certain limit. Magnification, of course magnifies unsteady air just like it does your object, but it does help in a couple of ways: (see preceding Chapter 5)

1) Use about **30x per inch** aperture to reach the maximum resolving power on a good night; on a "perfectly steady" night you can use up to **50x per inch** aperture; this allows the "image scale" to be large enough for the human eye to actually discern ITS OWN limiting resolution;
2) Up to 30x per inch aperture (about 105x on a 3.5" scope) will <u>darken</u> the background sky, thereby increasing contrast of your brighter object against the less bright sky or increase the contrast between a bright planet and a dark faint detail.

Every telescope is different; there are some telescope that I have used that have "knocked the socks" off of Mr. Dawes, splitting doubles far beyond this empirical value; on the other hand, I have seen telescopes that have fallen way short of even CLOSE to the limit they should achieve.

Invest in a good reference book and go out and find your limit for yourself....that is what amateur astronomy is about: self-discovery! If you have a computerized telescope, most directories have the best double stars; find ones that fit your Dawes' limit as determined above....and see if you can see some sky between them.

Let us know what you find. I would be most interested to see how good YOUR scope performs on the "acid test" of all telescope functions!

* * *

Chapter 7

Seeing Conditions and Transparency
You be the Judge of the Night Sky!

Every time winter rolls in frosty nights our way, questions surface and telescope performance concerns arise. Under these crisp, clear winter skies of North America, the stars almost reach out and grab us as they twinkle against the inky background of distant space. The deep and dark of these cold nights would seem to indicate excellent observing, right?

- *"Why can't I make out Cassini's division at 75 power?....I used to do it all the time. I'm worried about my telescope!"*

- *"I looked at Jupiter and could see the four moons [Galilean satellites] but for some reason all I could see was two brown bands! It was all washed out. What's wrong with my telescope?"*

- *"While looking at Betelgeuse last night I could see the nice red color okay, but now I'm really sick. I think something's happened to my scope! Shooting straight up (it was nearly overhead) I could see a bright flare of light, making the star look elongated. Should I send the telescope back?"*

Do these sound familiar? Have such experiences made your heart sink to your stomach out in the dark and cold of night? And to make it worse, these are actual concerns sent to me from people who spent a night in the freezing cold to "enjoy" the wonderful hobby of amateur astronomy.

In every case, EVERY telescope (and the dozens of others from users this winter) was just fine! It just so happens that, from where they were observing the sky was TOO CLEAR. Did you LOOK at the stars before observing? Those "twinkling stars?" **THEY** are to blame (or at least what you SEE of them); read on....

"Too Clear?!" You say. There's no such thing. Yep, there is for high power viewing and maximum resolution. And it usually happens in months when the days are warm and the nights cool down very rapidly, sometimes dropping as much as 30-40 degrees in just a few hours after the end of dusk. Fall and spring are the two worst times for these temperature plunges.

The term "too clear" refers astronomically to really good "**sky transparency**," our ability to peer through our own atmosphere filled with water vapor, gases, pollution, dust and even light toward the dark of space. Without the atmosphere, every night would be perfect, even when the moon was out! On a perfectly transparent night, the dark sky would be rated a "6", meaning the faintest star that can be seen by the naked eye would be 6th magnitude.

The term "steady image", known as "**seeing**" astronomically, is totally unrelated and - indeed -can be thought of as "inversely proportional" to the "transparency." Simply put....the BETTER the transparency (like in winter months) the WORSE the "seeing" or image steadiness is typically. And in reverse, when we have heavy inversions set it (like in summer), the seeing is remarkably steady....but the transparency of the sky might be a "3" (only 3rd magnitude stars can be seen).

Imagine it this way: "transparency" relates to how DEEP you can see in space through the obstacle of the earth's air. The better the transparency, the fainter the object or star that you might be able to see.

On the other hand, "seeing" has absolutely NOTHING to do with transparency; it is a gauge of how PERFECT, or steady, the image remains while you view it.

Now before going into detail, remember the "*really-nifty-Doc Clay-rule:*"

1) If the night is perfectly transparent, it is a low power night, but not necessarily a high power night;
2) If the night is perfectly steady, it is a high power night, but not necessarily a low power night;
3) If the night is transparent AND steady at the same time, it is a PERFECT night, both for low- and high-power viewing;
4) If the night is cloudy, it's *Miller* time.

SKY TRANSPARENCY AND WHAT TO EXPECT

In the old days when I was an astronomer, sky "transparency" was judged both visually and photo-electrically by a scale of 1 to 6.6, with 6.6 (the hypothetical limiting magnitude of the faintest star visible to the eye before light pollution) being perfectly deep clear. On the other hand "1" would represent ONLY the brightest planets and stars (magnitude "1" or brighter) being visible. Such a case would be in the summer months when smog or fog "capped" the night air preventing all but the brightest objects from being seen. As we will see, such nights are the very BEST for "steadiness!"

There are two charts (one that has appeared on this web site previously, but it is worth using again). Both are found under the GUIDES/Limiting Magnitude. Click onto the charts and each will print out a large quality print for you to use outdoors; slip them into a sheet protector or laminate and it will allow you to judge your transparency from any location on the Transparency Scale. The **Hyades** is excellent for fall, winter and early spring, and the "little dipper (**Ursa Minor**) is excellent for Spring, Summer and Fall since it is circumpolar (mid-northern latitudes) and can be seen nearly all night; however, it should (like the Hyades cluster) be used as a guide ONLY when the stars are reasonably high in the sky.

In transparency - and particularly for astronomy - there is a factor knows as "*atmospheric extinction*" which causes starlight to diminish as the object decreases in altitude toward any horizon. It is caused by the "stuff" in our air, and particularly troublesome during high humidity and smoggy conditions. This same thickening atmosphere as one nears the horizons, ALSO affects "seeing" since air closest to the ground, and miles more of it to look through, is going to be less steady than looking straight up through the thinnest route out of our atmosphere.

As I mentioned, "transparency" is a measure of how clear the sky is; this will allow you to see fainter stars and deep sky objects, but will NOT necessarily allow you to use more than about 25x per inch aperture, and then only for deep sky objects. Typically speaking (there are exceptions) the *CLEAREST NIGHTS are the most UNSTEADY*.

This is because of meteorology, not astronomy. All day long, whether cloudy or raining or snowing, the earth's crust and all upon it soak up the radiation from the sun in wavelengths that we cannot even see. Objects on the ground will HOLD that energy so long as: 1) the air around them is near the same temperature as the heat-soaked objects; or/and, 2) the object continues to receive an equal amount of radiation (heat) that it is losing into the air over the same period of time.

Okay, what happens at NIGHT? Same thing that happens in the desert on a hot summer day when your are driving down Route 66 at mid-afternoon: the GROUND is much warmer than the surrounding air and heat (the ground) always dissipates into cold (the air). When on the desert highway, we see this as a wavy image (mirage) as if water was standing on the road in the distance. At night, we see this rapidly rising air as "twinkling stars."

This nighttime convection is called attaining **THERMAL EQUILIBRIUM** (I won't use terms like that anymore) and continues until the ground is the exact SAME temperature as the air.

This condition (happens every night to a degree) DOES NOT affect transparency, although it is the most detrimental factor involving "seeing."

So, the first thing you will do each night you go out is to judge the "transparency" by seeing which of the stars you can distinguish from the charts. It is fun and interesting to keep track; for example, in Arkansas the month of OCTOBER is by far the best combination of "transparency" and "seeing" of any month, and also just happens to be the most weather-friendly month as well!

Transparency is usually worse in summer (inversion layers of air, stagnation of air) and best in winter (cold blast of arctic air bring pure, arctic-filtered and clean air southward).

Once you have checked out how clear the sky is, let's move on to see if you can make out the Great Red Spot or the Polar Region of Saturn with some really impressive high power.

ATMOSPHERIC SEEING CONDITIONS

We have seen that really clear skies don't normally indicated that the night is also a really good night for high power planetary work or splitting that double star you've been promising to do but keep putting off. Also, we understand now that it is the EARTH that is responsible ("....baaaad Earth!") for giving us nights when we THINK we should be able to see Cassini's division on Saturn ("...*but I saw it last night*!") but it is invisible except in fleeting glimpses.

Seeing can be rated on a scale of "1" to "5", with "1" good only for lowest power viewing and "5" perfect steadiness. Remember that I said that in summer stagnation and "inversion layers" cap our atmosphere to prevent really "transparent" nights? This is GOOD for seeing....because the air below the inversion is trapped, the heat cannot move rapidly in these upward currents. Consequently, the air immediately above your telescope is very, very "steady."

Early at night is the worst time for steadiness; just as soon as the sun sinks low enough in the sky, the ground begins "giving up" the heat it has basked in all day long. Those **heat currents rise rapidly** upward into the sky. Let's look at how we see them:

1) <u>NAKED EYE</u> - you can use the twinkling of stars to get a quick handle on the night's seeing conditions, using an imaginary scale. Look toward your most unobscured horizon (where you can see stars closest to the distant horizon) and mentally divide the sky into four (4) equal parts from that horizon to overhead. Each one of those "parts" we'll call a ZONE for seeing evaluation. "Zone 4" is from the horizon to 1/4 way up to the Zenith (directly overhead); by comparison,. "Zone 1" IS overhead. Now look at the brightest stars in each of your quadrants starting with the horizon; are stars in Zone 1 twinkling? how about Zone 3, a little above the first zone....are those stars also twinkling? Now Zone 2, nearly overhead...

Now it's easy: if ONLY the stars in Zone 4 (lowest to the horizon) are twinkling rapidly then your "seeing" is AT LEAST a "4" on a perfect scale of "5"; if however a bit higher up in Zone 3 the stars are ALSO twinkling like those in Zone 4, your seeing is NOW "3"; if the stars are twinkling rapidly all the way to the Zenith (Zone 1, then your seeing is a pitiful "1", the worst it can be).

Remember...this has NOTHING to do with the "transparency" of the sky!

If, on the other hand, NONE of the stars twinkle all the way to the horizon, then your "seeing" is possibly a "5." To determine whether it is that good requires that you move from your naked eye to the telescope!

2) <u>TELESCOPIC SEEING</u> - so far we have determined that you might have a really "steady" night of at least "4" and perhaps a perfect "5." Let's check: turn your telescope to a very bright star that is close to overhead. Make sure to center precisely in your field of view. Check you air steadiness through two steps:

1) focus precisely the star until the image (the "Airy disk") is as small as possible; the smaller the telescope, the LARGER the center bright "point", or Airy disk, will be. Once focused, you should begin to see clear "diffraction rings" surround the Airy disk that appear to be thin rings of light; if you are using a Newtonian telescope, you will see four spike-like streaks of light from the secondary mirror holder. Under steady conditions you will note that **a)** the Airy disk does NOT MOVE nor does it change size; and **b)** the concentric rings remain fairly steady, both in motion and brightness. You should see perhaps three rings in most telescopes, perhaps more.

2) now the tough test. Put the same star out-fo-focus until you see a medium-sized disk (not too big) of light with a center dark portion, looking much like a donut and its hole in the center. Put this de-focused star in the dead center of your FOV; well collimated optics will show the dark spot exactly centered on the bright disk. If your star moves toward any edge of your FOV, the dark disk will shift in the OPPOSITE direction! So keep centered. Once de-focused, look carefully at the bright disk. You will see alternating rings of dark and light, very, very fine. Examine all of them and see if they appear to move, or oscillation. If the image moves like a squirmy amoeba, then your seeing is terrible and you can forget high power. It's galaxy viewing tonight. On the other hand, if the image is very steady and uniformly bright your seeing can be excellent!

It's really simple to evaluate. Always remember: there ARE going to be nights when you want to give up and blame the telescope for poor performance when you might not have the contrast or detail you are used to seeing. Indeed, it is probably the "seeing conditions," and not your scope. If you find yourself in this situation "...take that lemon and make lemonaide!" as they say. Develop a mindset that "...tonight, I'm going to find those galaxies that I have been looking for!"

A CLOSING WORD OF CAUTION

The best "seeing" and the deepest "transparency" cannot offset a telescope that is suddenly rushed out of the house and plunged into the cool (or cold) night air. Here is the rule of thumb that we have always used, no matter HOW BIG or SMALL the telescope: put the scope in the observing location (in shade if the sun is still shining) for a TOTAL of **20 MINUTES PER INCH** of aperture of telescope. A three-inch needs ONE HOUR to adjust (thermal equilibrium - heck, I used the big word again!), while a five-inch requires TWO HOURS. But the wait is worth it.

This is why professional observatories (even with controlled ventilation systems) open the domes at before sunset and commence through dusk to allow the scope to equalize with the night air. The longer the night goes on, the more "married" are the temperatures of your telescope and the air around it.

Now that you understand at least some of this (isn't it interesting that most of this has absolutely NOTHING to do with your telescope itself?), you can probably have a better grasp of why the scope that you love so much sometimes can be ornery as a pet cat, tuned to your beckoning call one minute, only to ignore your very existence the next!

Heat currents rising from hot pavement give false impressions of water on the highway; this same rising hot air into cooler more dense air gives rise to atmospheric problems for observers early in the evening.

Not only does the air need to equalize for proper viewing, but so does the telescope need to become as near the same temperature as the ambient air as possible.

Chapter 8

Star Testing Your Telescope

Using the Airy Disk to Determine Optical Excellence

This was in response to an inquiry about star testing and folks concerned about everything being "perfect".

Star testing a telescope is easy, but should never be done under poor conditions. It will lead to unnecessary worry unless done under perfectly steady skies. Sometimes when things are going so well, we tend to "look for" problems that we think "should" exist....the old "too good to be true," nature of human-folk.

Let's talk about the star test for a second, though. First, in focus, there is probably a condition that may arise with "too many diffraction rings" from an Airy pattern around a bright star, but I would like that luxury. With my telescopes on a very clear night and, say Arcturus or Vega, I can see up to seven rings plus the central star "disk." Smaller telescopes reveals as many as five on a good night. Diffraction rings are good....if you were NOT getting them, or if they met/combined/mingled then you would have a problem. But you should never expect to see Airy rings around stars, even on better than average nights.

The Airy pattern is one of the most stringent of all star tests.

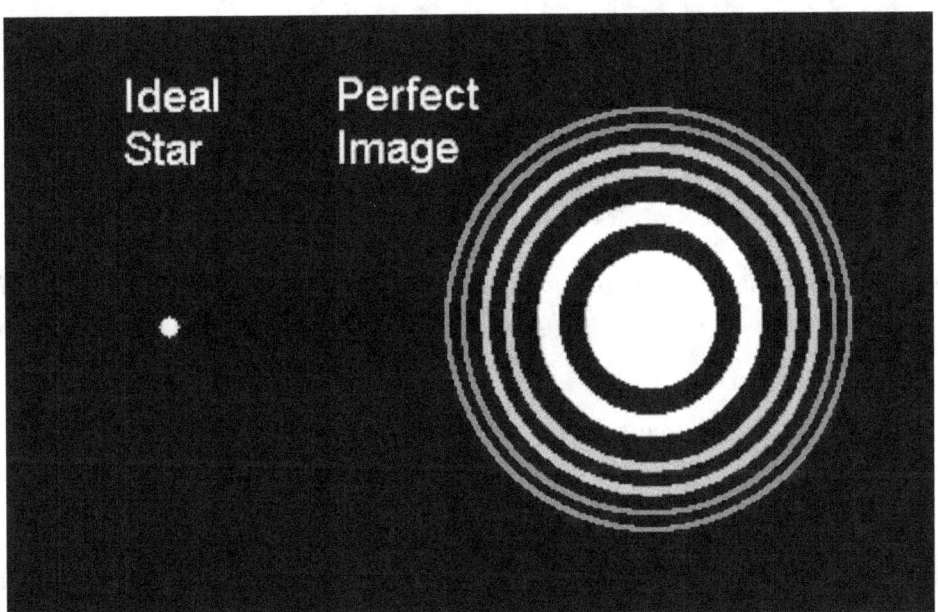

Using an ideal sky (no atmospheric seeing turbulence) or a point source in a laboratory will provide a small compressed Airy Disk (left), while the same source out of focus will show a perfect concentric pattern as seen on right (courtesy UWGB)

Now let's talk about your out-of-focus star disk test. Intra-focus and extra focus images

55

on a very clear night (always choose a bright star nearly overhead on a very steady night) will reverse themselves in nearly all catadioptic/compound telescopes as you move in and out of focus. It is common for one to appear "sharper" than the other. Use about 150x per inch for the test. If the out-of-focus star image (a disk) appears to wiggle or oscillate, forget the test,....the seeing is too poor.

In quality telescopes, the Airy disk should present a bright central "disk" surrounded by a series of very faint and concentric rings. If the rings merge together around "the circle" or if some appear abnormally large, then there may be a question of inferior optical figure.

Also look for **optical astigmatism** while moving inside and outside of focus; when the star is very slightly inside of focus and shows an ELLIPTICAL rather than circular pattern, and then as you move to OUTSIDE of focus the same elliptical disk appears, except rotated 90 degrees, then you have astigmatism with something in your optical system: main optics, eyepiece, etc.

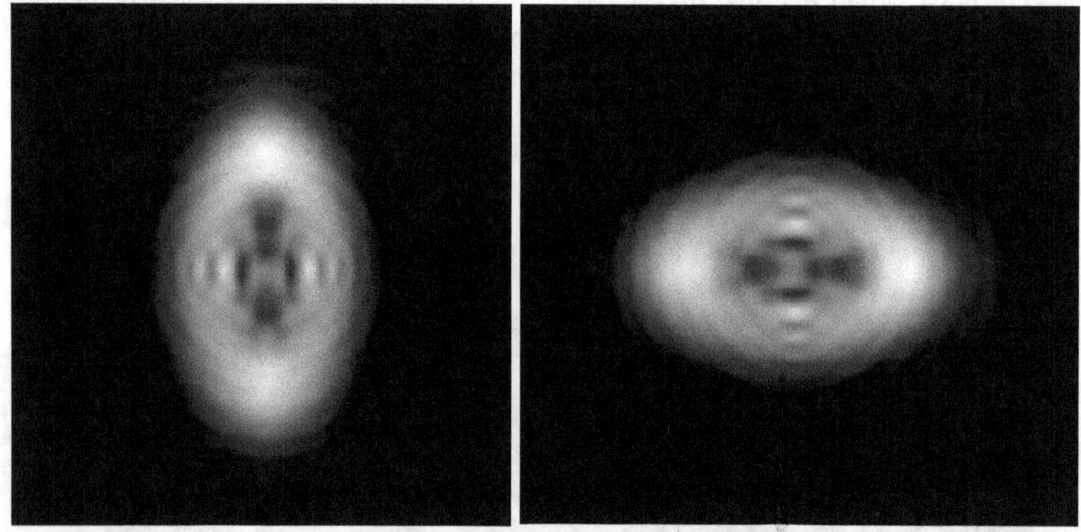

Figure above demonstrating serious astigmatism, common in some telescopes with poor quality manufacturing and particularly in binoculars marketed to the astronomical customer

Unfortunately, optical astigmatism is far more common than we would like. You can determine the component of your optical system (eyepiece, diagonal, reducer, Barlow or even the main optics, is at fault through process of elimination. Rotate your eyepiece with the star image defocused as shown above; if the ellipse rotates with the eyepiece, then you know the problem would likely be confined to that accessory; same with the star diagonal, and – once other accessories are all eliminated – then you might have to face the reality that your telescope optics suffer from astigmatism.

Cheap optical instruments and in today's market, large "astronomical binoculars" are particularly subject to this defect and should be tested for such upon arrival. Never purchase an optical instrument without a written guarantee from your dealer or individual that the item can be returned if inferior optics are found.

Another common optical problem with all telescopes is that of "pinched optics" in which usually a major component of the optical path (refractor lens, Newtonian mirror, Schmidt corrector plate) becomes torqued improperly at one or more places along its perimeter.

Such is the case with the optics above, this being a pinched primary mirror of a Newtonian reflector. At the three points in its mirror cell where it is being restrained with too much force, you can clearly see the disruption in the Airy rings. In some cases where an optical component is being pinched only in one location, the entire central condensation)disk) of the pattern may be greatly shifted to one side. Refractors with lenses being held too tightly in their circular cells by a too-tight retaining ring will show gross distortions throughout the pattern.

With a bright star at high magnification in your scope you should see very, very fine and close concentric rings within the out-of-focus star image, with the donut hole of the secondary (if you are using a compound or reflecting telescope) right in the center of the disk; an offset secondary shadow indicates a collimation problem....OR it might indicate that your scope is simply NOT CENTERED on the star. You will note that the "hole" appears to shift to one side as you move the star closer to the edge of the field.

NOTE that if the central disk is offset in the Airy pattern, this may not mean that you necessarily have "bad optics", but rather that your collimation of your optical system may be someone off and a slight alignment may be necessary.

The tiny little rings that you should see should be essentially un-disrupted from being cleanly separated from the others; if they appear to merge periodically, that is okay.

However, if you see major distortions on the disk, like a splash in a puddle or such, it is an indication of some degree of optical flaw. I doubt you will see this.

Always make sure your telescope has "equalized" to the outside air for AT LEAST two hours before doing any star test. If your telescope is a different temperature than the air, you will not ever get good results.

Important note: small telescopes have a distinct advantage (see the previous chapters regarding seeing and transparency) over large ones in average-to-poor seeing. In short, the greater the aperture the greater the resolving power; this means the larger scope can resolve AIR TURBULENCE even better as well, not just a festoon on Jupiter, or a faint star. This is why small scopes show considerably more detail on nights of poor seeing than large ones at the same star party.

There is a very good reason for this: the air is boiling with tiny vortexes call **EDDY CURRENTS.** These tiny swirling whirlwinds are typically about 0.43 arc seconds across. A 10-inch telescope of good quality has a resolving power of 0.46 arc seconds, just higher than necessary to resolve the air turbulence. A 12 inch telescope might resolve them clearly and that very added aperture is enough to render better images in the 10-inch than in the larger aperture. In the small the scope, the LARGER is the Airy disk (the central image of the in-focus bright star), appearing more like an actual planetary disk than with larger scopes. This, obviously is simply a result of the greater resolving potential of larger scopes.

You've determined that you have a good scope....***hide it***. There are other enthusiasts out there who would die for it!

* * *

Chapter 9

A Practical Guide to Eyepiece Selection
Making the Right Choice Before you Buy

One of the most frequently asked questions of all - particularly from "new" owners of telescopes is "....*what other eyepieces do I need*?" Likewise, there are frequently-asked questions regarding the best choice to use for a particular observing application, like Jupiter compared to a faint comet. Indeed, there are vast differences in the type and design of eyepieces to optimize your viewing enjoyment of nearly all celestial objects. This Guide will help you determine which eyepieces that I recommend from five decades of experience, to supplement your standard eyepieces that normally come with the telescope (normally the 25mm Plossl, which still remains one of my favorites).

In addition, there are tried-and-true rules that concern such issues as staying away from many "hyped expensive-type" eyepieces when a less-expensive alternative may be just as good or even better for some applications. I also want to warn everyone against "over-buying" both lots of eyepieces and filters. The astro want ads are packed full of used ones and the buying of eyepieces can not only be expensive....it can be addictive..

You will see references to both field of view and magnifications in regard to eyepieces; for the beginner I very much favor and recommend the Plossl design as the best quality for the dollar spent. However, you can easily determine within a few minutes (') arc for field and a few "power" for magnification by merely looking at the closest FOCAL LENGTH eyepiece that is listed to what you may have that may NOT be a Meade eyepiece. I have learned through my aging astronomy career that - if two eyepieces, made (or distributed) by different companies are of the SAME design (i.e., 6-element Plossls), then the ACTUAL fields of view will be nearly identical; the only difference might be a "stop" or small diaphragm placed in the barrel to improve contrast and image quality in less expensive eyepieces.

Be very, very wary of such promotional goodies as "120 degree field of view!", "Argon-filled", "purged and sealed", etc. The gimmicks go on and on. In fact, those being sold with such claims toward superiority are typically the ones that I intentionally avoid. They may be good, but if they need unconventional marketing to get the high price they are asking, I never even consider trying or buying them.

CONCERNING ZOOM EYEPIECES –

I have gone on record many times stating that I do not prefer "zoom" eyepieces and the reasons are stated below; this does NOT mean that they might not be suited for you, particularly if you really enjoy the ease of observing and not fiddling around in the dark looking for things. But for discriminating views of very subtle detail, the zoom eyepiece falls short....no matter WHAT brand you buy or HOW MUCH money you spend for it. Regardless of the claims. The shortcomings of zoom eyepieces (from my use of virtually all that are on, or have been on, the market) are:

1) field of view - the zoom inherently provides a more restricted actual field of view at a given focal length (say, zoomed to 8mm) than would a good eyepiece of 8mm focal length;

2) light loss - the zoom requires extra lens elements to accomplish the variance of magnification as well as maintain parfocal (focused the same at all settings) integrity; the extra lenses merely absorb more light through refractive indices and allow less light through than would a simple eyepiece; thus your images will be slightly fainter;

3) contrast - by far and large, the zoom fails on the contrast tests at all magnifications when compared to a simple eyepiece;

4) resolution and spherical aberration - at higher magnifications I have routinely seen very poor resolution on double stars and planets with zooms; in addition, at LOW magnifications, most seem to have inherent spherical aberration: a problem whereby if the CENTER of the field of view is in focus, the edges will be slightly out-of-focus.....if you focus the edges, then you lose focus toward the center; and last but not least:

5) cost - for the price of a quality zoom, a user could buy at least two good quality Plossl eyepieces of his or her choice and obtain the attributes noted above.

NOW ABOUT BARLOW LENSES –

I am an advocate FOR using a good quality Barlow/telenegative lens at the eyepiece of any telescope for two reasons:

1) the Barlow provides double (or in some cases 1.5x, 2x, 3x, 4x and 5x) the magnification of any eyepiece without robbing you of that precious eye relief (your ability to remain a comfortable distance back away from the front lens);

2) if properly chosen modern Barlow lenses provide excellent optical quality that are equal to the quality of the eyepiece itself; of course, cheap Barlows will adversely affect the quality of even the BEST eyepieces

3) the Barlow puts the eyepiece HIGHER toward the top rear of your telescope, thereby reducing you need to "scrunch" your face down into the back of the scope and against the small finder.

Of course, the Barlow adds either two (achromatic) or three (apochromatic, or "color free") more lenses to an already-taxed optical system. Remember that every piece of glass that you put in the path of oncoming light from your celestial object, the more light that is **"lost"** by refractive absorption or by simple reflection of the shiny glass surfaces.

The Barlow essentially allows you to double (or multiply) the power of any eyepiece, but you DO NOT reduce the eye relief by the same factor of "2", but only about 20 percent at most. Therefore if you own only the 26mm Plossl eyepiece PLUS a good Barlow, you essentially have TWO eyepieces: a 26mm and a 13mm.

It is VERY important when choosing eyepieces to remember that fact if you have - or plan to buy - a good Barlow. If you already have, say the 26mm and want (or already have) a good Barlow lens, it would be foolish to turn around and order a 13mm or a 12mm, or a 15mm eyepiece, when you WILL HAVE essentially just that combined with the Barlow. Do some math before you buy; one Barlow can save the investment of three or four eyepieces if you are choosing wisely.

A tip: stay away from any amplification in a Barlow GREATER than 2.5x, and be cautious with those; the 2x is ideal and most from reputable manufacturers are very good. The 3x and 5x Barlows that are so popular right now are the bargains on the "astro flea market" of tomorrow; they simply offer too much amplification and light loss from your image. EXCEPTION: Many of the 3x to 5x Telenegatives that are on the market today are not being used visually, but rather in concert with rapid-fire webcams to obtain some of the finest high resolution planetary photography ever made, all in the hands of amateurs. Using, say a 4x amplifier, at prime focus with no eyepiece of an f/10 telescope converts the system to an f/40 planet killer.

So the idea of a Barlow can greatly help you reduce your selection decision for eyepiece focal lengths and at the same time keep some of that "burning money" in your pocket!

MAGNIFICATION – *See Previous Chapter for full discussion*

Not to dwell on it here, but the magnification of your telescope and various eyepieces is computed below in a rather "overall" form. For eyepieces not listed, the "POWER" or magnification of the telescope is determined by taking the FOCAL LENGTH (in millimeters) of the telescope and dividing INTO that focal length the focal length of the EYEPIECE. The result is exactly the magnification of your scope. As an example with the 26mm (focal length) eyepiece:

Focal Length 350mm 350mm divided by 26mm = **13.5x** (27x with the 2x Barlow)
Focal Length 1250mm 1250mm divided by 26mm = **48x** (96x with Barlow)
Focal Length 1900mm 1900mm divided by 26mm = **73x** (146x with Barlow)
Focal Length 2000mm 2000mm divided by 26mm = **77x** (154x with Barlow)

You can see an interesting situation with the ETX 125 and the f/10 SCT (the last two); both have the same focal length even though one is a 5" and the other is an 8" scope! This is because of the Maksutov design (ETX 125) having an f/15 FOCAL RATIO and the Schmidt-Cassegrain using the f/10 FOCAL RATIO provide for nearly the same focal length. Although the "power" is the same in both, the "light gathering" and resolution is of course greater in the larger telescope using the same magnification.

EYEPIECE SELECTION GUIDELINES:

The following breakdown had to have some limits since there is a plethora of eyepieces flooding the market today. For my guide which follows, I have simply chosen the category of object being viewed with eyepiece options. My opinion is that it is always better to have THREE eyepieces of excellent quality and performance than a suitcase full of ones that give only marginal viewing.

Other eyepieces can be fit into my charts; for example, the data you need for say a **10mm Plossl** would be nearly identical to that for the **9.7mm** listed; a 5mm of another brand would fit nicely somewhere in between the 6.7mm and the 4.7mm version. This is assuming, of course, that the eyepiece design is similar to that used in this reference.

Virtually EVERY focal length, from 2mm (can you imagine the eye relief on that one!) to nearly 60mm super wide angle versions are available on nearly every page of popular astronomy magazines. What do you buy? How do you know if you are getting the best for the dollar? DON'T GUESS. ASK...... Or in this case "read."

Popular web sites such the many telescope forums and discussion group sites are excellent platforms for user feedback on particular types and brands of eyepieces. Star parties or astronomy club observing sessions are an excellent source of information and first-hand opportunities to actually <u>test</u> the performance of particular eyepieces that you might be considering. NEVER BUY AN EYEPIECE THAT YOU HAVE NOT HEARD OF, and/or one that is not recommend by someone you trust. There are some "generic" brand eyepieces out there on the market which I have tested and are excellent at about half the price of others more visibly advertised. But by the same token, I have learned to avoid brands that suddenly surface and appear to be assembled out of someone's garage.

As a general rule for beginners and those who enjoy CASUAL stargazing with the fine APO refractors available today, the Maksutov and the Schmidt-Cassegrain telescopes, the ORTHOSCOPIC or PLOSSL designs are the very best and any variation on that theme for your selection. Nearly all major brands now focus on the Plossl design, with "wide field" and "super-dooper wide field" versions available in some focal lengths.

NOTE that from my own individual testing I have found that the optical figure of large format (i.e., two-inch) eyepieces is significantly better than that of smaller eyepieces; the reasoning is logical: it is much easier to accurately figure to perfection a large glass surface than a smaller one. But my rule is still simple: you need ONLY ONE two-inch eyepiece – your very lowest power affording the widest field of view. That will be your favorite eyepiece of all and get the most use.

But remember this rule: the more pieces of glass in the eyepiece....the less light your eye gets. Sometimes simpler is better.

DOC CLAY'S EYEPIECE SELECTION AND PRACTICAL USE GUIDE –

Following are individual breakdowns on each of the various telescope sizes eyepiece selection and use. USE THE FOLLOWING INDEX CODE to select and eyepiece for observation or for possible purchase for a particular purpose. You will see these reference codes under the appropriate section for your telescope. If you DO NOT see a code by a particular eyepiece listed in your section, then this means that the EYEPIECE has no suitable application, or that it merely duplicates the performance of another one already listed.

NOTE: Asterisk (" * ") BEFORE A CODE indicates that this selection is recommended when only coupled with a good 2x Barlow lens, and in many cases is preferred over a straight eyepiece with no Barlow!

CODES TO IDENTIFY PROPER EYEPIECES FOR YOUR OBSERVING NEEDS

WIDE FIELD
w1 - lowest possible magnification
w2 - widest practical field @ low magnification
w3 - BEST RECOMMENDED

DEEP SKY

d2 - medium power to enhance object contrast against sky
d3 - higher power for smaller objects (i.e., planetary nebulae)

MOON
mw - allows the entire moon in field
m1 - excellent for low power views
m2 - scan the terminator shadows with this
m3 - ideal for highest power viewing

PLANETS
pw - aesthetic views, planet and its satellites, star fields
p1 - medium power, ideal for tracking Jupiter's moons
p2 - highest power on average steady nights
p3 - highest practical power on the very best nights

DOUBLE STARS
x1 - medium power for most bright doubles within reach
x2 - max power to reach Dawe's limit on nights of very best seeing

COMETS
c1 - widest good field for tail, etc.
c2 - good medium - details in head, nucleus

SUN
sun1 - can get the entire solar disk in for sunspot counts
sun2 - ideal medium for sunspot/flare/granulation details

Simply find the proper code in the following section listed by the appropriate eyepiece (REMEMBER: even though the focal lengths for the eyepieces are given, ANY similarly designed eyepiece CLOSE to this focal length will provide the same or very similar results! REMEMBER!! If the CODE has an (" * ") by it, then it means that this eyepiece is selected for use WITH BARLOW!

IN THE FOLLOWING LIST: Eyepiece Size, by mm / MAGNIFICATION / Field of View ( ) for YOUR particular telescope / Recommendations in the above detailed CODE.

COMPACT TELESCOPES (60-80mm)
25mm MA - w2 / d1 / m1 /pw / c1
plossl eyepieces (4-element, excellent light transmission) EXCELLENT INVESTMENT:
56mm (2-inch) - n/a
40mm - n/a
32mm - n/a
26mm - 13.5X / (3.9 degrees) - w2, mw, c1
20mm - 17.5X / (3.0) - w3, d1
15mm - 23.4X / (2.3) – pw
12.4mm - 28.3X / (1.9) - m1
9.7mm - 36.1X / (1.5) - d3, p3*, sun1
6.4mm - 54.7X / (57' arc) -m3, p1, x1*, sun2, c2

super wide angle plossl eyepieces (6 elements - moderate light transmission) EXPENSIVE:
13.8mm - 25.4X / (2.7 degree) (not recommended)
18.0mm - 19.5X / (3.5) (not recommended)
24.5mm - 14.3C / (4.6) (not recommended)
32mm - (2 inch) - n/a
40mm - (2 inch) - n/a

ultra wide angle plossls (8 elements, less light transmission) VERY EXPENSIVE:
14mm (1-1/4" / 2-inch) - not recommended
8.8mm (1-1/4" / 2-inch) - not recommended
6.7mm (1-1/4") - 52.3X / (1.7 degree) - d3, m2, p2, c2, sun2
4.7mm (1-1/4") - 74.5X / (1.2 degree) - m3*, p3*
DOC'S 3-in-the-box Selection: 25mm, 10mm, Barlow

IN THE FOLLOWING LIST: Eyepiece Size, by mm / MAGNIFICATION / Field of View ( ) / Recommendations

MEDIUM-SMALL APERTURES (90-105mm)
plossl eyepieces (4-element, excellent light transmission) EXCELLENT INVESTMENT:
56mm (2-inch) - n/a
40mm - 31X / (1.5 degree) - w3
32mm - 39X / (1.4) - w1, mw, c1
26mm - 48X / (1.1) - w2, m1, p1
20mm - 63X / (49' arc) - w3, d2, sun1, pw
15mm - 83X / (37') - c2
12.4mm - 101X / (30') - d3, sun2
10mm - 129X / (24') - m2, p2
5mm - 195X (15') - m3, x1, p3*

super wide angle eyepieces (6+ elements - moderate light transmission) EXPENSIVE:
19mm - 91X / (44' arc) - d2, c2, mw, sun1
22.0mm - 69X / (57') - w3, m1, p1
27mm - 51X / (1.4 degrees) - w2, d1
35mm - (2 inch) - n/a
40mm - (2 inch) - n/a

ultra wide angle (8 elements, less light transmission) VERY EXPENSIVE: Highly recommend Televue *Naglers* and *Panoptics*
15mm (1-1/4" / 2-inch) - not recommended
10mm (1-1/4" / 2-inch) - not recommended
7mm (1-1/4") - 187X / (27' arc) - x2*, p3*
5mm (1-1/4") - 266X / (18') - p2, x2
Doc's 3-in-the-box Selection: 32mm, 15mm, (or 13.8mm) 7mm, Barlow

IN THE FOLLOWING LIST: Eyepiece Size, by mm / MAGNIFICATION / Field of View ( )/ Recommendations

MODERATE APERTURE (125-180mm)
plossl eyepieces (4-element, excellent light transmission) EXCELLENT INVESTMENT:
56mm (2-inch) - n/a
40mm - 48X / (55' arc) - w1, d1, c1, mw
32mm - 59X / (52') - w2
26mm - 73X / (42') - m1, sun1, pw, w3
20mm - 95X / (32') - d2, p1
15mm - 127X / (24') - d3, m2, p1, c2, sun2
12mm - 153X / (20') - m2, x1
10mm - 196X / (15') - p3*, x2
6mm - 297X / (10') - p2, m3

super wide angle plossl eyepieces (6 elements - moderate light transmission) EXPENSIVE: NOTE: NO advantage to 2" size in this aperture range!
13.8mm - 138X / (29' arc) - d3, m2, p1, c2, sun2
18.0mm - 105X / (38') - p2, x1

24.5mm - 78X / (51') - m1, d2, p1, sun1
32mm - (2 inch) - n/a
35mm - (2 inch) - n/a

<u>ultra wide angle</u> (8 elements, less light transmission) VERY EXPENSIVE: - Highly recommend Televue *Naglers* and *Panoptics*
15mm (1-1/4" / 2-inch) - not recommended
9mm (1-1/4" / 2-inch) - not recommended
6mm (1-1/4") - 284X / (17' arc) - p1, p2*, x1, x2*, m3*
4mm (1-1/4") - 404X / (12') - p2, x2, m3
Doc's 3-in-the-box Selection: 32mm, 15mm (or 13.8mm), 10mm, Barlow

IN THE FOLLOWING LIST: Eyepiece Size, by mm / MAGNIFICATION / Field of View / Recommendations

LARGER APERTURES - 200mm and up
<u>Plossl / wide field eyepieces</u> (4-plus-element, excellent light transmission) EXCELLENT INVESTMENT: NOTE: *you only need ONE 2-inch eyepiece....LOW power.*
56mm (2-inch) - 36X / 1.1 degree - w1
40mm - 50X / 52' arc - w2, d1, m1, pw, c1
32mm - 63X / 49' - w3, p1, sun1, mw (Televue Plossl)
26mm - 77X / 40' - d2
20mm - 100X / 31' - m2, sun2
15mm - 133X / 23' - x1, c2, d3
12.4mm - 161X / 19' - p2*
9.7mm - 206X / 15' - x2*
6.4mm - 313X / 9' - p2, x2

<u>super wide angle</u> Plossl/other eyepieces (6 elements - moderate light transmission) EXPENSIVE: Highly recommend Televue *Naglers* and *Panoptics* and *Explore Sci.*
15mm - 145X / 27' - d3, m2, c2, sun2
18.0mm - 111X / (36') - d2, mw
25mm - 82X / (49') –
35mm - (2 inch) - 63X / (1.5 degree) - w3, c1, m1, pw
40mm - (2 inch) - 50x / (1.7 degree) –

<u>ultra wide angle</u> wide field (8 elements, less light transmission) VERY EXPENSIVE: Highly recommend Televue *Naglers* and *Panoptics* and *Explore Scientific*
18mm (1-1/4" / 2-inch) - 143X / (35' arc) –
9mm (1-1/4" / 2-inch) - 227X / (22') - x2, p2
7mm (1-1/4") - 299X / (16') - p3*, m3*, x2*

CLAY'S 3-in-the-box Selection: 32mm (SWA), 15mm (or 14mm/13.8mm), 6.4mm (or 6.7mmUWA), Barlow

REGARDING 2-INCH EYEPIECES

Some of the finest eyepieces you will ever use are made in the 2-inch barrel format. These include the incredible *Televue* Naglers, Panoptics, Ethos as well as the comparable *Explore Scientific* eyepieces.... all cutting edge optical technology. These provide the ultimate in viewing pleasure, like a porthole to space, but do so with a very high price tag. It is very important to note that there is absolutely NO advantage to the 2-inch design EXCEPT for your very lowest power providing he widest field of view. There is NO advantage to a high power 2-inch eyepiece.

Thus, when selecting your three- or four-eyepiece set for your telescope, put your MOST MONEY in the lowest power; that is the magnification and eyepiece that you will be using the most and the one that will provide the most exciting and rewarding views. If your telescope can accept a 2-inch eyepiece (not all can), then by all means invest in ONE low power wide field top-quality eyepiece. You will be thanking me every time you scan the Milky Way or exploring the Andromeda Galaxy.

Enjoy....and remember: RESIST....investing in eyepieces is addictive and there is no intervention group as yet.

* * *

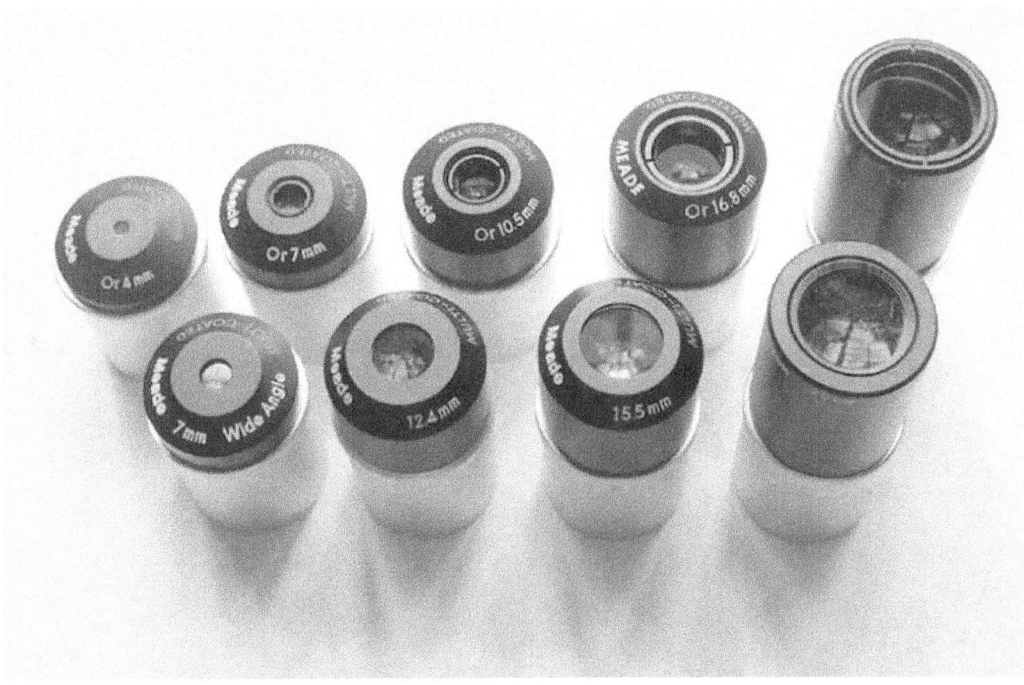

Chapter 10

Telescope Accessories: Choosing and Spending Wisely

There is not room in a hundred books for all the discussions that need to be addressed concerning ancillary equipment for your telescope, your observatory and your observing projects.

Astronomy, frankly, has become a very expensive "hobby" for many. Indeed the study of the night sky has evolved for tens of thousands of non-professional astronomers from a casual interest early in life to the pursuit of professional-quality astronomical research in the field.

The difference between the advanced "amateur" astronomer and the profession is quite simple: one gets paid for his or her work and the other does not.

Likewise as interests in applied study of astronomy rather than just looking around grows, so must the influx of equipment to support the study that one is interested in: there are specialized cameras, dew prevention equipment, computers and programs, specialized computer-driven focusers, wireless communication equipment, photo-electric photometers and spectrographs all of which are purchased out of the pockets of non-professional astronomer.

This is not to mention that most such scientists also work a "day job" and support all of this out of their own income, and still must support a daily life and typically a family as well.

Trust me when I back this up with nearly 60 years of experience: you can really get carried away with telescope and astronomy accessories....."do-dads" as I call them. It becomes and obsession of sorts, much like being an amateur photographer in pursuit of the perfect camera, or – more applicable – a gambler with a betting addiction.

So where do we draw the line on purchases? What is needed and what is not? What are the perfect items to match with each individual's observing pursuits? We have already discussed eyepieces. Eyepieces are the number one obsession with amateur astronomers. Every manufacturer touts theirs as the very best; you can see wider, farther and more clearly with each new innovation introduced in terms of expensive eyepieces.
Whereas eyepieces (oculars) once could be purchased for around $19 for a quality one, they are now well over one thousand dollars for advanced wide field designs. Is all this really necessary?

Now, that being said, I will digress slightly and tell you a story – a test that I conducted – about quality modern age eyepieces. In this story I will name names....nothing wrong with that: I want to make sure that YOUR money goes into the right places.

On a lark in late 2016 I decided to "go retro" and take some of the most expensive and high-end eyepieces from the 1980's and 1990's and compare them to what is offered today as the best of the best.

In my arsenal of eyepieces were the quality Brandon Orthoscopics and the early Televue Plossl eyepieces from that era, as well as a few Meade Instruments Research Grade Orthoscopic and Plossl eyepieces. At the time, these were the premium standard for all visual observing.

Mind you that my days of visual observing have been over for two decades, except for the entertainment of school groups, family and friends. Thus it was an ideal time for me objectively "look back" and just how far eyepiece design has come and to determine if the newer, far more expensive eyepieces are really worth the premium prices of today.

My older eyepieces include the Televue 55mm Wide Field Plossl (the best and widest field of its time), the Meade RG Series and a lot of wide angle Plossl eyepieces that were revolutionary for the flat, wide fields they produced some 30 years ago.

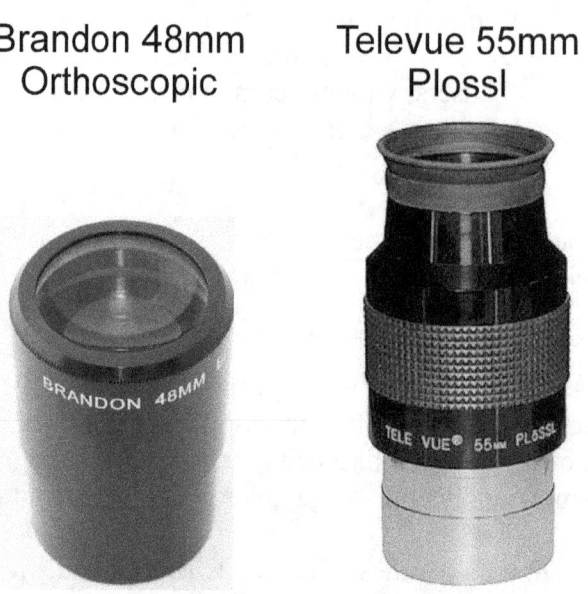

Two older eyepiece designs (wide field) both 2-inch barrels
Courtesy Brandon / Televue

One thing quickly noticed about most of the newer high apparent field eyepieces is their size and weight – about that of a traditional hand grenade, and will do about as much as one of those on your wallet. They are large and they are heavy, but boy do they work as advertised.

Seeing is believing – let me see if I can explain the advancement in optical design without illustrations and over-embellishing the descriptions. The quality of all of the eyepieces that I have is the best that can be. Some of these are 2-inch barrel designs (see my discussion on eyepieces in a previous chapter) and some are the standard 1-1/4 inch barrel for standard focusers or back on telescopes. Both older and modern wide field eyepieces come in both sizes, and in fact, the 1-1/4 inch high apparent field design of both Televue and Explore Scientific have equally impressive fields of view....so size does not matter in this case.

**Explore Scientific
28mm 68-degree**

**Televue 82-degree
20mm Nagler**

**Two modern ultra-wide-field eyepieces
that are unequaled in performance
(both 2-inch barrels**
(Courtesy Explore Scientific / Televue)

As I have mentioned, your eyepiece arsenal needs only ONE two-inch barrel design in my opinion – a high quality wide apparent field of view LOW power one....the rest can be the slightly less expensive and far less heavy smaller design.

Comparing these four on such fields as the Orion Nebula (for contrast) and The Beehive Cluster, Messier 44, (for star quality and limiting magnitude) I was in for a shocking reminder of just how far we have come in terms of optical technology.

Both the Naglers and the Explore Scientific eyepieces presented what appeared to be a porthole into space. Not like "looking at an image of" an object, but rather like literally looking out of a window of a spaceship at the objects. Sharp, clear and bright all the way to the edges, these eyepieces will "Wow" anyone no matter how skeptical.

Older design eyepieces were quite different – a difference that I actually had forgotten. While the two deep sky objects were bright and crisp with excellent star images across, the image to the brain/eye system was entirely different.....I was looking "down a long tube" and at the end of that tube was the Orion Nebula or the Beehive. A very striking contrast to the newer designs.

Enough about eyepieces. This discussion is a mere example of accessory choices.

Finderscopes and Locating Celestial Objects

Only a few words of wisdom regarding finder telescopes and finder mechanism (such as the red dot/reflex type of finders):

A finderscope on your telescope is just exactly what its name implies: a finder to get you started. If you could look down a long piece of darkened conduit with no optics and aligned to your main telescope to sight a bright star, that would actually suffice as a finderscope. There is far too many emphasis put on replacing the standard finders that come with commercial telescopes with "bigger and better" ones. Sure, these look great at star parties and up against someone that has the lowly stock finder, but how many times do you actually use it?

With modern GO TO telescopes a finder simply gets you pointed to your first alignment star. Once done, it is out of the equation. How big, how powerful and how technologically advanced does this device need to be to get that job done?
On the other hand a big Dobsonian telescope that is a "push-to" telescope does need a finder from time to time.

The standard fare for finders is either the 6 x 30mm or the 8 x 50mm finders. The latter is plenty large enough to actually show many of the brighter deep sky objects and certainly suitable for finding stars.

I have found that the pathetic 8 x 24mm finders on very small telescopes, mostly small catadioptic beginner telescopes, are fairly useless. The straight-through are virtually impossible to use because you cannot position your head against the telescope enough to actually even look through the finder; upgrading to a right angle finder of the same size has some improvement.

Illuminated finders are nice, but not totally necessary....sometimes it is very difficult to see the crosshairs to get your object centered enough to be in the field of view of the larger telescope.

But here is a nifty and no-cost trick for you if you do not want to invest in an unnecessary upgrade to your telescope system: hold your hand loosely over the light end of your flashlight and slightly above the front lens of the finderscope; this will illuminate the area coming into the finder and thus creating a glow against which the crosshairs are easily seem. Center your object and turn off the flashlight....now wasn't that simple?

Red dot type finders are very popular, lightweight and look very impressive on your tricked-out telescope. However, they are less effective in terms of accuracy and provide no magnification whatsoever. Thus the slight offset of parallax between the projected red dot and the actual field of view of your telescope is enough to render these devices – in my opinion – ineffective as a finding device.

Condensation, Dew Shields and Dew Prevention

As with every other aspect of your observing experience, there is an overwhelming supply of devices to keep your telescope and equipment from dewing over during the night. Many a deep clear night has been rendered unusable because of the formation of dew or frost.

There are two ways to reduce the formation of dew:
1) dew shield(s)
2) dew heating strips

Both work to some degree, but neither are a complete failsafe system to prevent dew on a really moist night.

Dew Shields:
I like the idea of using dew shields, and my suggestion is to ALWAYS use a good, hard construction, dew shield at all times, not just when you might expect dew to form. Outside of dew protection (remember, a dew shield does NOT prevent dew….it only protects optics from it, and typically does a poor job of that as well), there are two excellent reasons why I recommend using a dew shield:

1) the dew shield adds front weight to the telescope, thereby offsetting the addition of heavy eyepieces and equipment to the tail end of the telescope, thereby reducing the need for tube counterweights when using cameras and large eyepieces. NOTE that any time you add a dew shield you must always rebalance your telescope…if you do not, then do not expect your telescope to perform properly.
2) a dew shield's biggest asset is protection from dust, pollen and blowing debris…even insect poop. On nights when the pollen is particularly bad or the bugs are flying around, the shield will act as a barrier between your expensive optics and the outside world.

Outside of those two positives, a dew shield also looks really nice on your telescope at star parties and will ALWAYS assist in preventing curious hands from leaving fingerprints on the optics of your telescope.

In terms of dew control, they actually do very little…dew forms, not settles. It forms on surfaces that are at least one degree below the dew point and it does not matter if there is a dew shield out there or not…if the air is saturated and the temperature drops below the dew point, a dew shield affords very little protection.

The dew shield, to be at least partially effective, must be at least 1.5x the diameter of your telescope's optical tube assemble; if your 8-inch Schmidt-Cass has a tube that is nine inches, the dew shield must be at least 13.5 inches in length for protection.
All dew shields must be painted non-reflective black or have black flocking installed.
Dew Prevention Systems:

Much better solution to the dew problem is a complete dew prevention system, a series of wrap-around dew "strips" that you place on the optics-end of your telescope typically by means of Velcro. Small wires from these straps run down along your telescope and into a control box, or control system in some cases, that regulates the amount of heat that the straps will give off.

A simple dew strap with wire lead that goes into the control box. Courtesy Kendrick

There are many systems offered on the market and most work effectively; recommended for consideration are (in no particular order): *Kendrick, DewNot, AstroZap, Dew Buster, Thousand Oaks,* and many private labeled through telescope distributors. They can be as simple – or as complex – as you want. But my emphasis here is to not over-buy; always get exactly what you have determined you need and no more.

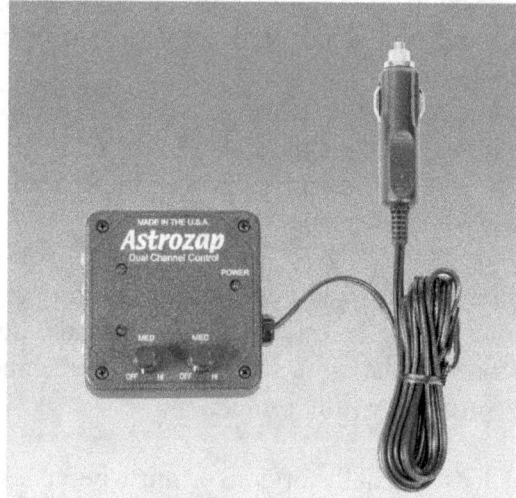

The dew system control box from which multiple strips can be routed. Courtesy AstroZap

Although effective, there are some drawbacks to using these systems and there are some tips to maintain the control without overheating the telescope components. the dew control system typically has multiple outlets from the control box that allow you to run wires to up to six optical devices to maintain warmth. The idea is to keep each component protected at a temperature just above the dew point. You can essentially supply dew-preventing warmth to: telescope, eyepiece, finderscope, guidescope, hand controller, focuser….anything you can wrap a Velcro fastened strap to. With all these wires routed – and many people do protect everything – there becomes a traffic jam of wires that must be carefully placed and routed to prevent tangles and dangerous binding of the telescope when in slewing motion. Many a telescope drive has been ruined by being stopped dead in slews because of wires entangled.

Many users of these devices use them incorrectly….NEVER wait until dew forms and then kick up the power to the dew strips to full blast

Routing the wires: get all the wires assembled just as you are going to use the telescope; color code them so that you can quickly assemble and route every time you set up your telescope. Once selected; temporarily bundle the wires leading from the dew control system into one group and hold together using wire ties. Once done, move your telescope to ALL sky positions manually to determine if the telescope will move freely to all areas and move so unimpeded by any wires that might get tangled. Make adjustments if necessary (it will be) and when you have them properly routed around the telescope hardware, use permanent wire zips and ATTACH them along safe places on the telescope and mount so that the final wire bundle moves WITH the telescope the same way with every slew.

Maintaining temperature control: there is only one effective way to use a dew heating system – turn it on and leave it alone. NEVER WAIT until you see dew! At that point it is probably too late anyway because on large optics some time is required to gently warm the glass to eliminate dew. The idea is to PREVENT dew, not eliminate it. Therefore take my advice: when you get started for the night and ready to observe, turn ON the dew system then….do not wait until you think the air is becoming moist. Turn your dew system on LOW and leave it there. Typically doing this is perfect and works nearly every time; if necessary as the night progresses, you might turn up one notch….but never all the way. The gentle heat from the padded dew strips will ultimately give off heat currents which can actually be seen as distortion in the views that your telescope provides. Keeping the temperature to the lowest possible is always best.

IMPORTANT: Concerning Condensation on Your Telescope

There are two ways that you can end up with damaging moisture condensation on your telescope, both optics and equipment. Remember that near urban areas much of the condensation in the form of dew or water vapor ALSO contains industrial pollutants that come along for the ride: corrosive acids and bleaching alkalines. Each can eat the coatings right off the from glass and the anodizing off of your fine equipment. Please read carefully to assist you in preventing excessive moisture both INSIDE and outside your telescope.

Dew control while at the telescope has been discussed above, but that is only the observing session side of the problem. Bringing in your telescope and storing it after a damp night can be very destructive if not done properly as can not properly using accessories at the telescope.

Condensation while moving the telescope: This occurs when taking the telescope from one temperature extreme to another – moving from COLD conditions into WARM conditions. Condensation can form instantly, for example, if you move your telescope out of the cold winter night air into your warm house after your observing is finished. Instantly you will have a telescope covered in water. The same is true in summer when moving the telescope out of a very cool air-conditioned house into the hot back yard….moisture is going to form quickly. MILDEW is the number one enemy.

Follow these suggestions to minimize condensation, particularly when bringing in the telescope:

1) Examine from front glass, or all optics of an open tube design, for moisture using a flashlight. If you see fog/dew on the glass, then DO NOT CAP the telescope with its dust cover…leave it open and cover with a cotton pillowcase that will breathe.
2) Make absolutely sure with any closed tube telescope (refractor or catadioptic) that the back end (focusing end) is completely blocked with a plug or eyepiece…if not and you bring it in to warm conditions, moisture will rapidly enter the OTA and condense on the inside of the optics, something that will lead to mildew within only a few weeks.
3) The mechanical portions of the telescope and exposed surfaces will likely be covered in condensation and wet. Use a terrycloth towel to gently wipe off all NON-OPTICAL surfaces until dry; you can later buff with an old soft cloth if necessary.
4) NEVER wipe the optics when damp….let them air dry and then clean as described in this book. Never, ever wipe your optics, whether it is eyepieces or your main telescope optics.
5) Do NOT open the plug on the rear of the telescope for at least 8 hours to allow the system to acclimate to the surrounding air.
6) If you covered the front optics with the cotton pillowcase, leave that in place for at least 12 hours. Any droplets that you see can be "wicked" dry using the point of a Kleenex, but do not wipe nor rub at all.

Condensation while at the telescope: This is going to happen ultimately and you must follow some similar rules for the sake of the optics of your telescope:

1) Examine the front glass with a flashlight; if you see dew formed and have access to a hair dryer, then gently on "warm" setting, warm up the glass until the dew disappears.

2) If you do not have access to a dryer, the immediately cover the front glass with a cotton pillowcase and NOT the dust cover; do not remove until the entire telescope has warmed.
3) While using the telescope NEVER leave the rear opening without a plug or eyepiece for more than 8-10 seconds maximum; doing so will allow the damp night air to rapidly seep into your telescope; keep plugged at all times or keep accessories in place until acclimated.

Some Serious Equipment – Your Focuser

If you have focusing that is rough, the image shifts with every tweak of precise focuser, or you simply cannot focus to precision without shaking the telescope, then you have a sub-par focuser. Frankly, nearly every commercial telescope made, whether advertised with "deluxe dual-speed micro-focusing" or "internal Crayford no-shift focusing", will be soon discovered to have poor focusing ability and some degree of image shift.

In many cases the focuser is incapable of supporting the load that the observer wishes to add to the focusing end of the telescope. Heavy eyepieces and large diagonals, off-axis guiders, cameras and filter wheels – these are heavy items and typically a standard rack-and-pinion focuser is simply not going to be stable enough to support this load and maintain focus. The internal focusing systems such as those supplied with advanced RC designed telescopes and modern catadioptic instruments are excellent in this regard, since there is no weight whatsoever outside of the focusing hardware.

Without hesitation I can say that the single-most important upgrade that you can make to your existing telescope system is a quality Crayford-style focuser. They are manufactured to fit virtually any brand and any type of telescope design, from Dobsonians to advanced RC's and worth every penny of investment – provided that you choose the right one.

Like every accessory, there are robust-looking focuser for aftermarket installation throughout the world wide web. Please be careful in your choice; many of these fall apart within weeks of use, the fine adjustments do not work as you would expect and they fall short of expectations to say the least.

There are several quality USA manufacturers of robust and precise focusers who provide the quality that is needed for years of your output at the telescope. Do not waste your money trying to save money.

An upgrade Moonlite focuser which allows the ultimate in precision and accuracy. These focusers are robust and beautiful, capable of supporting the weight of heavy equipment. The Moonlite is the finest aftermarket focuser available
Photo courtesy Moonlite Telescope Accessories

* * *

Choose one that is adaptable to your needs: they can be purchased with electric DEC motors, stepper motors, rotators, dual speed fine focus, adjustable weight load bearings and much more. Drawtube lengths for these focusers are carefully designed for the type of instrument that you have; for example, modern SCT telescopes will not allow long in-travel and thus the focusing distance for such must be limited; on the other hand, refractors have plenty of focusing distance and may need more focusing distance.

With these add-on focusers for the compound (catadioptic) telescopes, the limited focusing range is perfectly fine, since these popular telescope have internal coarse focusing that can get you quite close. For exacting focus to view or photograph however, you need the delicate touch of a dual speed with no image shift, or the precision of motorized focusing. Thus, you rough focus with the main telescope and fine tune to precise focus every time. **In terms of preference of dual speed manual over electric**: I highly recommend electric handbox or computer controlled every time.

Never consider a system without talking to others who are using them. Look at the used equipment listing for other aftermarket focusers and see which ones are typically for sale….those are the ones you want to avoid.

There are many high-end and expensive aftermarket focusers out there; I urge you to read much and choose wisely. You want the finest in performance and the most adaptable to all applications as your interests grow and change. In my tested opinion (Arkansas Sky Observatories uses seven of these) the best on the market and worth every penny are those made by Moonlite Telescope Accessories.

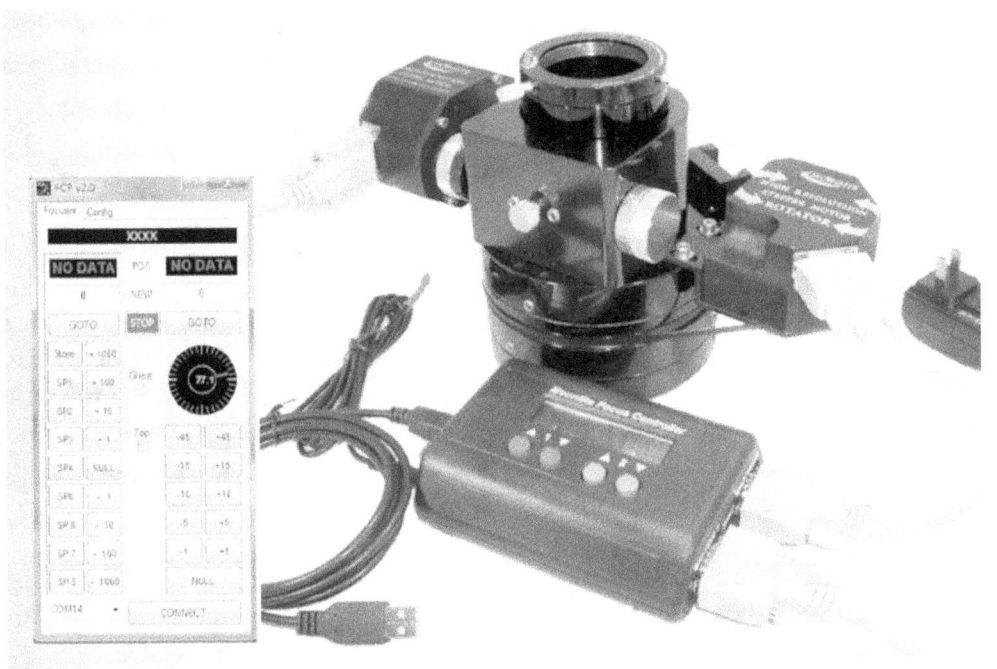

Note that the focusing system above is "not your grandpa's focuser." This unit has not only computer controlled (or handbox controlled as shown) and accurately repeatable precise focusing at a micro level, but also a built-in computerized image rotator by which the entire assembly rotates with a computer-integrate worm gear assembly for precise image orientation. Please consider that this focuser started out exactly as the one pictured on the previous page: a basic manual focuser. This is shown to demonstrate the diversity and adaptability of this Moonlite product: start with the basic and build as your proficiency grows! Your choice of many beautiful anodized colored finishes too.

Nearly all modern focusers are equipped to handle both two-inch and 1-1/4 inch standard eyepieces, with simple adapters for both as standard equipment.

Precision adapters are available through all aftermarket focuser suppliers which will couple their product exactly to your telescope, no matter what design or model.

Star Diagonals and Visual Backs

If you are going to use two-inch accessories (eyepieces, star diagonals, camera adapters), you are going to have a telescope that is capable of accepting these large accessories. Some smaller refractors and SCT/catadioptic telescopes are not capable of using anything but 1-1/4 inch accessories. My advice to you on those instruments is to locate the very finest in 1-1/4 inch accessories and not even worry about the larger offerings.

Star Diagonals – the days of old "prism diagonals" are almost over, but these cheap Chinese offerings are still supplied with smaller telescopes by manufacturers. One of the leading cause of poor images see at the telescope is the lack of quality of the star diagonal. Nearly everyone uses a diagonal to view the heavens visually and there is a great difference in cheap prism diagonals and excellently made first surface **mirror** diagonals.

Mirror diagonals are now available in both 1-1/4 inch and two-inch sizes; a minimal amount of searching on the Internet will reveal countless choices.

I have seen far too much worry and concern and too much money being spent on mirror diagonals advertised with "99%" reflection and transmission, and ultra reflective coatings. The hard fact is that almost ALL of the mirror diagonals made today are coming out of the same two optical houses in China and nearly all are made to the same specs. Those that are made to more exacting specifications are excellent, but in reality the human eye cannot see the subtle differences in the $300 price differential. And keep in mind that these accessories ARE going to be used visually….they are not intended for use with cameras or precise CCD instrumentation.

In fact, the best views are had by taking out the diagonal altogether and putting your eyepiece directly into the focuser and looking straight through the telescope, thereby eliminating yet another piece of glass. But that does become uncomfortable.

Obtaining a replacement mirror diagonal is a good idea if possible, because the improved over the glass prism is indeed evident. However, stop at the "dealer private label" range and do not over-invest in these devices; your eyes will not be able to benefit from the measured laboratory differences in specified reflectivity ranges.

And, remember that when using any diagonal, your image that you see is inverted and reversed. Make an attempt to always be able to determine North, South, East and West with the diagonal in place so that you can match your views to those of your star charts and computer monitor programs!

Visual Backs – if you decide to use two-inch accessories you are going to need a two-inch visual back or adapter. These are not as easy to come by as two-inch diagonals, but some of the larger dealerships private label these as such and there are several manufacturers – Televue and Astro-Physics among the leaders – that actually manufacture quality two-inch visual backs that will screw directly onto the threads of modern catadioptic telescopes, thereby replacing the smaller 1-1/4 inch visual backs altogether. They are expensive for a machined block of metal, but worth every dime if you choose to make the transition to the larger accessories.

The Astro-Physics screw-on adapter for conversion to 2-inch accessories
Photo courtesy Astro-Physics

Telescope Counterweights for Accessories
When you start adding more and more sophisticated accessories, most of which is heavy, to your telescope, you are creating more of an imbalance and thus the telescope performance is going to degrade both in GO TO accuracy and long term tracking for astrophotography.

It does not matter what type of telescope you own and operate….all of them must be balanced for good performance and for the longevity of the instrument.
Typically, adding accessories to the eyepiece end of the telescope results in the need to balance via one or both methods:

slide the optical tube, if possible, forward or backwards in the mounting rings or cradle as you would in any ring-mounted telescope such as a refractor; and

add weights to the front of the optical tube assembly (OTA) that will offset the heavy weight of the accessories on the eyepiece end of the telescope (that would be opposite for the use of a Newtonian reflector where the eyepiece end is at the front, or top, end of the telescope.

With most commercial telescopes today, outside German Equatorial mounts outfitted with clamping rings, sliding the optical tube forward and backwards is not an option. Even with some ring-mounted telescope where tube sliding can be done, this is not always a choice or a good practice when certain equipment is permanently installed. So the option of some type of TUBE WEIGHTS as they are called is a very good idea.

Please refer to our chapter on Telescope Balancing for a full discussion, but at first pretty much any metal object, such as a hammer head or a bag of BB's will be just fine to determine exactly what weight is necessary; getting an idea of weight requirements can be quick and temporary with wire ties, Velcro and clamps, but eventually this ugly arrangement will want to be replaced with something more attractive to you most expensive investment.

Meade sliding "2-D" Tube Counterweight Set
Courtesy Meade Instruments

Sliding tube weights are available from all dealers and come in various creative forms; some slide, some screw into various points and some can be mounted on rails and then semi-permanently fastened for security.

As noted in the chapter on Telescope Balancing, there is a problem that arises when using sliding counterweights, or any weights mounted directly to one side of the optical tube assembly: they may take care of imbalance of accessories on the tail end, but one side of the OTA is now out of balance with the other side, 180-degrees around the OTA; unless a heavy guide telescope or piggybacked camera is mounted on that opposite side, there is still going to be a balance problem with just one sent of weights mounted in such a way.

In some situations and applied uses, sliding counterweights can be added and solve imbalance perfectly the first time; the sliding weight set seen above is a "two-dimensional" weight set; some manufacturers – ADM and Losmandy in particular – offer excellent "3-D" weight sets that add a perpendicular offset of the added weight which can assist in perfection in balancing.

An excellent and useful accessory from ADM telescope accessories
Photo courtesy ADM

As seen above, this added dimension allows for the accessory tube counterweight to not only slide parallel along the OTA, but also to increase the "torque balance" by running the weight out along a threaded rod which can greatly assist in balance. In both the 2-D and 3-D models of this system, a rail is needed (supplied) that will perfectly match the length of the existing OTA and mount via screws already on the OTA.

WARNING:
when adding any screws into the OTA, make sure that the LENGTH of the bolt or screw is such that it does not go so far into the OTA as to touch any of the optics! Many telescopes have been badly damaged by inserting screws directly into the optical glass.

A third, and very versatile, option to weight use is to use small weights that are not only useful and easy, but also attractive for permanent use on the telescope; these weights, available through www.scopestuff.com are certainly far more desirable than a rusty hammerhead secured through a plastic wire zip onto the OTA. These weights are drilled through so that they can be stacked in succession for added weight or a threaded stud inserted for attaching to standard threads of most telescope accessory holes.

The photo on the following page, courtesy ScopeStuff, shows this innovative and convenient method of balancing perfectly, not just on the OTA but easily adapter to various mounting points on the fork or German equatorial mount as well.

Screw-in stackable accessory counterweights from ScopeStuff

*   *   *

Computers, Smart Devices, and Programs for your Telescope Control

Nearly all telescopes today are encoder-equipped and have high torque motors for GO TO slewing acquisition and tracking of celestial objects.

Some of these come standard with their own programmed hand controllers, such as Meade's *Autostar*, Celestron's *Nexstar* and Astro-Physics' *GTO*. These devices have the capability of on-board libraries in the hundreds of thousands of objects, with ultra-precise GO TO pointing accuracy.

No longer are setting circles even equipped on telescopes; the encoders and programming is far more accurate and convenient. Computerized telescopes make the novice stargazer look like the proficient professional to his neighbors the first night out of the box. No longer is a flashlight and good night vision necessary to read the fine details of the etched setting circles; no fumbling around with the locks and clutches of the telescope to release every time a new object awaits to be viewed. With the computer activated and the telescope properly aligned and initialized (it has to know exactly what time it is, where you are located, and how the telescope is pointing when it starts up), the modern telescope handbox displays your object of choice, gives information about it and allows you to command the telescope to GO TO that object.

When finished, you need not even look at the hardware of the telescope; you simply select another object for the next GO TO.

The diagram of my ETX and it controller is seen below, showing the comparison of data from the difficult-to-read setting circles compared to the vivid digital display of the

Autostar handbox. In addition, the use of computer control versus the "old way" is far faster and much more accurate!

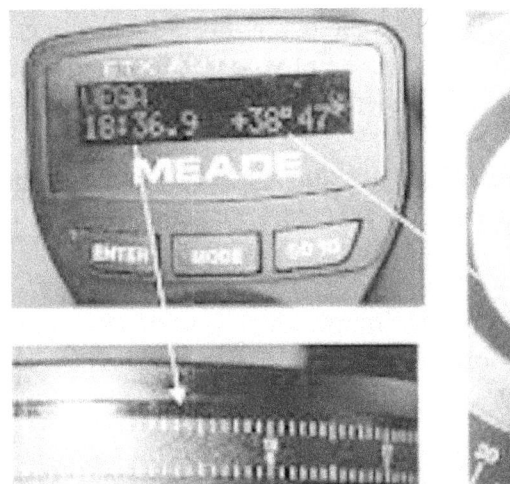
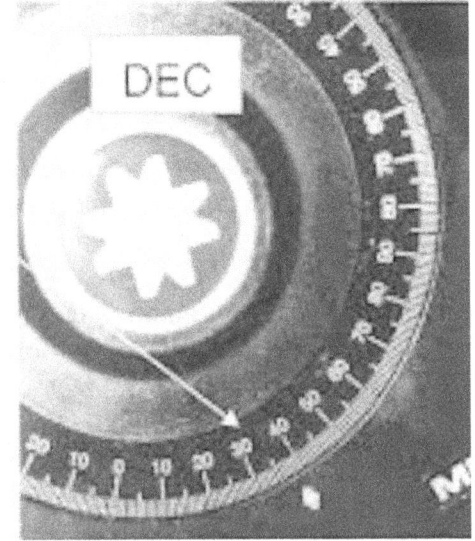

The next generation above the internal on-board computer control of any telescope is that of personal computer (PC), laptop computer or smart device control.

Computer and Laptop Control of Your Telescope: To operate your GO TO capable telescope by means of a computer, you will need the proper cables to connect the telescope to the computer. Or, if you want to be even more advanced, you might equip your telescope with a wireless "bud" such as a Bluetooth device for wireless reception at the computer, and no wires are needed.

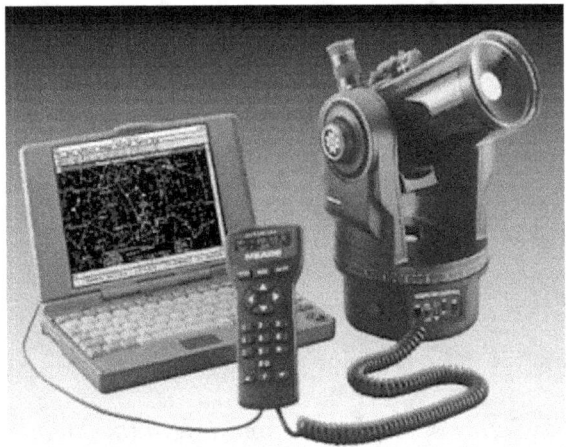

In either case, you have just advanced your telescope possibilities to the next level Your object database will expand as will your referencing ability at the telescope.

Using a wired connection requires that you are VERY careful to obtain the exact connecting hardware necessary for your telescope and your computer; typically a "serial converter" is required at the telescope end of the cabling and for this I recommend ONLY the BELKIN serial port converter; the cable will have a pair of similar connectors on each end, one for the telescope output and the other to plug into the adapter at the computer; check your telescope manual of on-line Help for the proper connectors and cabling. Many dealers have knowledgeable staff on hand who can assist in determining what is precisely needed with no guesswork....let them do this for you.

If you are operating your telescope permanently, a desktop PC is fine and preferred; such a computer will allow you to also connect your CCD camera, focusing control and other devices for control entirely from that one central command.

For portable use, of course a laptop is preferred. Similarly, a powerful laptop computer can control many devices all at once from one monitor out in the fields of your dark sky site.

Wireless Control of your Telescope: Modern technology certainly has not forgotten the telescope and the knowledgeable astronomers who use them.

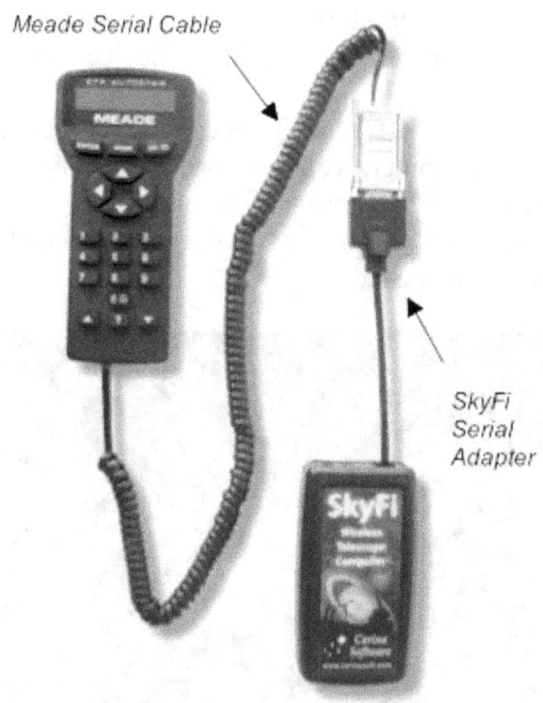

Showing SkySafari's wireless adaptation of the Meade Autostar

Not only can you control the hand controller of your telescope via wireless technology but you can also control your entire telescope wireless over your smart phone or pad without need for a handbox or computer. There's an App for that.

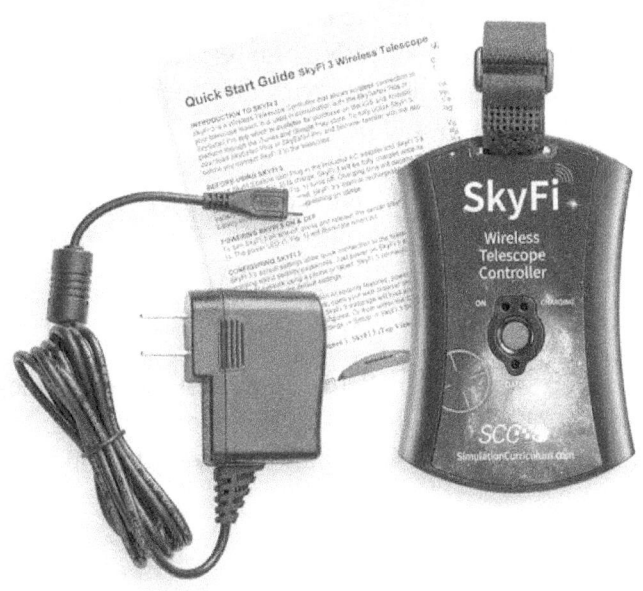

The SkySafari kit that allows effortless wireless control of your telescope; note that the wireless bud for your output on the telescope is needed and the DC adapter is not necessary when operating in the field. Courtesy SkySafari.

Although one of the most effective and creative devices every concocted for the modern age of social media and smart devices, the limitations of these wireless devices is in the database that you might choose through via Apps for your smart device. Various tested and acceptable telescopes, PC Sky Planetarium Programs as well as smart device Apps are discussed briefly following.

COMPUTERIZED TELESCOPES: I very much hesitate to start listing manufacturers of computerized GO TO telescopes….the list has new entries every year, and many drop out in the course of only a few short years. Nonetheless, I will acknowledge from my personal experience those who provide telescope systems with on-board computers for GO TO operation – those which I feel have the quality and reputation to support:

Astro-Physics GTO – absolutely the finest telescopes and mounts available; very high end, but clearly worth every cent. The highest recommendation I can possible give.

Celestron Nexstar – making fine telescopes for many decades and the pioneer in compact, portable instruments. Their NexStar GO TO system is fine and quite user-friendly although in my opinion not so much as the Meade product.

iOptron – primarily a supplier of moderately priced computerized telescope mounts (not complete systems), this company offers a very nice series of GO TO mounts with computer-adaptable keypad that is similar in many respects to that of the Celestron telescopes, although somewhat difficult in terms of viewing of display.

Orion Telescopes – around for many years providing privately-branded production telescopes, Orion has a large selection of reflectors, refractors and catadioptic designs, all with very nice GO TO mounts with keypad virtually identical to that of the Celestron Nexstar, with a database of nearly 43,000 objects.

Meade Instruments – in my opinion by far the most user-friendly of all computerized controllers, the Meade *Autostar/Audiostar* allows for personal input of user object data and has a database of more than 37,000 objects (Audiostar) to over 300,000 objects (Autostar II for advanced telescopes.

Planewave Instruments – a high-end maker of fine RC-type telescopes, utilizing several larger computerized mounts (buyer's choice) to include: Parallax, Mathis Instruments, Micron, Astro-Physics and others. All fully computerized, the operation will depend on the mount selected.

Vixen – a Japanese manufacturer of excellent quality, they offer a full line of smaller – but somewhat expensive – GO TO telescopes, utilizing the Sphinx mount and *Starbook* computer control technology.

Many other manufacturers of mounts with computer capability can be found by doing a computer search on this subject. However, I urge anyone considering a new telescope to attend a local star party and examine first hand the array of possibilities in these telescopes and talk to users about the pros and cons of each.

<u>SOME PLANETARIUM / SKY PROGRAMS:</u> Just like telescopes, there is an increasingly innovative list of suppliers of excellent computer control software and planetarium sky programs for telescope operation. When choosing, make sure that you get one that is very user-friendly, one that can allow input of TOURS and USER OBJECTS that you will want to personally archive in a sky library, and one that will function with your particular telescope.

The following examples, are my first choices, but there are many others which I unfortunately have not had the opportunity to test and assess. And – just like computers – new programs are always being introduced, and old standards are always being updated and improved!

Cartes du Ciel – this is a very popular FREE telescope control software and planetarium program which has far-reaching advanced applications, and the ability to incorporate current star catalogs such as UCAC4 and others. You cannot beat "free."

Deep Sky Planner by Knight Sky – used by many and has complete and accurate telescope control. Perhaps not as adaptable as some others, but a very good program.

GUIDE by Project Pluto – by far my preferred telescope and planetarium program for telescope operation. Intended mainly for advanced observers, this program is still suitable for all; the nice advantage of GUIDE is its ability to be updated very easily with User Objects, comets, asteroids, and new objects direct from links available on-line.

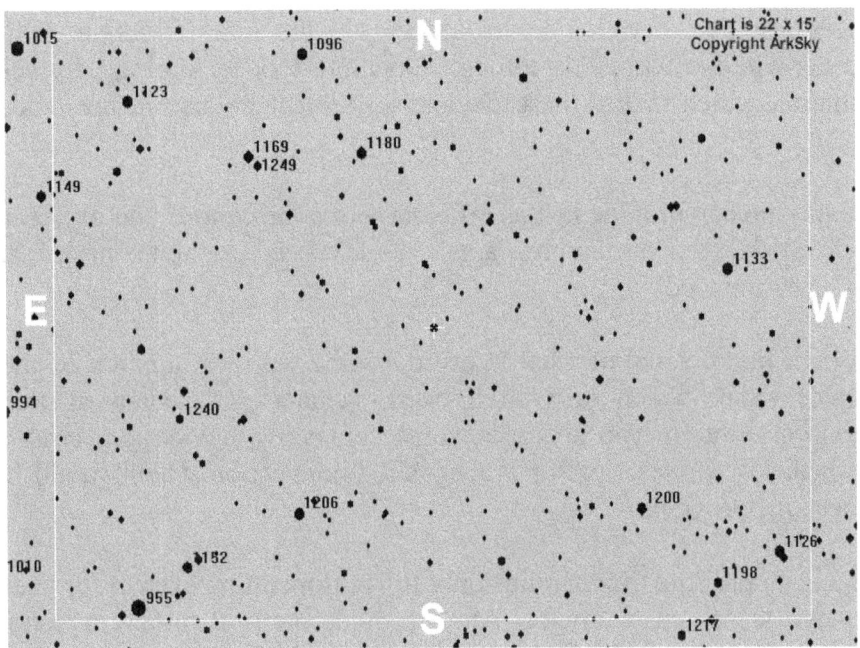

A screen shot of GUIDE display from ASO; in this example the background is grey with black stars

One very fine feature of GUIDE is its ability to have the extensive toolbar which allows your selection at a fingertip of color combinations for display, limiting magnitudes, GO TO objects and much more. No cute horizons, no constellation drawings….just incredible ease of use.

Starry Night – outside of research use at my observatory, my preference for teaching and sky demonstrations is Starry Night software, allowing full telescope control and easy input of current objects such as asteroids and comets. Many versions of this excellent program are available, from beginner/novice to Starry Night Pro for advanced observers. Beautiful realistic horizons an optional mythological constellation depictions.

Stellarium – this is a very popular and sometimes free software that allows both planetarium use and telescope control.

The Sky – this is a mainstay resource for many advanced observers, and this is required software for the Software Bisque line of computerized mounts. This is an excellent all-inclusive (allows astrometric and photometric measures and other applications), but is many times not as user friendly as other software packages.

APPS FOR SMART DEVICES: We are in a new generation of technology, pretty much ruled by social media and "smart devices" such as the iPhone, tablets, and pads. Nearly all of these can produce effective results in terms of telescope operation and display of any computer.

These devices can be connected to telescopes via the many creative APPS that are available, either wired or wirelessly through Bluetooth-type technology, for very efficient telescope control, which makes these devices so perfect for use in the field, or at star parties.

A good iPhone App can provide just as efficient telescope control and object information database as many PC planetarium programs. For advanced use and complete automation, they are not recommended.

Follow is a brief list of some of what I consider to be the most applicable and complete Apps available today which provide both sky charts with database and complete telescope control. Note that in all cases, some connective device is necessary for your telescope, whether a wireless bud or proper wired connection which would be obtained separately from the program.

You MUST obtain the App that pertains only to your operating system for your device as noted. All of these smart device Apps provide very detailed depictions of the constellations, and can be used with beautiful mythological graphic representations of the 88 constellations, or those can be turned off to show the sky in its natural appearance

Programs for Remote Operation of your Telescope: Following are the tried and true PC sky programs as well as a few Apps (the list of Apps is growing by each day).

For iPhone/Apple Devices:

   Sky Safari – by far the most advanced and my favorite of all smart device Apps; as with many PC programs, this one comes in several versions, but all have complete telescope control and fantastic database reference and sky maps.
   Starmap – also with several versions; this one has the capability of ASCOM output for accessory control.
   Luminous – honestly, this one has at present a limited user base, but is one of the highest rated sky Apps with complete telescope control among those who have tried it.
   Orion's StarSeek – also among the highest rated Apps for iPhone, this is distributed under private label by the major telescope manufacturer.

For Android, etc.:

   Sky Safari for Android - by far the most advanced and my favorite of all smart device Apps; as with many PC programs, this one comes in several versions, but all have complete telescope control and fantastic database reference and sky maps.

Virtuoso – a very fine and highly rated Android App that allows excellent sky charts, information and total telescope control.

Sky Portal – efficient and very clear, user-friendly; ratings are not as high on this one as they are for Sky Safari.

* * *

The eight-inch Meade LX90 computerized telescope
A perfect telescope for the advanced amateur
Courtesy Meade Instruments

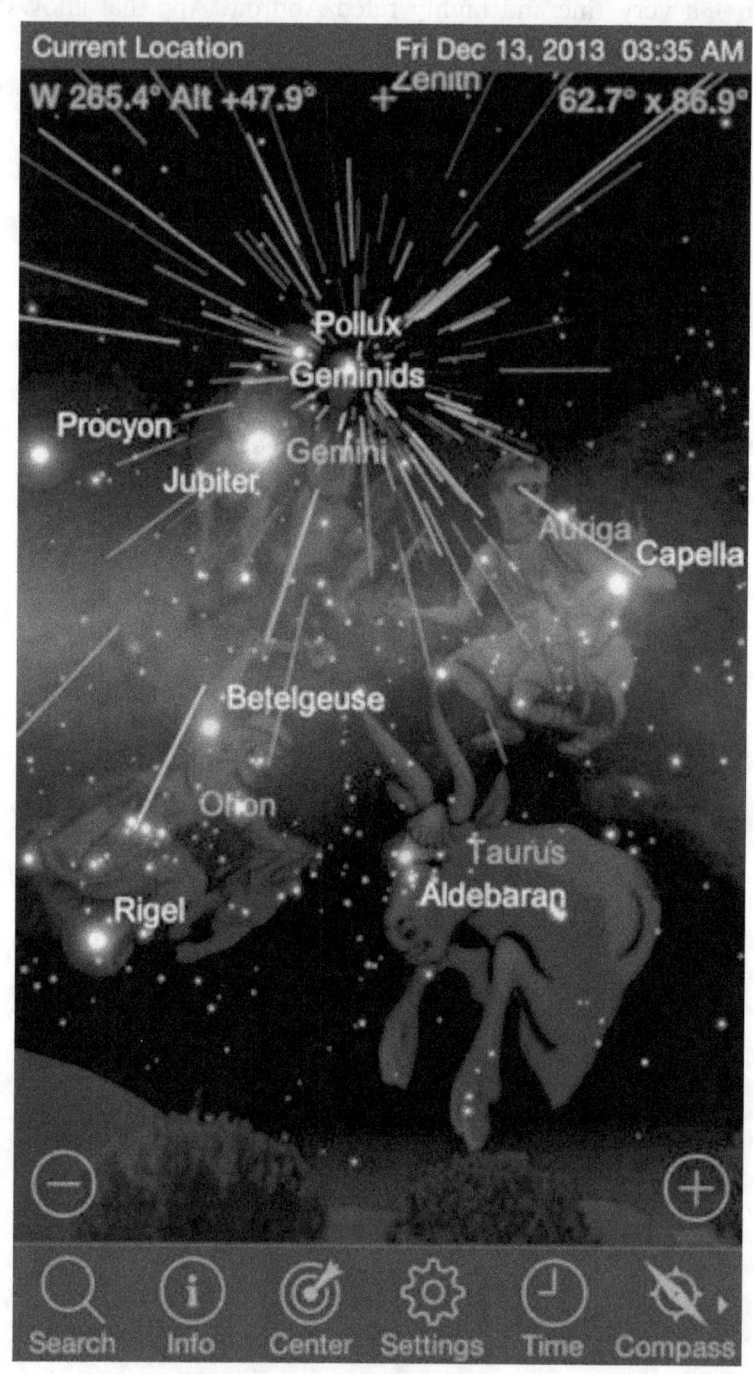

Screen shot from SkySafari Pro 4
Showing the constellations and the radiant of the Perseid Meteors
Courtesy SkySafari

Chapter 11

Limiting Magnitude on Your Telescope
The ASO Telescope Limiting Magnitude Determination Chart

The GO TO telescopes are incredible instruments, both in electronic innovation and in their optical performance. I have published a field test on this web site in which the results of some stringent parameters were tested with my ETX 125 5" Maksutov. Many amateurs who have not had the opportunity to REALLY use such a telescope may not understand that the optical design of these telescopes is pretty close to perfect and their performance reveals just that.

Lately, I have had a lot of correspondence from modern telescope users and other astronomers regarding "limiting magnitude" when using the telescope visually. Visual limiting magnitude MUST be differentiated from PHOTOGRAPHIC magnitude, in that the camera's film (and CCD imaging) has the ability to accumulate light, like a "sponge" slowly soaking up water.

In addition few amateur astronomers realize that there is a significant difference in their abilities to see certain COLOR stars, normally with very reddish stars appearing more difficult than blue or white ones to discern when very faint

The telescope's (or the eye's) ability to "reach" a limiting magnitude (say, 11.8 with a 3-inch or 12.8 with a 6 inch) is termed the visual "threshold," that point when the very faintest of stars.....sometimes can be seen....and sometimes can't! If you can, indeed, get even a momentary glimpse of that faint star you can honestly attest to the fact that the magnitude of THAT star is your limiting magnitude.

There is considerable discrepancy in the literature as to the limiting magnitude of ANY optical system, including the human eye. In a telescope it will be restricted by

1) the type of telescope, i.e., refractor, reflector, catadioptic;
2) the type of glass the image must pass through (including eyepieces);
3) the transparency of the dark skies in which the scope is used;
4) the visual acuity of the observer; and,
5) the quality of the optics, including coatings.

There are formulae available in all the books that I will not bore you with; from that formulae, I have prepared a MEAN value, an average of sorts, of all of them and offer the list below. My 32 years in astronomy has shown me that this list is, indeed, VERY close to actual performance.

Under the darkest conditions (see below)

```
HUMAN EYE - 6.5
   2.5"    10.5
   3.5"    11.4
   4.0"    11.7
   5.0"    12.8
   6.0"    13.2
   7.0"    13.6
   8.0"    13,9
   10"     14.4
   12"     15.0
   14"     15.6
```

and so on....conditions vary, scopes vary and observers vary. And so will a limiting magnitude from scope to scope. BUT NOT BY MORE THAN 0.4 magnitude under the identical conditions at the same instant with two observers of equal visual acuity.
Now, that being said, let's find out what we can see.

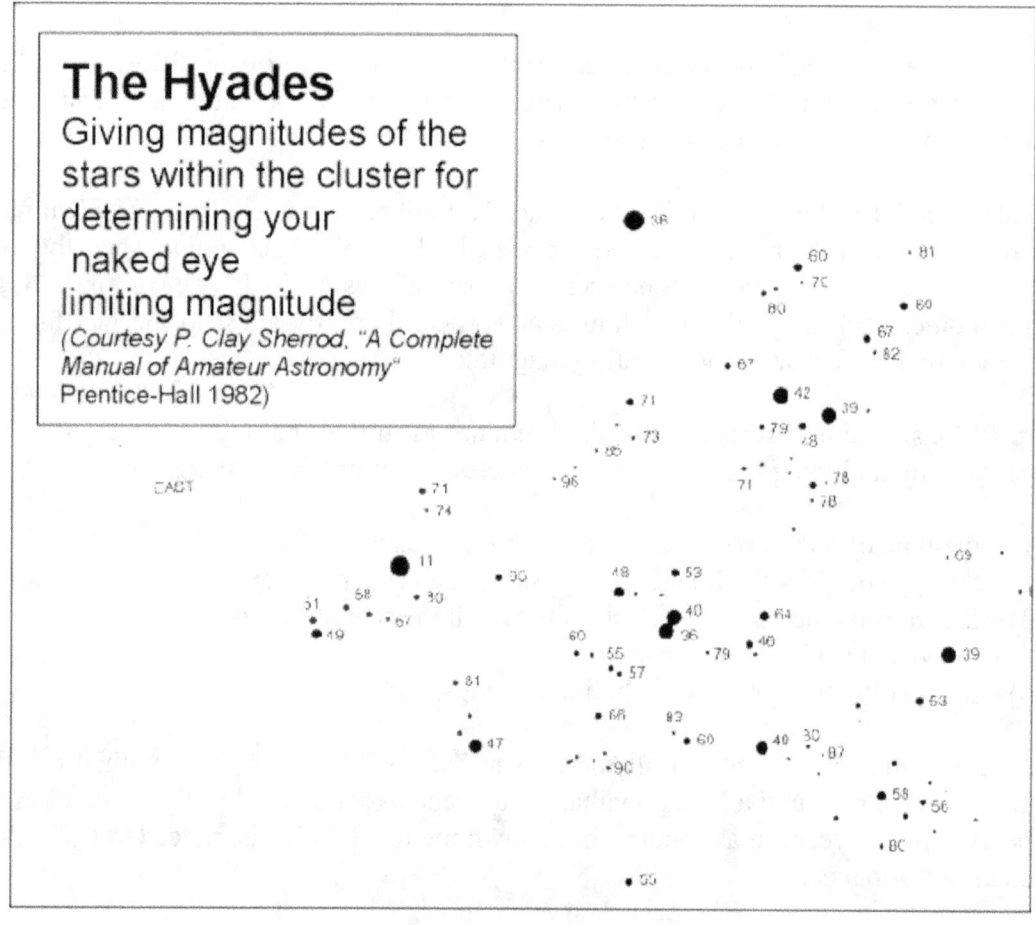

First you must determine how visually acute YOU are. I noted above that on dark skies, you might be able to glimpse a star magnitude 6.5. That's pushing it under perfect conditions and age is most definitely a factor. The Native Americans of North America used the Pleiades as an eye test as well as Alcor and Mizar in Ursa Major; in the latter, if the "rider" could be seen atop the "horse," the young man was deemed suitable to become a warrior if all other tests (no drug tests back then) proved likewise. Many times skilled observers can see to magnitude 7.1 with the naked eye on high desert dry mountains and from peaks such as Mauna Kea in Hawaii, high above the earth's vapor layer.

We, too, can use the "asterisms" of the Hyades and Pleiades in the constellations of Taurus to determine our visual (eye) and telescope's limiting magnitude.

1) The first chart is a star diagram of the Hyades. (from, "*A Complete Manual of Amateur Astronomy*," P. Clay Sherrod, 1982; permission by the author) giving the magnitudes of stars from bright Aldebaran to the very faintest. In addition to your eyes - try your BINOCULARS on this chart to determine their limiting magnitude as well.

With both charts, be sure to use a red flashlight to view or sketch on the chart to maintain your night vision; even when using the red light, you must wait about 2-3 minutes after illuminating the chart and then turning off the light to commence viewing to allow your eyes to adapt.

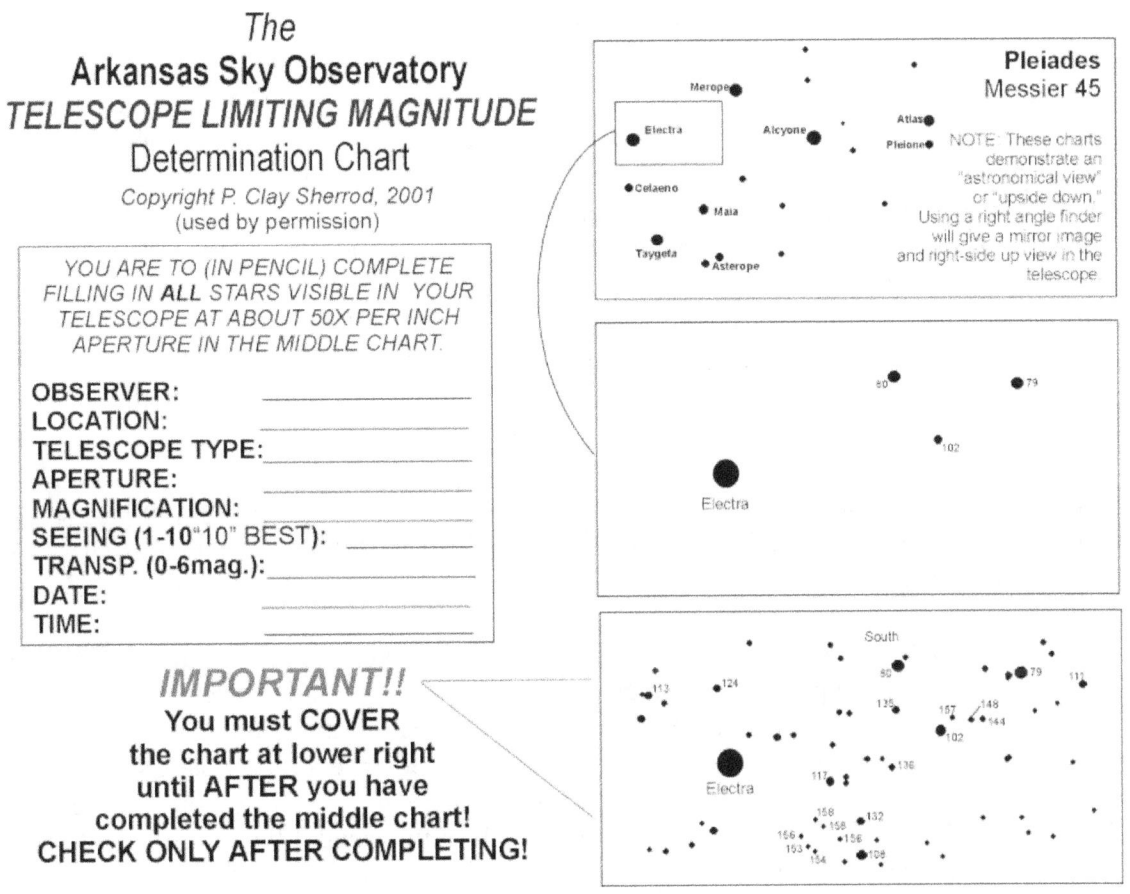

2) The second chart is a challenge. This is a detailed three-part chart of the Pleiades star cluster, centering on the star ELECTRA. The first Pleiades chart shows the familiar

pattern provided by the "8" stars of the "Seven Sisters." All of these should be visible to you on a dark night. Like the naked eye test ONLY try this on the darkest of all nights and ONLY when the clusters are nearly DIRECTLY OVERHEAD!

The second Pleiades chart is your target: locate *ELECTRA* and center it at high power (about 50x per inch aperture); identify the three stars that comprise the triangle just EAST of Electra; they should be visible in your telescope. This pattern including and surrounding Electra will be in your field of view at 150x.

NOW, without looking at the lower right chart (the one that gives you what stars are really there and their magnitudes) sketch in with a pencil EVERY STAR you see through your telescope; take your time....take some breaks. It should take at least one hour, and it's worth it.

When you're done, compare what YOU have drawn with what I have provided. The numbers given (the largest being the faintest stars) are without decimal points; if you have indicated a star of "135" and no larger number, that is your limiting magnitude!

There is a distinct advantage in using the Pleiades in that nearly all stars are "O" and "B" type whitish and bluish stars and more visually attainable; also, it eliminates any error that could be introduced if you were, say, "red sensitive."

IMPORTANT NOTE!! These charts were developed by me in 1976 for an asteroid identification program using large professional telescopes. Those telescopes always photographed and displayed "astronomically correct" images, that is they were "upside down" but correct right and left. Unfortunately, nearly all telescopes today utilize right-angle attachments which give an upright, but mirror-image view. Consequently, you will be required to orient yourself to the star pattern surround Electra....it is VERY important that you identify the small triangle composed of the "7.9, 8.0 and 10.2" magnitude stars in RELATION to Electra. Another alternative would be to LOOK AT THIS CHART IN A MIRROR, AND TURN IT UPSIDE DOWN. That would provide the correct orientation.

A third and best alternative would be to use a "visual back" that accepts eyepiece onto the rear cell of the telescope and view directly through the instrument, making your view exactly that of the charts.

If you have the terrestrial erect image prism for daytime viewing, you can use that and simply turn the chart upside down. However, this prism has optical elements that will REDUCE your limiting magnitude by as much as ONE-HALF magnitude!
Good luck, and report back here your results! If enough observers report back soon, I will extrapolate the curve for different telescopes and provide you with a sample. Not only is this project FUN, but it will educate you very well on the capabilities of your pride and joy - that wonderful friend, your telescope!

Chapter 12

SUMMERTIME.....WINTERTIME
And the Livin' Ain't So Easy for your Telescope!
Temperature Precautions for your Telescope

for many telescope users. Through this cold weather, perhaps you have learned a lot more than you wanted about the sensitive nature about your electronics, optics and mechanical attributes of your telescopes.

There is a lot we wish we did not have to know:

1) cold weather drastically affects the mechanical performance of telescopes because of the heavy green grease and lubricants that gum up in sub-freezing temperatures; the only solution to that problem is to "de-grease" ("Performance Enhancement Guide....Part 1");

2) the required 12V power during cold months is adversely affected in very cold weather, usually beginning below 27 degrees F and getting worse the colder it gets; after battery power gets too low, motor function and GO TO accuracy can become next to impossible;

3) All computerized scopes are most definitely temperature sensitive at the same cold range. It must be kept warm, and it must be supplied with no less that a constant 10V before your display begins to talk to you in "Martian" and the keypad commands to the telescope becomes inoperative.

But we're through all that right!? Summer is coming, and the "...livin' is easy." Easy perhaps until the sun bakes your control panel or until temperatures get so warm on your precious telescope awaiting nightfall on some August afternoon, that the very glues that hold your secondary baffle in place, or until your circuit boards no longer can "resist" nor "capacitate"....

Welcome to the perils of summer and hot temperatures. We are in for another bout of temperature-related telescope frustrations and failures. I want to alert all users to the dangers that can cause irreparable harm to their telescopes, the circuitry, and particularly to the computerized telescope – Much of these precautions are based on both experience and personal testing that I have just completed on standard computers, cameras and telescope control systems during the hottest of conditions

You are likely to be surprised...even if you are a seasoned "veteran" user of the telescope and computer control or other servo device.

YOUR TELESCOPE IN SUMMER HEAT - Part 1:
Optical Tube Assembly, Mount, Accessories
Following is a quick checklist of (P) potential heat-related **problems** with the telescope and mechanical equipment itself and (S) the **solution** to prevent the situation from

occurring; and/or (R) the "**remedy**" if that particular heat stress DOES result in down time or damage to your telescope.

P1 - Leaving the telescope in the hot sun uncovered.

S1 - Simply don't do it....it is never a good idea. The only solution to this (like setting up for a weekend star party where your scope must be left outdoors) is to cover the scope with a CANOPY, not a tarp. A tarp or even a sheet (NEVER COVER YOUR SCOPE WITH PLASTIC as this traps moisture which can ruin the scope and its circuitry) will still build up a significant amount of heat under it, particularly if tied at the bottom. Use a CANOPY instead, and THEN cover the telescope with a sheet beneath the canopy; since the canopy does NOT rest directly on the scope, air is allowed to circulate and continually cool in the shade of the suspended canopy

P2 - Leaving the telescope in the hot sun COVERED.
S2 - See above....NEVER DO THIS!

P3 - Internal Heat in the telescope OTA
S3 - A big problem taking out and setting up your scope early, particularly with the Maksutov design and only a bit less with the LX Schmidt-Cassegrain, is that a closed optical tube assembly (OTA) will build up the day's heat. This can cause a pro of significance:
1) Long cool down time to observing, and 2) Heat damage to telescope

    R3 - use the Canopy discussed in S1, above to keep the scope as cool as possible during the day;

    R3b - wait until dusk or slightly before to take the scope out into the evening air and
allow AT LEAST one hour for the smaller SCT telescopes and TWO hours for the larger ones to reach thermal equilibrium.

    R3c - you can use the "CHIMNEY TRICK" to accelerate cooling of the OTA. DO THIS CAREFULLY to prevent the accidental entering of debris or dust into the OTA! All you need to do is to orient your telescope so that the LENS (front of the scope) IS FACING TOWARD THE GROUND! Remove the end cap of the scopes or the prism screw mount from the any compound telescope (closed tube) until the rear of the scope is OPEN; in the ETX telescopes you will need to flip your small mirror as you would to "view" straight through the scope (you will be able to see the secondary through the opening in the correct position). Place a clean lightweight cheese cloth (or similar lace-like fabric) OVER this opening to block dust and debris.....and then merely wait about one-half the time! Your scope will cool internally TWICE as fast, as the port acts like a chimney, rapidly dispelling warm air otherwise trapped inside the tube!

P4 - Excessive Heat Damage to the OTA
S4 - Safeguard as in S1; however, likely damage that can occur is in the cements that are used within the OTA, particularly that holding the secondary baffle tube to the meniscus lens of the Maksutov telescopes. This thin metal baffle is held in place ONLY by a thin ring of adhesive that becomes like jelly when the temperature gets excessive in the scope.

[NOTE: This does not apply to the Schmidt-Cass. Scopes but internal heat should always be avoided at all times]

R4 - If your baffle DOES become loose and begins to slip, you will know it; before you can actually SEE this with the eye, you will notice flares from images of bright objects like Mars and bright stars, flares that you did not have before. If you notice this, un-focus (turn your focus know counterclockwise) and look at the out-of-focus image....you will see the "disk" pattern of the star; if it appears oblong rather than circular, then your baffle tube has slipped. It will be necessary to carefully UNSCREW the entire end cell that holds the meniscus lens from the OTA (it may take some work to get it moving at first and be very careful not to damage the pretty blue tube!). Once out, you can get a replacement adhesive ring from Meade OR you can merely use a very good temperature-resistant glue (something that never actually "hardens" but remains flexible as the temperature changes) VERY sparingly against the flat edge of the baffle that mounts to the glass; use the MIRROR itself (the secondary) as a circular template to place the baffle....let dry for about two hour in a protected environment and then reattach via screwing the cell and its lens back onto the OTA firmly (no need to do any alignment whatsoever).....and that's all there is to it!

P5 - UV Fading of Telescope Exterior
S5 - Even if kept indoors and covered, light from windows and even in shade can eventually bleach the beautiful colors of your modern scopes. I highly recommend (in addition to keeping out of direct light) using TURTLE WAX's "Scratch Gone" (or similar) soft wax for new car finishes. This comes in many colors and that is the key to buying it....if it is Turtle Wax soft and it has a choice of colors, you are getting the RIGHT product. It has a UV block and is excellent for both the BLACK PLASTIC (and metal on the metal parts) and any dark tube. Use this very sparingly (just moisten your rag with it) and rub gently and even on all surfaces....then buff. It resists fingerprints, UV sunlight, and makes your scope look like a million bucks!

P6 - Telescope Electronic Circuitry (OTHER THAN AUTOSTAR)
S6 - During tests of hand controllers on 90+ degree days in mid-April 2001, I also exposed the mounting of an two commercial SCT telescopes (less tube assemblies) to the same temperature/sun exposure as their hand controllers. These conditions would closely simulate about a two-hour observing session of the sun with the sun at its highest point during the day. The temperature on both days reached at least 92 degrees F. in my observatory for a prolonged period of two hours of direct sunlight. ALTHOUGH SOME MAJOR PROBLEMS AROSE the hand controllers, there were NO adverse conditions resulting to the internal telescope circuitry or electronic function within the drive motors/encoders whatsoever.

P7 - Protecting your Optics from Direct Sunlight
S7 - It is easy if you are readying for a star party to inadvertently allow the direct sun's rays to enter your telescope assembly; you should always (for your safety and that of the scope) point the telescope in any direction away from the sun. If outdoors, REMEMBER YOUR FINDERSCOPE; direct sunlight will MELT the crosshairs within about 8 seconds flat as the sun might inadvertently pass across the field of that little scope!

P8 - Accessories Left in Sunlight
S8 - Your eyepieces and ALL accessories should be kept in shade. NEVER put your accessory case out in direct sunlight, not even for 10 minutes. It is like a little greenhouse and WILL RUIN your eyepieces, particularly any of them (and many do) that have cemented components; if you carry a flashlight in your accessory case, it too can be damaged by excessive heat.

P9 - Transporting your Telescope Cargo in Your Vehicle in Summer
S9 - NEVER, NEVER lock up the telescope in the trunk of your car, even while the car is moving at 70 mph down the freeway toward vacation. The inside of your trunk does not understand the chill factor of the speed and sunlight bearing down on the metal trunk lid will eventually heat that interior upward to 150 degrees F. I always recommend, if possible, transporting the scope and accessories in the back seat and all non-temperature-sensitive parts (i.e., the tripod, wedge, charts) in the trunk.

YOUR TELESCOPE IN SUMMER HEAT - Part 2: Your Computer Control – *EXTREMELY IMPORTANT - PLEASE READ*!

Just when we thought we were getting our computer control keypad away from the environmental perils of winter's cold....here comes summer's HEAT. And, boy if you thought you had "cold temperature problems" this past winter, please take notice here!

P10 - Battery Power and Hot Weather
S10 - not a problem...unlike winter months, your internal batteries (or your external DC power station) actually will provide a somewhat enhanced output during summer, provided they do not get excessively hot.

P11 - Prolonged Exposure to keypad of Direct Sunlight for Very Short Periods of Time - MALFUNCTION!

IMPORTANT NOTE: The following descriptions are what prompted me to write this very important guideline to summer safety for the telescope. As I was training the motors on a "Supercharged" scope one afternoon I began to run into some serious malfunctions related to the hand boxes that at first I could not identify. After examination and further testing the next day in 92-degree heat with both controllers, it was learned that direct exposure to sunlight is VERY detrimental to your computerized keypad and will result in AT LEAST the following two situations in less than 10 minutes of exposure:

[DO NOT ATTEMPT TO DUPLICATE!!! Permanent Damage to YOUR handbox Will Result!!!]

1) the handbox LED display will no longer function, although the commands of the handbox can still be entered; what you will get will be a LED display of filled "boxes" all the way across both lines of your display screen; this will NOT go away until the handbox has cooled for about 30 minutes, but your problems are NOT over;

2) not all commands entered into a hand controller once the display goes blank (#1 above) result in proper functions being obeyed; again, about 30 minutes is required before the handbox cools to where some commands can be rightfully accepted;

3) speed settings on your handbox (the number keys "9" fastest, "1" slowest on Autostar become totally inoperative and everything runs at the fastest speed when the controller becomes overheated; NOTE: letting the handbox cool down as above DOES NOT always result in your speed function returning; and so,

4) a total RESET (obviously requiring all new user data entry, owner info., site specifications, telescope type AND "Train Motors" will be necessary) must be done to put the controller back into normal operational mode once it has become overheated by exposure to the sun.

I used both of my "test keypads" and was able to (thank goodness) successfully revive both after the alarming results of only 12 minutes in sunlight. It is important to note that this is NOT a "maybe this won't happen to me" situation....it will. I ran the test throughout two successive days on a hot muggy April week here in Arkansas, and believe me, I will do it over and over again.

So, I go back to my 50-year-old rule that I have preached to all telescope users, even way before microprocessors took the place of *Magnusson* Setting Circles, before Maksutovs began outperforming my old *Unitrons*.....

....NEVER subject your telescope to any condition - permanent or temporary - that you would NOT subject your 6-month-old infant child. They are BOTH just as sensitive....and certainly BOTH are just as temperamental and difficult to please!

* * *

Chapter 13

Controlling Heat Currents and Thermal Equilibrium
IN AND AROUND YOUR HIGH RESOLUTION TELESCOPE

I have noticed many questionable heat-related issues that will affect your observations and imaging of very fine planetary detail or detrimentally affect your ability to see the faintest stars on a given night. Equally, heat currents, particularly in early evening, are the curse of astrophotographers, causing bloating of star images and lack of detail in deep sky objects.

In planetary imaging, the trend now is stacking..... our main priority is rapid-fire patrol-type images that are taken (sometimes as many as 2,000) in fractions of a second and combined for one quality image, eliminating those fuzzy images where no detail can be gained.... yet even with such short and rapid imaging, we too suffer from sever heating effects that appear to be non-atmospheric in nature.

There are essentially TWO types of bothersome convection currents that can be seen in all telescopes:

a) those caused by atmospheric effects as the nighttime air cools rapidly and heat is dissipating upward from the ground, roofs, pavement, etc.;

b) those created locally within your observing environment which may or may not be able to be controlled.

The 16-inch telescope that we use for the hourly 120-image shoots is in an observatory that actually has temperature controlled walls; the roof is an 8" insulated double-slide (appears like a large pagoda) with a special thermal-resistant plastic resin material similar to automobile body fiber material.

Even with such precautions, the affects of heat currents is quite detrimental:

1) I have noticed considerable heat currents directly over a very large DC inverter using that I run for powering dew preventing apparatus for the computer and the telescope/guide scope/imaging camera;

2) In the corner where our image monitor (a 21" computer monitor with its own computer and image integrator) is "no-man's land" when it comes to imaging over that;

3) I open the roof at about 10:00 p.m. if I plan to image or observe a high magnification by midnight; this helps tremendously;

4) We have two power vents in the roof and one floor air vent and a "Florida-style" louvered window that we force air through DURING observing and imaging, which helps

tremendously; I have also found that leaving the front (north) door open to allow additional circulation helps.

5) The dew strips (absolutely essential here in Arkansas on every night) cause considerable turbulence. Since we image every hour; the main scope dew strip (A large Kendrick) is turned off 10 minutes prior to imaging; then turned back on again until the next sequence.

The best practice in terms of using a dew control system, or "dew strips" is to regulate the heat in a practical manner. For example, do NOT turn the unit on to its highest output ever....there is a tendency to do that when you have damp nights or there is a chill in the air. The best practice is to simply turn the dew strip(s) on to low power before you even uncover the telescope and leave it there for the duration....unless the dew point is excessively high and the humidity reaches near 100%.

These are some of the many things that we CAN control; however:
If a tremendous cold front roared through your area last night, dropping the temperature at 6:00 p.m. last night from 76 degrees to 27 degrees by 4:00 a.m. this morning (yes, this happens often in Arkansas), the rapidly rising heat from the ground is going to dissipate throughout the night and likely result in poor steadiness throughout the night. When the temperature drops so rapidly there is NOTHING that can be done to offset the tremendous amount of heat that is generating from the ground, the rooftops around and the highway to our south and west.

Indeed, during "rush hour" in late afternoon, although a mile distant, observers can see rapidly deteriorating seeing with the increase in commuter traffic.

Telescopes should always be allowed to cool and equalize at least **one hour** prior to viewing for planetary OR for imaging certainly. I have a trick that I highly recommend to those attempting very high resolution work, either visually or electronically with any closed-tube catadioptic or refractor telescope:

1) turn the telescope where the front (lens end) is facing pretty much down to the ground;

2) remove the visual back, eyepiece or camera connectors from the rear of the telescope to reveal the opening going into the telescope and leave open;

3) cover this opening with a very fine mesh cloth to prevent debris from entering the telescope;

4) allow heat to convect upward from this hole for at least 1/2 hour like a chimney.

This will remove the internal heat three times faster than mere cool-down. Also, if there are many people, dew strips, and heated accessories close to the telescope, it is a good practice to exercise this routine about once every hour as well for about 8-10 minutes.

Observatory Temperature Control for Optimum Performance

Arkansas Sky Observatory is presently nearing its 50th orbit around the sun; in that period of time, I have gained a considerable working knowledge - including some rude awakenings - about the affects of temperatures and humidity on telescopes and telescope equipment.

In those years, it becomes obvious that the temperature and atmospheric conditions inside the observatory DURING USE is of very little importance, if those conditions were ignored while NOT IN USE during daylight hours. The maintenance of a good observatory is far more important during daylight than it is at night. Use the night time to enjoy God's Creation....use daylight to protect your right to do so.

Here is a summary of my thoughts on observatory temperatures:

OVERALL CONSIDERATIONS
When telescopes that are going to be used for:
1) imaging
2) high resolution (i.e., planetary or double star) viewing or imaging
3) serious research applications

the ambient air temperature and humidity within the observatory confines (the actual room in which the telescope is being used) must be as close to that of the outside ambient air as possible.

Going one step farther: the actual telescope itself, including the AIR INSIDE the optical tube assembly as well as the primary optics (the main lens or mirror) must also be within ONE DEGREE of the outside ambient air for useful applications to be successful.

To use telescopes that are not so "acclimated" will result in differential heat currents which are highly problematic from these sources:

1) air "boiling" inside the optical tube assembly, even if the OTA has one end open; some form of ventilation can be used to allow faster cool-down times; boiling air within the OTA will result in images that are soft, out of focus and very low resolution, much as would be seen if the telescope were looking through a thin layer of plastic sheeting.

2) surface temperature of the optics exceeds that of the outside air: this will result in "soft images" and no focusing efforts will improve upon this until the mirror or lens is within less than one degree of the outside air. Mirrors are more subject to this influence than are lenses.

3) mechanical contraction and expansion: this is the number one cause of

the requirement to constantly focus through any given night, particularly as the night air cools rapidly from a warm day OR in the case of a telescope being confined inside of a closed observatory during daylight hours and being forced to operate at a warm temperature when the observatory is first opened. This is a direct result of the metal components of any OTA, and even the dovetail mounting brackets that are used. Aluminum tubes are very prone to differential contraction and this is the number one frustration with users of SCT or Maksutov telescopes, since closed tubes require longer to acclimate than open tube designs. Modern telescopes are now being equipped with Carbon Fiber or Kevlar tubes and fittings to minimize expansion or contraction during nighttime hours.

4) differential air temperatures within an observatory environment and that outside the environment; this is compared to looking down a hot highway on a summer day and seeing "water" on the surface of a dry road.....this phenomenon is particularly troublesome in summertime when a cool observatory is opened to warm night air.

5) the one that we do not have any control over: temperature inversion, which is the cause of "star twinkling"; on nights when stars overhead are twinkling, very little serious astronomy can be done by any observatory, no matter how clear the night might be; this is air mixing in refractive "zones", where warm air and cold air intermingle to change the refractive values of the air, and thus create multiple "lenses" in our atmosphere which change constantly; the result is blurring of your celestial target, or constant changes in focus. Many night with demonstrative star "twinkling" will result in absolute frustration in a user's attempt to focus, with focus good one minute, and terribly off the next.

OBSERVATORY TEMPERATURE MAINTENANCE IN COLD WEATHER

For the purposes of this discussion, "observatory equipment" refers to:
1) Telescopes and mountings
2) Cameras and CCD equipment inside the observatory
3) Computers that remain inside closed observatories
4) Books, charts, wires, metal surfaces

It is far easier and safer to maintain instruments and observatory equipment in cold conditions than in hot. I strongly recommend insulation in all observatory buildings.....not for comfort or to maintain temperature during observing times, but to keep extreme temperatures and humidity from ruining observatory equipment. Condensation is far less likely to form inside your observatory during daylight hours if the building is insulated. In fact, most damage from rust, mildew and electronic failure as a result of condensation that I have seen has been a result of equipment being stored in an un-heated and un-insulated observatory building.

When fall, winter and early spring days bring very cold temperatures, equipment within the observatory is typically unaffected by such cold when not in use. When temperatures drop below 20 degrees F, it is best to turn off computers however. No special attention needs to be given to equipment maintenance. Part of the reason for this is that, as temperature drops very low, dew points drop to a point where condensation inside of the building - which will always be a slightly warmer than the outside air - is likely not prone for moisture to form on any items in the building, even if un-insulated.

However in storage mode (daylight or nighttime), when humidity exceeds 75%, an un-insulated building will result in condensation onto cool surfaces quickly: these surfaces include: 1) computer surfaces and screens; 2) telescope optical tubes; 3) metal surfaces of telescope mountings; 4) particularly a metal pier and its base; 5) inside walls and panels of your observatory.

This problem is FAR more common in summer than in winter. In winter I suggest a small 1000 watt or less space heater or 200 watt light bulb being left on during closed periods to minimize humidity. If you are able to keep the inside temperature to within ONE DEGREE of the dew point on all exposed surfaces, you will prevent condensation from ever forming within your observatory.

* Without insulated walls and flooring, this is virtually impossible. Insulation is imperative if you wish to protect your astronomical investment *

Because most observatories are not ventilated, prolonged periods of exposure to such condensation WILL eventually lead to not only rust, but dreaded mildew which many times cannot be removed from metal surfaces and certainly not from optics. Like plant roots, mildew exudes tiny amounts of strong acids as a means to soften surfaces on which to attach; these acids will eat through the surface of your paint, anodizing, and optical coatings and will ruin electronics in 3-4 days unattended.

If such condensation is found in your observatory after being closed for a long period - either in summer or winter - do NOT turn on the power to the instruments, but carefully dry by opening up the building, turning on a space heater and using a hair dryer on exposed parts if possible.

More about humidity condensation is discussed in HOT CONDITIONS.

To Acclimate the Observatory Prior to Observing in Cold Conditions

In cold weather, always open the observatory up for an extended period prior to your planned observing. Do not open prior to sunset or you will be introducing more warm air than releasing in many cases....if it is necessary to open the building up (for example, if there is moisture on some surfaces), open when the sun is at a very low angle OR if domed, rotate the shutter opening AWAY from the sun for best results.

The observatory must be opened about 2 hours prior to observing schedule for optimal results with equipment.

If blowing dust, pollen or debris can be a problem, leave the telescope capped and covered with a cotton sheet, and cover computers and screens appropriately.

A thermometer should be located in the observing room on a wall that is not subject to daylight heating or direct sun (north, if in northern hemisphere); a similar thermometer outside in shade should be located, remote if possible. When the two are within a few degrees of one-another, your observatory is optimized for viewing.

Planning Your Winter Night's Observing Period

One tip that I always promote is to simply plan ahead a bit if possible on wintertime observing. If, for example, the moon is a first quarter and the sky is going to be filled with moonlight until it sets at about midnight, then plan to go out about 11 p.m. local time to begin your observing. Check the weather forecast and if all seems good for the night, simply open up the observatory AT DUSK, cover your equipment with sheets (never plastics!), get some sleep or watch television and leave the observatory OPEN until you go out there at 11 p.m. That way, all equipment is equalized to the outside ambient air.

During some extremely cold nights, there may be periods where equipment operates sluggishly and this may not be particularly good for your equipment and pocketbook.

1) if your telescope begins to slew difficultly, and if you can actually hear the sound change as if laboring or you see it visibly moving slower than normal, do NOT continue to operate the equipment. This affect in some telescopes will begin to get troublesome at about 15 degrees F., and if you do not use all temperature low viscosity grease in the bearings and gearworks, you are asking for a repair ticket on your mounting, electronic focusing control and other moving parts.

2) if your CPU fan begins to hum louder than normal, this is a sign that the computer may be struggling from excessive cold....this is rare and typically the temperature would need to get into the single digits or lower to become

problematic.

3) if your CCD camera continues to frost over, the night should be over.

4) if any of your electronic readouts (such as the hand control for the telescope, a digital clock, computer screen characters, etc.) become jumbled and unreadable, the temperature is beginning to affect them; this can occur even as warm as 35 degrees F.

NOTE that I do not suggest using LCD computer screens nor laptop devices in the cold observatory when the temperature drops below 10 degrees.

OBSERVATORY TEMPERATURE MAINTENANCE IN HOT WEATHER

Unlike cold weather, in hot summer there is MUCH care and consideration necessary for all telescope and related equipment inside your observatory. Temperatures INSIDE a closed observatory with no insulation nor ventilation WILL easily reach within 150 degrees for facilities located with 38 degrees north and south latitudes. When temperatures inside your building reach 95 degrees, you must begin to think of keeping it at least at that temperature or below that mark for three reasons:

1) at about 102 degrees, some electronic components begin to "get soft" and can become damaged or technically challenged once activated; electronics will be absolutely ruined at high temperatures;

2) although the optics of your telescope will likely not become permanently damaged from summer heat, they will retain this heat LONG into the night and likely not be able to acclimate to outside air temperatures whatsoever during the following night; this will result in deformed images and inability to properly focus your telescope;

3) camera equipment will not perform properly, with a dramatic increase in noise-to-signal interference; likewise, it is very easy to damage delicate CCD cameras and digital cameras in excessive heat.

Your computer equipment located in the hot building in summer months should NOT be left running; the CPU fan cannot keep up with the temperature of the CPU and the increased outside temperature of a closed observatory on hot summer days. If you must, for whatever reason, leave the computer running, I suggest always using a small desktop fan directed at the back of a PC unit, or at the keyboard of a laptop at all times during daylight hours.

If the temperature inside your observatory reaches 110 degrees, you MUST either open the roof, door, or both. Ideally some type of ventilation should be provided at all times.

Using an Air Conditioner and Condensation

For extremely hot days, the problems with condensation are typically not an issue since the air temperature is far above the dew point and moisture will not form on surfaces. However, if you are running a small cooling unit in the observatory during the day, you MUST be very careful prior to opening the observatory AND taking the caps off the telescope; if your air cooling unit has gotten the inside temperature and the surface temperature of the equipment down TO the dew point, then you will immediately have "fog" (condensation) form on the surfaces of all cool equipment. Make sure to turn OFF your cooling unit at least ONE HOUR prior to opening your observatory for an evening's session.

Also note that on very humid summer days, such as when rain is expected or if the dew point exceeds 50 degrees, you may likely produce condensation if your cooling unit is working "too well." Always make a habit to check the surfaces frequently (I always use the section of the pier very near the floor level) for moisture....if you have some developing, it is time to turn on a small space heater or light bulb (see COLD, above) similar to winter months, simply to lower the dew point INSIDE your observatory environment.

To Acclimate Your Observatory for a Summer Observing Session

If proper observatory maintenance is maintained as described above AND if proper ventilation and insulation is in place, it is typically easier to acclimate your observatory equipment for a nighttime summer session than it would be for one in winter. Once again, wait until the sun sets until the observatory is opened....in summer, your twilight will last up to 1 hour or slightly more in some latitudes, so this will provide plenty of time for acclimation to take place.

Remember, that - if you have properly ventilated and insulated your observatory - the inside temperature should be about 12-15 degrees F. cooler than the air outside while the sun is up; this means that the air outside will begin dropping quickly after sunset and at some point be at the same temperature as your observatory inside. The trick here is to not keep your observatory SO COOL during daylight (if using an air conditioner) that the outside air will never reach the temperature of your equipment and thus the equipment must actually acclimate UPWARDS to match the warmer air outside.

Warning about Warm, Humid Days

Warm and humid days can occur in summer, winter, spring or fall. Humidity WILL get inside your observatory and the only way to control it is either through the use of a small space heater, a de-humidifier, or a 200-watt light bulb left burning continuously. If you own an observatory, you also without a doubt have invest a LOT of money in your equipment that is inside that observatory.

You are the keeper of that equipment and only you can prevent mold and mildew and direct water damage through constant maintenance. Keeping an observatory is much like keeping a prized dog: you love them and they love you....but they only love you if you feed and pet them. Pet your observatory.

Watch the local forecasts regularly and look for high dew points and high humidity. Fog and condensation, and even rain, will form when the air temperature reaches the dew point temperature. This can happen inside and outside of buildings, and will happen in your observatory on a regular basis if precautions and care are not exercised.

When you see that there are going to be prolonged periods of high humidity, rain, high dew points and general cloudiness, there are several precautions that you - as a responsible observatory owner - must take always:

1) turn off and unplug all electronic equipment, including your telescope and dome control if you have it;
2) provide for some way of warming the interior of the room as described previously;
3) cover your telescope and other equipment with a soft thin cotton sheet - *NEVER* cover with plastic, as the plastic traps moisture, assuring you of condensation, damage and certainly mildew.

Remember that you have invested in your observatory as a TOOL to your pursuits of astronomy as they increase your knowledge of the night sky. The observatory is not a panacea of all things irritating to your time at the telescope and away from the comforts of your warm bed. It is only a tool.....and a good one, provided that you are a good maintainer of that building.

If YOU would be uncomfortable living inside your observatory, day or night, when it is NOT being used in your explorations of the sky.....then something is not right with things internal; it might be improper ventilation, improper temperature control, too much humidity, too many bugs, or a cold draft coming from a wall or floor with no insulation.

My acid test of a good observatory room is simple: If you can take your favorite astronomy book inside during each of three specific days:

1) a hot sunlit summer miserably oppressive day;
2) a gloomy rainy and foggy day; and,
3) a cold, bitter, wind-blowing day.....

....and sit down on your observatory stool and actually READ your book without distraction.....

....then you are a good maintainer of your astronomical pursuits. Enjoy.

Just like heat currents rising from hot pavement creating this water mirage, the heat of the daytime rapidly rises after sunset to create waves of rapidly moving air which makes the air very unsteady for viewing. If the stars are twinkling wildly, you have unsteady rising air.

Chapter 14

Power Supplies and Voltage
For Computerized Telescopes

It was not until the mid-1990's that commercially available telescopes were equipped with high torque, low voltage direct current (DC) motors. Nearly all telescopes prior to that came equipped with a heavy power cord to supply voltage to one or more Hurst or similar AC synchronous motors.

Not only did this restrict telescope use to places where AC power was available (or the use of unreliable DC inverters), but there were performance and safety issues with using AC voltage in the moist night air as well. While the durable motors held up well, the wiring and connections for these soon became frail and leads would break, solder would snap and power to the telescope became a constant concern.

From the old Cave Astrola telescopes, Criterion Dynamax to the revolutionary Celstron-Pacific Schmidt-Cassegrain orange telescopes – all utilized the synchronous motors.

A telescope is like a clock and needs pretty much the same precision; in fact, the telescope is like a clock that is running slightly fast....23 hours and 56 minutes per revolution rather than a typical 24-hour civil time clock. Sidereal Time – the faster of the two – is what creates the progression of our seasons each year.

Gone are the days of the AC-powered telescopes and with them are the problems with electrical power, wiring and safety concerns and accuracy of tracking.

Today we have in our modern telescopes – computerized or not – inexpensive yet powerful tiny DC motors which can be regulated to all sorts of tracking rates and powered by simple, inexpensive and portable voltage supplies.

Please study the following guidelines carefully for selection and operation of power supplies for your telescope – it can mean the difference in a pleasant experience or a very expensive repair bill.

POWER SUPPLIES FOR MODERN TELESCOPES

We must start this discussion with a firm warning: TEST each and every power supply that you use with a telescope, whether it is supplied with the purchase of your telescope or you buy one separately. As much as I resist negativity toward telescope suppliers, I have found that the "power lumps" that are provided by pretty much all manufacturers for powering their telescopes are nothing short of useless.

Not only that but they can be dangerous in terms of the safety of the electronics of our modern sophisticated telescopes. Power **surges** are the number one cause of failure in telescopes sold and bought today. Using **improper power** (such as incorrect voltage or reversed **polarity**) is number two.

Types of Power Supplies

Other than manual "push-to" tracking and finding objects, there are only three types of power supplies that we are concerned with for modern telescopes: AC-to-DC converter power; battery power; and, battery pack power. The difference in the last two will be discussed fully.

AC Power Lumps

Power supplies that are standard equipment with your telescope purchase are far better than they used to be. Today it is rare to see a telescope manufacturer or dealer sell any power supply from AC input that is not fully **regulated**. Regulated power (sometimes referred to as "switching power") simply means that the output of the converted 110 volts (or 220v) does not waiver more than +/- 0.5 volt from the specified output.

If it looks like this, never use with a telescope. The common "Wall Warts" are fine for powering flashlights and portable vacuums, but unreliable voltage output makes them unwise for telescope use

* * *

If the power lump is advertised to provide "12 volts" of power, then regulation will provide no less than, say, 11.5 volts and no more than 12.5 volts, the target being somewhere around 12 volts. This range is perfectly acceptable for all electronically operated telescopes.

Only as recent as 2010, many power supplies were being sold for telescopes that – when first powered-up – could surge from their purported "12 volts" all the way up to 31.8 volts or so. Likewise, these unprotected, unregulated, power sources were subject to spike, or surges, throughout a night's use, sometimes dropping voltage down so low that the telescope control would simply shut down, freeze up during very cold weather, or surge to voltage so high that the circuits would blow and the telescope mounting rendered useless. The voices of telescope users worldwide quickly resulted in major manufacturing specification requirement changes for the largest commercial manufacturers and nearly all sold today are regulated fairly well.

Nonetheless, the dependability of the power supply – as well as the knowledge of the person using it – can make or break the future of your equipment. Consider that you would never even consider powering a laptop or smart phone by way of a third party untested power source. Now consider that telescope electronics are far more sensitive to power shifts and disruptions than either of those two devices.

Pyramid makes, in my opinion, the most reliable and stable power sources for quality astronomical telescopes; at below $50 US, the unit above is heavens above other sources in terms of very precise regulated output

*   *   *

Choosing a lifetime power supply should not be difficult: most telescopes require 12 volts DC with amperage requirements UNDER 5 amps. The unit shown above is perfect for that specification, as are others from other manufacturers. There are some telescopes manufactured that require 18 volts DC – so always look for the exact specifications for your mounting prior to purchase.

The key with any power supply is regulated voltage – the Pyramid, Radio Shack and other suppliers' units are regulated so precisely that you will see no more than 0.1 volt variance in output throughout hours of use.

Typical of all commercial power supplies is a wide range of amerpage – from 3 amps to 35 amps. That is your choice....more amps is not necessarily better, simply because most telescope, under full slew, will draw less than two amps. You cannot buy "too many amps" in your power supply.

The best rule of thumb on any power supply is to stick a quality digital volt meter to it. Put the tips of the meter into the proper output points (the PLUG has a sleeve and a TIP receiver, these discussed later) and simply watch the output. If it flies all over the place and the voltage changes constantly, it is something you do not want to use. After about five seconds in place, the voltage should remain quite constant with very little, if any fluctuation.

ALWAYS REMEMBER POLARITY of your telescope power requirements!

Power cables for your power lump will require connection to the voltage unit (as seen in the Pyramid on the previous page, you have "OUTPUT DC" which can use either RCA-type plugs direct into the front (red is polarity positive, black is negative), OR those caps unscrew for direct bare wire wrapping around the posts and then re-secured by the caps for direct contact to the output points. More about power cables in a minute.

Battery Supplied DC

The single most reliable in terms of eliminating unknowns (other than reversed polarity) is the old "car battery", actually a **Marine Battery** as shown below.

Available at any auto parts store or large big box retailer, the 12-volt battery is dependable, provides incredible stable and constant voltage output and can be used for weeks under normal telescope use.

For portable use, there is nothing quite as reliable as the stand-alone marine battery and the units will last you a lifetime if kept properly changed.

Proper charging is the key to dependability and longevity however; this means that you will need also a trickle charger for a constant charging while not in use (sometimes the charge may take as much as 30 hours). Cloudy nights and periods during full moon are excellent for battery charging time.

Note that the voltage output on these –as well as the Jump Start batteries described below – will be around 13 to 13.8 volts. This voltage is just fine for operation of "12 volt" telescopes.

To use these requires only a pair of battery ("alligator") clamps for the battery end, and fused power cable to the telescope (1 amp in-line, slow burn) and the proper plug end for your telescope with the correct polarity. *ALWAYS REMEMBER POLARITY!*

The downsides to the big battery is that it is quite heavy to tote to the field – its biggest advantage over the other two sources of power – and that you must remember to maintain the charge and condition of the battery.

Jump Start Battery Packs

How did we ever do without these? For use at star parties, field trips, quick setups, dark sky sites, there is no better unit than the now-popular Jump Start battery units available in every big box retailer across the world.

These units will run a telescope for as much as four nights with regulated 13 volt output without the need to recharge. The beauty and true advantage of such power sources is that they contain their own integral AC charging device….just plug into the wall when not in use (extension cord optional) and it will quickly charge fully for your next night out.

Now that I have bragged on these portable and efficient units, here is a very serious word of caution: NEVER, ever use a "Combo-Power-Unit" or whatever name they attach to it to power a telescope. Never. The single greatest cause of power surges that can ruin delicate electronics in a telescope is the use of a multi-tasked power source such as those with the following:

Lights…whistles….bells….USB port chargers….radio….AC outlets….flashers…..cell phone chargers….and the list goes on.

You are NOT buying a power source for all that. You are attempting to power your telescope, period. Those multi-task Power Combo units are fine for your dew heater strips, your laptop, or other devices, but never – ever – have anything running off of the same power source as your telescope. The same rule applies to an expensive CCD camera. Independent and isolated power output only for those two devices.

At least three major telescope suppliers private label these power units and they are quite expensive. In my opinion they should not be offered as a safe source of power for telescopes. In addition, why are you going to need a spotlight at a dark observing site? A flashing red light? How irritating would it be on a quiet dark night under the stars if the fellow next to you turned on his radio? Do you really need to charge your smart devices at the star party?

But in terms of reliability, grab-and-go, and dependable power, a simple Jump Start unit like the EverStart (Wal-Mart) shown on the previous page is excellent and highly recommended. You will need a cigarette-lighter-style output plug at the power source combined with an in-line fused power cord to your telescope, available from many telescope accessory suppliers.

Battery 'Packs"

Many commercial telescopes come with either an external small batter pouch or a built-in battery compartment for using common "AA" or "C" dry cell batteries.

Do yourself a favor and forget that you have that item. On the external pack will be a power cable that will connect to the plug on your telescope; on the internal batteries from most manufacturers, no cable is necessary and you can easily forget that the batteries are being held internally in some compartment. Many a serious case of acid corrosion has been the ultimate demise of a portable telescope because of these batteries being left in the compartment and forgotten over time.

Not only that, but dry cell batteries – even fresh out of the package – must be used in series circuitry in order to accumulate voltage up to the necessary power to run the telescope, and because of the metal, contacts, and hardware in the path, at best a series of even the best batteries will put out under 11 volts, not nearly enough to power a GO TO telescope under cold conditions.

Expect unreliable and erratic operation and performance from your telescope if using such a batter pack. The best advice that you can get on powering in such a way is simple: Do NOT power a telescope with a battery pack of any type.

A DISCUSSION ON POLARITY

There is nothing as important to assure that you have set properly than the polarity of the line of voltage going into your telescope.

Polarity is simply the "pole arrangement" of the electric charge – positive and negative, like poles of a magnet – of each of the two wires coming out of a DC power source and into your telescope.

Nearly all commercial telescopes are universal – **Center Pin** (see discussion following is **positive** and outer sleeve is negative.

The "center pin" is the small tip that you can see protruding like a tiny finger inside the power port of your telescope's control panel; the sleeve is the outer sheath into which the power plug slides, like a housing surrounding the plug. The two are insulated from one-another by a plastic or other non-conducting material both in the plug and in the control panel receiver for the plug.

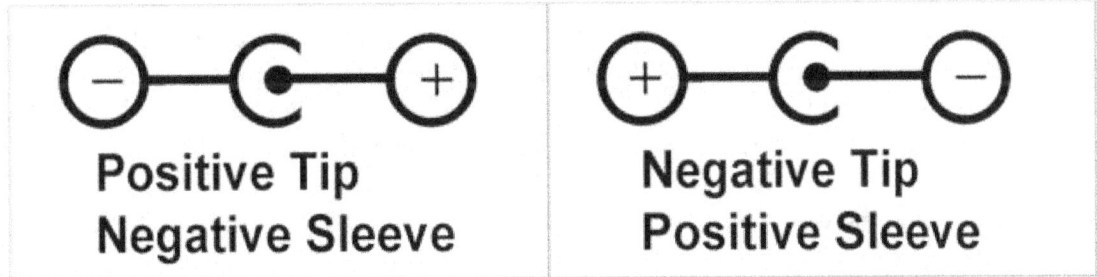

All power sources are required to have one of the two labels affixed to the backside of the power source. The one on the LEFT is the one that you typically want to see if you are using for your telescope power supply, this indicating that the center tip is indeed positive (+) and the sleeve negative (-).

That will also indicate that the wire going to the center tip on the PLUG end (that which attaches to the telescope) also must be the same wire that attaches to the RED output port on the power supply.

Reversing polarity with either crossed wiring, an improperly wired plug, or –more often seen – using the wrong type of cigarette lighter plug with wiring for center pin negative – will instantly render your circuitry useless and permanently damaged.

It is very difficult with the lighter plug to determine polarity unless you actually attach it to the power source and then measure the output at the telescope end. If you have the red tester probe inside the center pin and the black on the outer sleeve and the output is "+" voltage, then you have a cable with center pin positive and wired correctly; if, on the other hand, the digital output on the volt meter is showing (for example) "-12.3 volts" (minus sign), then the center pin is wired negatively and should never be used.

For the reason of confusion, I recommend **avoiding** any "universal" plug kits in which all sizes and types of control panel plugs are included with two prongs allowing reversing polarity if you wish. Although clearly marked with "+" and " - " signs, it is all-too easy to slip up and install a plug improperly. All it takes is one power cycle and your electronics are fried.

By all means, with your new telescope, always check first before you power up, even with the supplied power source if one comes standard. Check both the output with a volt meter and also check the polarity AT THE SCOPE END of the power line that runs from the cable....you must have adequate voltage and that power must have the correct polarity, usually **CENTER PIN POSITIVE** except when described by the manufacturer.

If using a marine battery or any device that uses clamps, connect the red clamp to the positive pole of the battery and that must run entirely to the center pin wiring for the telescope.

CABLES TO POWER THE TELESCOPE

Unfortunately in this profit-driven world, many telescope manufacturers do NOT include the standard power cable necessary to run the telescope from a power source. That component is option and it is up to the owner – or the qualified dealer who sells you the telescope – to match the proper power cord with your telescope requirements.

To properly match the cable, you must consider the following three requirements:

1) compatibility to power source (type of connection required)
2) compatibility to power panel on telescope (size of plug)
3) polarity of tip entering port on telescope

All three of these requirements are essential for safety and efficient operation of your telescope.

Unfortunately not all telescope control panel power ports are universal – far from it. Each telescope has consistency from brand to brand in that they all require some type of PLUG to connect the wire from the power source. Some of these plugs simply slip into the power port, while other have a retainer screw that can firmly fasten the power plug in place. Some power plugs are straight-line, while others enter the telescope at a right angle….the latter is preferred simply because they are less prone to be bumped and interrupt power service to the telescope.

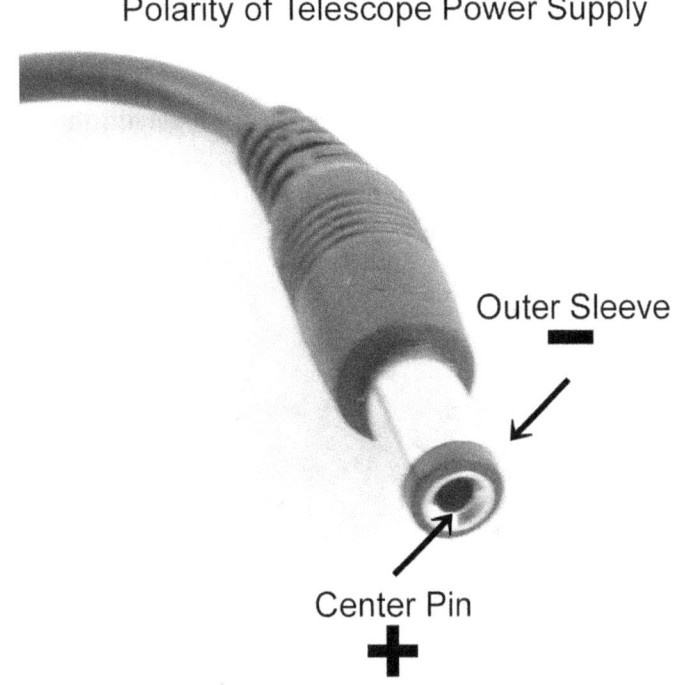

A typical power plug configuration is shown above. As noted, most telescope required center pin POSITIVE as this one is marked.

Equally important is the connection on the other end, that to the power source. You must assure polarity coming OUT of the power source as well as proper connection. The connections can vary from bare wires wrapped and secured around a terminal, to plugs, to a cigarette lighter plug.

The latter can be troubling if care is not made to assure polarity. Always MEASURE the output at the **telescope plug end** to assure both voltage and polarity.

The example below, a very cheap cigarette lighter style plug with the proper plug for a telescope is very confusing as to polarity; there are no markings whatsoever on either end to signify polarity. If this device was wired with center pin **negative**, your telescope electronics would pop and burn as soon as power was supplied. Unless you are confident in your testing abilities and equipment, avoid this type of plug.

On the other hand, the lighter-style plug assembly shown above (original in color) has TWO wires, the lighter grey one here being actually RED in color and positive while the negative polarity is the black wire. No mistakes should be made with this highly recommended configuration.

Most telescopes do NOT have an internal safety fuse installed in their circuitry which is a huge shortcoming since a simple fifty cent fuse may be the last line of defense between a power surge and your telescope budget.

IN-LINE FUSE APPLICATIONS

All telescope systems should be equipped with some type of safety fuse within the incoming power line. Protecting the system prior to a surge even reaching the front power port will save the life of your telescope.

Such fuses can be purchased as an integral part of the cigarette lighter type plug (previous page) in which they are containing inside the large plug that inserts into the power source, or they can be manufactured or installed within the power line wires between the power source and the telescope.

Showing the component fuse within the power source end of the cigarette lighter plug adapter; if the fuse blows, the front merely unscrews for fuse access within a minute

* * *

Anyone who can use a roll of electrical tape or shrink tubing can easily install an in-line fuse using the appropriate adapter (following page) to accept the simple fuse. The time spent in preparing this safety precaution is well worth the time spent and the total cost, including the fuse, will be under one dollar.

For all telescopes, I recommend a 1-amp slow burn fuse, which is ample to stop damage to the telescope, yet substantial enough to eliminate "fuse burn" during every observing session. Keeping a set of fuses in your telescope equipment case or next to the telescope at the observatory pier is always recommended.

The following photo shows an aftermarket-installed in line fuse on the positive wire coming from the power source. The small canister that contains the fuse simply twists to unsnap, the old fuse removed and the new one quickly inserted. An internal spring contact maintains firm connection of the power flow at all times.

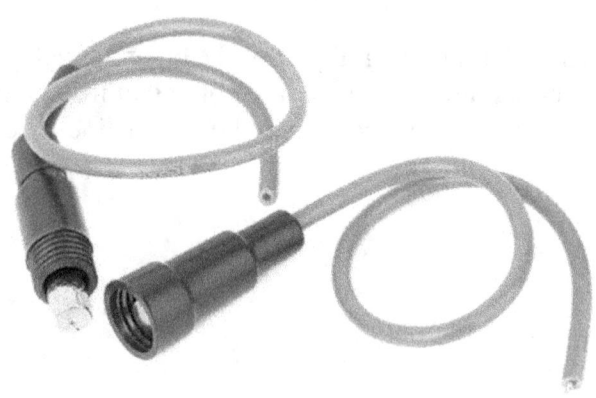

MULTIPLE POWER REQUIREMENTS

A short word about an important subject that was discussed briefly. Today's modern amateur astronomer (actually better labeled "non-professional") has a vast array of digital and electronic equipment that all require power.

TWO components must have independent power from all other devices:
- your telescope main power supply
- any CCD camera or similar that is in use

No exceptions to this rule: any other devices such as dew heaters, a laptop or smart device, lighting, guiding programs or equipment, dome control, etc., must NOT be operating on the same power source as the telescope and/or camera.

Other devices can easily create surges on the power going into the telescope or CCD and create a sudden influx of voltage damaging to telescope electronics. For this reason, it is advisable to always have two power sources on hand for your observing sessions: ONE for the telescope and ANOTHER for everything else. Never, ever run a dew heater from the same power source as is supplying the telescope.

Chapter 15

Lightning or Electrical Surges
Basic Guides for Protecting your Telescope
With discussions on Telescope Pier Construction

I will contribute what I know from experience. The older I get the more my "experience" gets compounded, so drag up a chair and read a while. Those of you who know it all, need not read any further.

First of all, you need to ask the telephone and electrical company nearest you if the "ground has proper grounding potential" for electrical discharge and lightning. If it does, then your problem is minimized a bit; if it does not then you must take very special precautions because you MAY create the best ground that your nearby area has via the pier that you are putting in the ground.

Two things to remember:
1) lightning does not always come from the "air"; and,
2) your will get more surges and strikes from the ground than you will from lightning aloft. Nearly all telescope electronic damage, and component (i.e., CCD cameras, computers, clocks) damage is a result of indirect ground surges during lightning storms than from aerial strikes.

Lightning may strike a tree, tower or even high ground two miles away from you attempting to find grounding. If it cannot find suitable discharge, being the fun stuff that it is, it will travel through the source of the strike and through any medium, particularly wet and rock-laden ground, to the nearest "best" grounding potential point. This point may be an iron-rich lump of sandstone beneath the surface of the earth, a mobile home metal anchor, a large structure standing on the surface, OR worst case: your pier filled with metal rebar.

If you live on top of a hill or in one of those grounding-poor areas such as Arkansas Sky Observatory is on Petit Jean Mountain in Arkansas (we have taken five lightning hits in five years until I finally figured out the cure), your observatory is going to be a prime target for both aerial and ground lightning surges, particularly if you do not design your pier properly.

1) My first rule of thumb is simple: install a complete shut-off system for the observatory, where all power can be disconnected from all equipment via a lever switch outside the building at the breaker box. Remember that lightning and surges are also very common through phone lines; if you have your PC plugged into a modem, or even a phone connected to a land line in your observatory, you are inviting electricity into the privacy of your sanctuary.

2) Second rule: even with the breaker totally disconnected from the electrical

source (pole/transformer from utility company), you can and will likely be in danger of a ground surge during lightning storms; these will enter your system via the pier and not from outside power sources nor through direct hits ("strike").

3) Third rule: knowing the second rule, never leave any wire whatsoever connected to the control panel of the telescope, even with all power disconnected; something so simple as the rs232 cable running from your PC to your coiled Autostar cable running to the Autostar resting on a nearby table or telescope base can act as an "antenna" for any surge traveling up the pier and looking for that juicy PCB behind the control panel for your telescope.

4) Fourth rule: NEVER install lightning rods nor tall antennae close to your observatory for the intent of diverting lightning away from your building. You will attract a lot of energy toward your location that might otherwise go somewhere else. Lightning rods should never be installed nor used to your observatory.

5) Fifth rule: know how to build your pier, construction material wise, for your location. If you are in a grounding poor region then you must take extra precautions to protect your telescope from ground surges which will "get your equipment" even when disconnected and unplugged. If your telescope pier is metal OR if your concrete pier has metal rebar in it you must be very careful to protect....since most piers are similarly made, then let's look at how to protect.

NOTE that virtually all telescope mount/pier arrangements are going to transfer part of the support of the telescope to any pier whatsoever by means of a piece of metal; this can be either a metal pier from the ground or floor level to the base of the telescope, or a metal wedge from the top of a concrete pier to the base of the telescope. So....the fact that metal is going to be in your framework is a given.

In any case, high quality *Guaranteed" (i.e., such as the Radio Shack replacement guarantee) surge protectors should always be used; I recommend using a major commercial "Utility Line" surge box (this attaches to your main meter via your Utility Company and there is a monthly service fee for this) for the overall building, and a separate small surge protector strip for EVERY outlet into which your equipment is plugged.

PROPER CONCRETE PIER CONSTRUCTION

This discussion is based totally on my vast experience of 40 years of doing things the wrong way and then knowing how to never do it that way again. So all of you lightning experts and electrical gurus can just sit back and listen because I have "been there and done that." I know what works and what does not work with regard telescope systems in regard to lightning protection. In some lightning-prone regions, proper protection can become expensive quickly and NOTE that surge protectors are always recommended, but they are certainly NOT assurance against hits. In fact, nearly all equipment damage that I have seen and have reported to me has resulted with the surge protectors (and fuses in line) not even being tripped whereas the electronics inside a telescope or other equipment are aptly fried.

A quick rule to remember for your observatory: any surge that can stop a digital clock can damage your PCBs inside the telescope, and GO TO telescopes have at least four for surges to nest in. If you keep having surges on your electrical line, or if you have lightning storms that repeatedly will require you to reset a digital clock, then you have potential danger for your computerized telescope and nearby equipment.

There is a precautionary step that can and should be taken with every concrete pier poured (more discussion will follow on the applications of this): whether you your using Type A or Type B piers (below), always install TWO to FOUR 3/4" J-bolts (preferably one at each corner of the base pier, a few inches from the top surface, sticking outward like "arms") extending out the SIDES (horizontal) for either immediately or in the future creating "terminals" for an outside "grounding cage" to divert surges totally out of your observatory. More on this later.

There are TWO types of concrete piers for telescopes:

A. Those that are built to slightly above ground level onto which your metal pier will be bolted; the metal in turn will rise typically 30" or more above the surface of your observatory floor and that is what the telescope or its wedge will rest upon.

B. The pier is designed to anchor to the ground AND to rise above the floor level and serve, in addition to being the ground anchor, as the riser pier between the floor and the base of the telescope or pier.

For lightning protection, I would always opt for "B" above.....concrete all the way to the base of the telescope.

** **PIER A** (substructure only, concrete base anchor for metal pier) : This pier will go deeply into the ground, preferably below the frost line for your region and should flare outward as would a pyramid base into the ground. THIS section is where the rebar or any structural enhancing steel should be placed, below ground. However, the more rebar you are using the greater ground you are creating for lightning and ground surges to "find", particularly when the ground is saturated. You must use some rebar or metal cage in the below ground construction, but use it wisely. Remember, "mass is good" in regard to the base, but it can be overkill.....you do NOT need three tons of concrete in the ground to support a 90 pound telescope. Period.

For most modern amateur telescopes and those in small college observatories, a below ground anchoring pier should be no less than 3-feet square, flaring toward the bottom and no less than three feet in depth into the ground. Ideally, this pier should come up about one foot above ground, this section out of which you will have nominally three (3) anchoring "J-bolts" of at least 1/2" diameter; 3/4" to 1" are better, and preferably about 12-15" deep inserted into the concrete during the curing process.

For a 3' deep anchoring pier, you should drive SIX (6) pieces of standard 6-foot rebar into the ground: one driven at an angle outward from top to bottom inside of each "corner" of the below ground pier, inset with enough space to properly assure engagement with the concrete being poured around it. The other TWO (2) pieces can be driven directly toward the middle of the anchor pier and straight down into the ground. Drive each of these metal bars deep enough so that not one bar extends to within 6" of your large "J-bolts" and certainly never should the rebar make contact with these bolts.

That is all the rebar (and concrete for most applications) that is needed for PIER A.

For PIER A, added surge protection can be applied when fastening your metal extending pier (above ground) by placing very large nylon or Delrin Fender washers around each J-Bolt and allowing these to rest between the metal pier bottom plate and the concrete below it. I also recommend using nylon bushings AROUND the extended threads of the J-bolts to isolate the metal threaded bolt from the inside metal of the drilled holes in the base of the metal pier, followed by a second top Nylon Fender washer before the large nut is applied to fasten the metal pier plate firmly in place. These added isolators seem minor....but they can make a huge difference in the drop of surge potential from the ground, through the concrete below ground, and into the metal pier.

** **PIER B** (complete structure including interior rising pier to telescope, all made of concrete): This is the recommend method for lightning prone regions. This method of pier construction offers advantages over a base pier (concrete anchor) with an extended metal pier above it. Using concrete throughout the pier has added benefit in both stability and vibration suppression AND in reduction of surge potential if properly constructed.

For this discussion, we will assume a "model pier" of 36" above floor level to the point at which a telescope base/mounting plate or wedge will be attached.

The ideal concrete structure would have an anchor pier very similar to that described previously as "Pier A;" however, during the pouring process, all concrete MUST be poured at the same time to prevent differential drying which would greatly reduce the stability of the overall structure. So planning a "PIER B" must be done very thoroughly before the mixing of any cement or before the concrete truck arrives.

The base section ("Anchor") in this example will be a minimum of 36" square with rebar placed exactly as previously described EXCEPT at the center; in this example, we will be using a total of EIGHT (8) rebar sections, each 6-feet long. The corner sections are driven as described about half-way into the ground at an angle away from center; however the center section (about a six-inch diameter circle minimum) will contain a total of FOUR pieces of 6-foot rebar, each driven into the ground until they extend ONLY HALF WAY to the anticipated TOP of the overall pier.

In other words: if your anchor hole extends 3 feet into the ground, and you want the anchor pier to be elevated above ground level by ONE FOOT to your floor level, that is a total of four feet for your anchor pier. Your pier above floor level is going to extend (in this example) 36" more, so you will have four lengths of rebar, each extending TWO FEET above your anchor pier. This allows the rebar to secure and strengthen the overall pier, both bottom and top sections, but also diminishes the proximity of the potentially-grounding metal away from the telescope system which will be sitting on top of the pier.

For the actual 36" pier that "shows" in the observatory, I always recommend simple Sonotube, available and any large hardware outlet, such as Lowes or Home Depot. Minimum diameter should be 8 inches. Ten inches is preferred, this being centered in the larger anchor pier below, and resting OVER the four extended lengths of rebar; those metal rods will be extending about two feet into the three-foot section of Sonotube. The tube is placed into position immediately after the anchor pier has been poured and is still quite fluid; you must immediately set up a framework to support the Sonotube in two ways:

1) to assure that the Sonotube is perfectly perpendicular to the ground and level vertically using a good square and level; and,
2) to assure that the Sonotube does not SINK into the anchor pier and shorten your desired height above floor level.

Two by four lumber frames, previously constructed prior to pouring is recommended for this. One frame can be designed to serve both purposes.

Quickly, the remaining concrete must be poured into the Sonotube prior to any curing of the cement already poured in the anchor pier. Once again, prior planning dictates that you MUST have a bolt pattern template (thin plywood with anchor bolts already inserted is recommended) already made and ready to place on the very TOP of the Sonotube to allow for your anchors for the final wedge or pier/telescope mounting plate. The anchor bolts used at the top of the final pier must NOT make contact with, nor come too close to, the rebar that you have driven into the ground and extending into the vertical Sonotube. Isolation here is a must.

Once the cement is poured into the Sonotube and your top template is in place,

RE-CHECK the very top assembly and surface to assure that you are perfectly level at the top; if not, make slight adjustments to your Sonotube and 2 x 4 frame and secure firmly to assure that the cement cures to as perfect to level as possible.

Once cured, do NOT attempt to remove the outer cardboard of the Sonotube; leave it in place, as it makes an excellent base layer for a beautiful paintjob using Epoxy resin (marine) paint, the color of your choice. Leaving the cardboard in place, if properly finished and filled where necessary, will provide a beautiful exterior finish to your pier.

You have just constructed a very safe (in terms of electrical surge) and rock solid telescope pier for nearly every small observatory application.

FURTHER SURGE AND LIGHTNING PROTECTION - The "Grid Cage"

Remember the four horizontal anchor bolts extending like arms from each corner of the Anchor Base Pier? You can actually divert surges and direct lightning strikes away from your delicate and expensive observatory equipment if you connect these like you would terminals on a automobile battery to an external "Grounding Cage." (a.k.a. *"Faraday Cage"*)

This method, highly recommended by telecommunications and electronics storage companies, is a bit on the expensive side,but will assure you of considerable added protection against surges. The idea is to assume that you ARE going to take a surge or strike at the observatory at some point and no matter what precautions you have already taken, powerful electrical potential is going to enter your building.

So....you are going to essentially "get it OUT of your building" using this cage method.

On each "terminal end" of those J-bolts coming out of the sides of your anchor pier you are going to use two hex nuts and Fender washers. Between each pair of nuts you will secure a length of #6 or #8 copper braided wire which will lead outward from your pier......these four lengths of braid will each join outside the four corners of your observatory to an external grid that you have buried in the ground using a common rented trenching tool. The grid can be pretty much any kind of metal, even long welded lengths of rebar, but preferably thick copper rod is best, the kind that is used by electric companies to ground your electric meter at the service point of your home or observatory.

The underground rod, no matter what type of construction, is all connected in a huge square outside of your building....the farther away it is from the building and the deeper in the ground you can put it the better. However copper is expensive and the cost will add up quickly. Nonetheless, the grid must be at least 6 feet distance from each wall and no less than two feet in the ground.

The large copper or steel square will be connected entirely from each corner; at each corner a separate copper braid from each J-bolt will be connected via a heavy duty grounding wire connector assembly (available at any hardware store). Once all this is connected, added protection can be assured by driving ANOTHER copper rod six feet or more into the ground straight down from each corner and either welding or connecting (via a wire connector) each vertical rod to the square cage once driven into the ground.

Once in place, bury the cage and forget about it.

You will sleep very well at night.....but only if you take every precaution I have just outlined.

* * *

Chapter 16

POLLEN ALERT!!
Protecting your Equipment from Nature's Sandpaper

Every mid-spring I start getting inquiries about *POLLEN* and its affect on coated optics and so it is well worth a reminder here....once pollen is allowed to accumulate and sit undisturbed onto your fine glass optics or the finish of your telescope tube or mount, its presence will remain there indefinitely, etched in place by the slow and dilute acids provided for by Mother Nature.

I believe that the best (and it is tried and true....) method to remove pollen is the procedure that I have listed from my Arkansas Sky Observatories protocol, discussed in a following chapter. This will cut right through pollen and sap.....that is correct: there are considerable amounts of sap in every granule of pollen...that and the abrasive burrs are what allows the pollen to stick to its intended surfaces.

Cleaning is one option, but a good rule of thumb with fine optics is to NEVER clean unless absolutely necessary; therefore, prevention is a better solution if possible. That being said, avoiding pollen in most places in springtime and many times in late summer would require that the astronomer simply not even use his or her telescope. That is not an option for those of us who take advantage of every clear night.

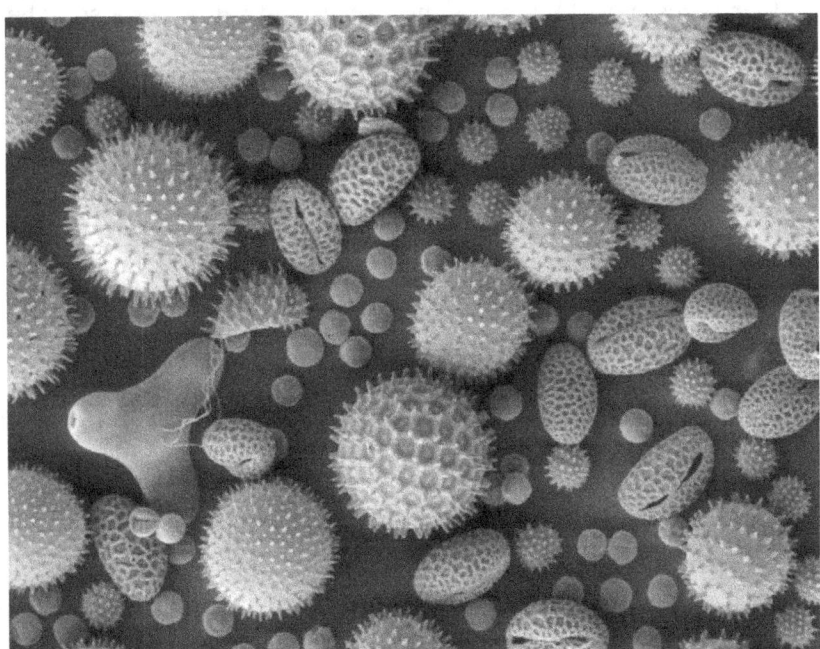

Pollen is one of the most destructive things that will get on your optics and should be treated as such. It also can damage the beautiful finish of a fine equatorial or fork mount. All one has to do is to look at a pollen grain under a microscope to see how incredible abrasive it can be on coated surfaces.....you really do not want this stuff on your telescope.

Protect Your Investment from the #1 Most Damaging Substance

We all have problems with pollen irritations in the springtime. Pollen is a pesky little package that - under a microscope - looks like something from the movie "Alien." The same factors that cause human irritations through pollen (itchy eyes, swollen nasal passages, inflamed glands....) are responsible for POLLEN to be the NUMBER ONE most damaging factor of telescope lenses, corrector plates and mirrors.

Right now, the pollen counts throughout the United States have been off the scale due to the unusual climatic conditions of the previous summer and winter. Don't look for things to level out anytime soon, either, as other pollens throughout late spring and early summer will merely replace those we are dealing with right now.
Pollen collection on your telescope surfaces will happen whether or not the wind is blowing; the pollen is so lightweight it is easily transported throughout the air; the only remedy is a good rain shower to cleanse the surrounding air, but that is only a temporary fix as the pollination procedure progresses.

Consider the following:
1) **Pollen is abrasive** - it varies, but most pollen grains have "spicules" (or spikes) that protrude like sharp points so that they may adhere to surfaces (such as the wings of a bumble bee); other types have "knobs" which serve the same purpose....there is even a "Velcro pollen" which attaches itself via a "hook" type extrusion to soft surfaces like the hair of a bee or the feathers of a bird.
2) **Pollen is coated** with a sticky emulsion, like the consistency of honey, that intentionally is there to act as a "bandaid" to get it to stick to ANYTHING that it comes in contact with.
3) **Combine the two** above, and you have ONE MEAN CRITTER if it gets on your mirrors, Maksutov lens or corrector plate of your expensive telescope. DO NOT ATTEMPT TO REMOVE UNTIL YOU HAVE READ THROUGH THIS THOROUGHLY!!

Attempting to remove pollen from the wonderfully-coated glass of our telescopes can be the most damaging and irritating thing that you can do to the telescope.
Merely "cleaning", even if you use the right techniques and solutions, will do nothing more that "grind" the pitch-fork spicules into the deep coatings....it is like using steel wool on your corrector plate. YES, it really is.....and pollen is sappy and it will smear when cleaned improperly, simply making things worse.

PLEASE follow the guidelines I have outlined below to protect your investment during this time of massive pollen production....this is NOT dust, folks...it is NOT dew nor fingerprints. It is POLLEN, and it can strike a death blow to your optics if not removed early, and removed properly!

1) protect the telescope from pollen during observing: either use a dew shield while your lens cap is off, or put the cap ON when you have extended periods between "looks;" during the spring of year, you can have a one-component layer of pollen across your corrector plate within 5-7 minutes unprotected;

2) make sure to GET THE POLLEN OFF immediately upon coming in for the night; if you allow the adhesive to dry onto your corrector plate, chances are you MAY not get some of it off; this is particularly true if you allow it to remain on during the hot summer months;

3) carefully remove the pollen and clean per the EXACT STEPS following. NEVER simply clean the lens....always follow these guides when it comes to pollen.

DR. CLAY'S DE-POLLINATOR (patent pending.....not really)

As soon as you are finished observing and the scope is indoors away from open screened windows or doors, remove the lens cap in a darkened room; look at the lens or corrector plate with a flashlight held obliquely (looking across lengthwise) the glass and see all the pollen. You must get that stuff off before it dries and sticks to the glass surface more permanently. Pollen sap remains fluid for a very short period of time and removing will leaves streaks if not done properly. However, once the sap solidifies, it is very much like dried shellac and can be nearly impossible to remove the pollen grains from the dried sap.

1) I use a medium-firm flat-edged artist's camelhair paintbrush that is 1" across to loosen particles of dust and pollen from the corrector plate. DO NOT RUB with the brush, merely very gently LOOSEN and fleck off all that you can....some likely will NOT come off in this step. Some will smear across the glass.

2) Once you have done that, use some "canned air" to blow off the loosened particles. CAUTION: You must be careful using compressed air cans on your optics, and some people never tell you this. Be sure to: a) NEVER shake the can; this stirs up the propellant and causes condensation to form which will blow water droplets all over your optics; b) NEVER assume that the canned air will get the particles off without BRUSHING FIRST....if you don't brush, well, your optics will be toast; c) use the canned air only in SHORT bursts, not a continuous stream, as long cycles will result in cooling of the propellant inside the can and result in liquified substances spraying onto your optics.

3) Now, use the optical cleaning solution and procedure described in the following chapter.
Kleenex tissues work just fine for this cleaning, provided that you do not "over-use" any one tissue until it becomes limp and begins to shed particles or tiny shreds of paper. Never use scented tissues or those with "softening agents"....only pure white, non-fancy Kleenex brand.

4) Daub the solution sparingly, but enough to wet the surface, without any rubbing at all at this point; this step is to further loosen particles and to begin to break down the tiny stains left by the pollen adhesives....YES, they will still be there.

5) Once the entire glass surface has been daubed down, USING THE SAME KLEENEX, very, very gently begin to wipe (not rub) in a slow circular motion, making sure that any particular area does not become dry at this point. Keep the glass WET at all times....do not allow time to dry.

6) Now that you have wiped the entire surface, get a FRESH Kleenex or two and apply a generous amount of solution TO THE KLEENEX (never to the optical surface) and very gently begin rubbing in similar circular motions, but GO BACKWARDS from step 5); you are actually allowing the detergent to clean off the pollen goo.

7) Repeat with a plenty of fresh Kleenexes, but rub even lighter, just smear some cleaning solution across the glass.

8) With a fresh Kleenex (actually several) that has been "Misted" with pure distilled water (not wet at all, just moist), begin to "buff" every-so-lightly about ¼ of the area of your glass at a time; if you see stubborn stains, or still some pollen debris, merely get another Kleenex and gently rub those areas to remove. Continue until the entire surface has been done.

9) NOW, take a bunch of Kleenex and use TWO at a time (so that the oils in your fingers do not penetrate to the glass) and dry-buff with the lightest possible touch, just like on a vintage automobile. Do this step twice to remove any streaking. Never let the same surface of the Kleenex rub across the buffing surface twice....tissue is cheap.
The alcohol is there in part as a drying agent. NEVER use more than a 3:1 ratio of water-to-alcohol, as this can dull your coatings. In the proper solution this cleaner is entirely safe and excellent for you glass and its coatings.

REMEMBER....as in so many things, prevention is always better and easier than the cure. Protect your optics when you are outdoors. On my portable telescopes I use a pillowcase or cotton sheet and place over the entire telescope when it sits idle for very long on pollen-laden spring nights. I merely remove prior to putting my eye up to the scope each time.

Sound like a lot of trouble?? Not nearly so much trouble and expense as replacing the SINGLE-MOST-EXPENSIVE piece on the telescope – your optics.

* * *

Chapter 17

ASO *fine optics* CLEANING SYSTEM: Part I
PRECISION COATED OPTICS
Lenses, Corrector Plates and other REFRACTIVE GLASS

PREFACE:

There are many, many variations of high precision, high reflectivity and high transmission coatings presently offered on the market for both amateur and professional scientists who use OPTICS in their respective lines of study. Smaller glass surfaces with high transmission coatings have always been seemingly easy to clean, since the smaller surface area is not as prone to spotting, sleeking and streaking of the cleaner used. On the other hand, large optical surfaces such as telescope lenses, corrector plates and optical glass "windows" are very difficult to properly clean without some residue being left behind as a result of cleaning.

The ARKANSAS SKY OBSERVATORIES' new protocol for cleaning optical surfaces includes:
1) judging carefully when cleaning is actually necessary;
2) preparation of the optical surface for proper cleaning;
3) a new solution that combines the attributes of all previous formulae and results in very fast, easy, and streak-free results if used properly;
4) the proper new technique that is highly recommended for cleaning.

WHEN TO CLEAN OPTICS:
Although we are attempting to obtain the best possible light transmission efficiency from our optics by cleaning them free of deposits, film and debris, lock firmly in your memory that *cleaning coated optical surfaces is the single-most damaging* action that will be done to them, short of actual physical damage or breakage.

No matter how careful, how delicate, nor what cleaning solution is used.....every time cleaned will result in a microscopically-reduced optical performance than before cleaning. Note that the coatings themselves - regardless who makes them and from what they are made - are nothing more than molecule-thick deposits of a very delicate film left on the optical surface from a vacuum process in which air is evacuated and the gases of the coating materials are gently and uniformly distributed across the glass surface after the vacuum container is void of air.

This system is devoted to the cleaning of large astronomical refractive optics: lens, corrector, and other optical glass; however the techniques discussed here as well as the new ASO *SuperPlus* Solution is excellent for the cleaning of eyepieces, eyeglasses, binoculars, camera lenses and all other fine coated optical surface.

So....the ground rule here is: *CLEAN ONLY WHEN ABSOLUTELY NECESSARY.* In most cases, dusting alone will lead to tremendous improvement in performance and overall light transmission.

PREPARATION TO CLEANING

DUSTING OPTICAL SURFACES: Large area optical surfaces are frequently plagued by DUST, POLLEN, GRIT, DEBRIS and even human skin and airborne hair.

If the surface of the glass is allowed to be exposed at a temperature BELOW THE DEWPOINT, these particulates will stick to the glass and will be stubborn to remove. However, for optimum performance, it is essential to, indeed, remove debris from the optical surface.

Your optical glass MUST be dusted when:
1) a flashlight held obliquely against the glass reveals a uniform and fairly thick layer of dust, etc;
2) when POLLEN is on the glass, as leaving pollen will result in "pollen sap" leaving a very difficult-to-remove stain on the surface;
3) ALWAYS prior to cleaning the glass with the solution and technique which follows.
Never clean optical glass without gently dusting first!

You will find in 3 out of 5 cases that merely dusting off the glass is sufficient to greatly enhance your performance back to optimum and that further physical cleaning is NOT necessary after dust removal. There can be a lot of smudges, stains, flecks and streaks on the glass before it actually begins to degrade your optical performance for all but the most exacting (i.e., high resolution planetary imaging, CCD spectrography and photometry, etc.) demands put upon your telescope.

To dust, use a SQUARE-CUT (not a tip-cut) very soft brush that is about 2" (50mm) wide with tapered bristles. I have found several excellent such brushes at Lowe's and Home Depot and other stores where quality painting supplies are sold. Look for the very soft and flexible "touch up" and/or "delicate trim" brushes....most of these are short-handled and have the bristles as an angled radius cut. Make sure that the bristles are incredibly soft; I use the "cheek method" for testing softness: take the brush out of its package and push the tiny ends of the bristles hard against the cheek of your face....if they do not "prick" then they are fine for optical use. Another tip on selecting a brush is the number of bristles....the MORE bristles on brushes just described, usually the softer and better the quality.

I start dusting by dusting the METAL SURFACES that surround the optics, ridding them of all debris first; just whisk away. Then start at the top of your glass and gently swipe the surface IN ONE DIRECTION....do NOT move back-and-forth with the brush. Stroke in only one direction. Do NOT rub....merely "pull" the brush across the surface and apply no pressure; let the brush do the work for your. Any particles that do not come off with such brush will be removed in subsequent cleaning with liquid if necessary.

The object of your dusting is to essential "move" all the particles to the bottom of the surface you are working on...once there you can brush them off the area and actually assist their removal by blowing gently against the areas being brushed.

USING COMPRESSED AIR: **DON'T**. Period. Dusting is easy, although it may take a little more time, and it is more effective. I have found that compressed air is virtually worthless for attempting to gently remove embedded particles on a glass surface and the chances of the liquid propellants within the can being expelled in liquid droplets against the glass is quite great.

The ASO SuperPlus Optical Cleaning Solution - *how to mix it yourself!*

There is NOTHING magic about the new concoction developed over a period of about five weeks here at the Arkansas Sky Observatory. *SuperPlus* Solution is quite simple, and indeed, there are many familiar components that are being used that have been touted in cleaning solutions before. Nonetheless, after hundreds of elixirs and hours later, this combination - in exactly the proportions given below - results in near-perfect results every time!

In striving for the "perfect cleaner" the following criteria were evaluated:

1) Streaking - the solution was required to dry streak free with minimal "dry rubbing" which can damage optical surfaces;
2) Spotting - the solution must dry spot-free with minimal rubbing;
3) Safety - the solution was required in all respects to be totally impervious to the optical coatings and totally safe for all variations of them on the market;
4) Simplicity - it needed to be something that anyone could mix up when needed with over-the-counter inexpensive components;
5) Sure-fire - it must work every time the first time....the less rubbing the better.

Experiments on all types of optical glass surfaces were conducted with EVERY cleaner offered by all makers and groups; the following SuperPlus Solution was derived as the "best of all of them" since all had some attributes that were worthy, with some extreme cases omitted. Interestingly although some of the solutions that have been previously offered were deemed very hazardous to the quality of cleaning and even the surfaces themselves, some components used within those solutions did HAVE MERIT and have been incorporated! You will be surprised perhaps at the simplicity of this.

HERE IS WHAT YOU WILL NEED:

Nearly all components should be available locally; suggested outlets for obtaining these are in parenthesis.

1) distilled water (supermarkets) – 1 gallon
2) "pure" isopropyl alcohol 94% or higher (pharmacies, drug stores....may have to be ordered) – 1 pint
3) coffee filters
4) "regular" *Windex*, the blue kind (supermarket) – 2 ounces
5) Kodak *PhotoFlo* solution (camera and photo houses only) – ½ teaspoon
6) Synthetic Cotton Replacement Pads (some finer pharmacies, medical supply companies....ask your local M.D.!!) DO NOT USE COTTON!

7) two "atomizers" or simple squirt bottles for dispensing liquids (Wal Mart or similar)
8) box of *KLEENEX* [brand only!] pure white, no additives tissue (supermarket)
9) quart mixing jars, very clean and sterile (try your cabinets!)
10) sterile eye dropper (drug store).

NOTES ABOUT THE INGREDIENTS:

What an how you combine these components, as well as HOW you use them will make or break your success in streak-free and perfect cleaning; please make note of the following:

Pure Isopropyl Alcohol - NEVER use "regular" isopropyl alcohol. Isopropyl is what you commonly see in stores as "Rubbing Alcohol." However, most on-the-shelf varieties is about 70% or less pure....the remaining 30% is impurities which WILL result in streaking and deposits on your glass. USE ONLY **94%** OR HIGHER proof isopropyl....this is found on the same shelf typically, in very large and well-stocked pharmacies. If not, simply ask your pharmacist to order some! Expect to pay about double the price of the "store brand."

Windex - Many cleaning formulae suggest Windex, indeed from one of the largest optical houses in the world. However, there has always been "something wrong" with Windex in that it leaves a ghostly film on optics. After much experimentation, I have found that it is the heavy impurities that are SUSPENDED in the solution that are responsible for the fog....you CAN get them out as you will see. NOTE that ONLY the blue Windex should be used. NEVER use any cleaner with vinegar on your optics.

Kodak Photo-Flo - If you have never used this before NOTE!!! This is extremely concentrated stuff and a tiny, tiny bit goes a very long way! We are talking DROPPER amounts here....NOT ounces. DO NOT USE MORE THAN RECOMMENDED....your results will be horrible.

Kleenex - *ONLY USE* pure white Kleenex, no other brands at all. Do not select Kleenex with "ultra softeners" or with scented oils. Only plain and simple pure white.

Cleaning Pads – I suggest using Intrinsic Synthetic Pads from Barnhardt Industries, http://www.barnhardt.net/intrinsics/ for best results. Nothing else even comes close to these "spa cleaning" pads. You will use these pads to APPLY and swab the optics with the cleaning solution....you will use KLEENEX (and lots of it) to remove the moisture and buff the surface to a slick luster.

HERE IS HOW TO MIX ALL THIS STUFF:

You are making TWO solutions:
1) **Solution 1** - Cleaning Solution: This is the active part of the cleaning and should be mixed very precisely in the quantities provided.
2) **Solution 2** - Rinse Solution: This is ABSOLUTELY necessary for most cleaning session; however, you MAY find that you do NOT NEED the final solution if your optics dry streak-free (which likely they will!).

SOLUTION ONE: Cleaning Solution.

You are going to have much more solution of each component than need for one gallon of final *SuperPlus* Cleaning Solution. Keep all left-over unused and unmixed components well sealed and marked for future use. NOTE that you will be making a gallon of this cleaning solution. It will keep indefinitely, but you will have enough to share with other telescope enthusiasts.

Step 1: FILTER THE WINDEX VIA THE COFFEE FILTER into a thoroughly washed and dried container; go ahead and filter the entire bottle as this is much simpler and more effective than attempting to filter one ounce.
Step 2: REMOVE one pint of your distilled water and discard.
Step 3: FILTER THE DISTILLED WATER using a second clean coffee filter into another clean gallon container. Yes, I know that distilled water is supposedly inclusion free, but trust me on this one.
Step 4: MIX...... In a quart jar, add the following (do NOT substitute nor change amounts!)
    a) the filtered and purified WINDEX - 2 ounce
    b) ALCOHOL – 1 pint
    c) PHOTO-FLO – ½ teaspoon...that's RIGHT, I said "1/2 teaspoon"....any more and you will be sorry. And I mean SMALL drops!! (even this amount is pushing the limit)
Step 5: MIX together gently but do NOT shake.
Step 6: ADD all this to your gallon (less one pint) of Distilled water. I chose to mix my solution in empty quart plastic alcohol bottles; if doing so, merely fill the bottle to within 1" of the top.
Step 7: Pour liquid into your MARKED squirt bottle for use.

SOLUTION TWO: Rinse Solution.

In 32 ounces of filtered distilled water add TWO drops (only!!) of Photo-Flo solution. No more no less. Transfer liquid into SECOND MARKED squirt bottle.
You are now ready to CLEAN your optics.

The ASO SuperPlus Cleaning Technique
You CAN do it right! The FIRST time!

tip #1
CLEAN OPTICS ONLY IN THE DAYTIME WITH THE OPTICAL SURFACE "LOOKING" OUT OF A WINDOW OR TOWARD A BRIGHT OPEN SKY BUT NOT FACING DIRECT SUNLIGHT!

tip #2
NEVER....NEVER...ATTEMPT TO SURFACE CLEAN LARGE OPTICS WHEN THE HUMIDITY IS ABOVE **65%** !! Streaking will result. If you attempt to clean your optics when the humidity is high, you will be very disappointed in the results.

tip #3
PLAN TO USE AT LEAST *ONE TISSUE PER INCH APERTURE BEING CLEANED*....ALWAYS keep a dry tissue to the surface for best results!
There is no solution that will result in satisfactory cleaning if your technique is NOT good when cleaning. Unfortunately with cleaning large glass surfaces, you must normally move quickly, but gently in order to obtain a streak-free and spot-free result. If you follow this technique, you can move a bit more slowly and deliberately AND achieve the same results.

** MAKE SURE YOU HAVE DUSTED OFF THE PARTICLES FROM THE GLASS PRIOR TO FURTHER CLEANING! (see above) **

STEP ONE - Turn your telescope so that you are FACING the corrector plate or lens head-on; you are NOT going to use so much liquid that you need to be worried about cleaning solution getting away from you and down inside the retaining rings of the optics. Make yourself comfortable....you may be here a while! I prefer placing the telescope if possible in a position where I can sit down to clean. You must have a small table or area within reach where you will have your Synthetic Cotton Replacement Pads, solutions and Kleenex waiting.
STEP TWO - Imagine your corrector plate or lens in QUADRANTS or quarters, like large sections of pie. You are going to begin at the TOP left and work your way down to the BOTTOM left piece of pie.
STEP THREE - Gently shake the container (Solution ONE - Cleaner) for just a brief moment and spray a generous amount of liquid onto the Synthetic Cotton Replacement Pad, NOT the glass surface. You want the Synthetic Cotton Replacement Pad WET, but not dripping; make sure you hold the pad only on ONE side and do not TURN to use the side where your fingers have been.
STEP FOUR - Begin in your upper left "quadrant" and gently daub (do NOT rub) this section until you have generously smeared the cleaning solution across the surface of ONLY that area. Never "push" the Synthetic Cotton Replacement Pad, only pull. Do NOT rub. The idea here is to ONLY move the liquid across the surface to break the adhesion of film and dirt deposits against the glass. *MOVE QUICKLY TO STEP 5.....*

STEP FIVE - Before the liquid begins to collect into large areas and before any drying takes place, immediately begin wiping the quadrant just soaked with KLEENEX tissue to dry it....to do this, you want to gently PULL the Kleenex across the surface in ONE DIRECTION ONLY...do NOT go back and forth as this will streak and will tear the tissue into endless amounts of clumps that will have to be removed from the surface. You will see the liquid rapidly drying behind you. Follow each swipe IMMEDIATELY with a DRY Kleenex tissue.

[reminder: **keep changing to a dry tissue constantly!!]**

STEP SIX - When entire quadrant is reasonably dry, buff gently with a totally dry Kleenex; repeat a second time with another Kleenex while gently "puffing" a bit of your breath against the corrector plate or lens to expose possible areas of streaking.

[reminder: keep changing to a dry tissue constantly!!]

STEP SEVEN - Repeat same procedure on remaining three quadrants with a bit of overlap on each.

[reminder: **keep changing to a dry tissue constantly!!]**

STEP EIGHT - Check each point where areas overlapped during cleaning and "touch up" using a fresh Synthetic Cotton Replacement Pad sprayed with a VERY SMALL amount of cleaner....you want this swab nearly dry, but just enough moisture to touch up defects in cleaning.

STEP NINE - Using your breath as a guide, gently "puff" against the glass while using a Synthetic Cotton Replacement Pad to buff the final cleaned surface to a high luster with not streaking!

STEP TEN - [[OPTIONAL]] - *USING THE RINSE SOLUTION*

This step is likely NOT necessary and should ONLY be used if there is any streaking left after the careful cleaning procedure outlined above. If there are problem areas, you should rinse your cleaned corrector/lens as follows:

spray a VERY SMALL amount of rinse solution onto the glass OR place some on a fresh Synthetic Cotton Replacement Pad.....you want only a tiny amount of liquid present to break the surface tension of the glass....remember, the glass is already cleaned from the CLEANING PROCEDURE. All you are attempting to do is to remove any streaks at this point.

You want to gently rub the Synthetic Cotton Replacement Pad across the entire glass area quickly but very lightly and follow WITH YOUR OTHER HAND a fresh dry Kleenex tissue to absorb any moisture remaining from the first pass. This should take care of streaking very quickly.

To finish and polish buff the entire surface with a fresh and dry Synthetic Cotton Replacement Pad to finish.

Breathe on the surface and fog the glass to examine for streaks or missed spots....polish lightly with a light buff using fresh white Kleenex immediately upon fogging.

Best of luck and take your time.....this solution and technique will work on all coated glass surfaces (NOT MIRRORS) and the solution is ideal as well for your binocular, eyepieces and camera lenses.

The key to success is: 1) take your time; 2) work in small areas; 3) use LOTS of dry Kleenex; and, 4) use ONLY the materials and techniques described.

*   *   *

THE CLEANING OF OPTICAL SURFACES: MIRRORS
Part Two:
Front surface mirrors, secondary mirrors and mirrored diagonals

Part One of the **ASO** Fine Optics Cleaning Procedures discusses the methods and protocol for proper cleaning and maintenance of refractive optics: lenses, corrector plates, binocular lenses, eyepieces, etc. The cleaning solution and procedure is a tried and proven method for all coated (enhanced or non-reflective) surfaces as well as glass surfaces which have no coatings.

The cleaning of front surface mirrored surfaces is much, much different than that of refractive optics; many times the reflective surfaces might be of deposits of enhanced silver or aluminum which may or may not be overcoated with some protective layer (usually a molecule-thick layer of *Silicon Dioxide* or similar) of transparent material.
Your first step in attempting to clean ANY reflective optics is to first ascertain whether or not your mirrored surface is indeed protected by such a coating, since the cleaning solution AND procedure to clean without damage is quite specific for protected vs. unprotected mirror surfaces.

Note that the following discussions include all reflecting optics: primary mirrors, secondary mirrors (both flats and curved), mirror diagonals and any ancillary optical equipment which uses a mirror in the optical interface.

DETERMINING MIRROR PROTECTION
A very simple rule for deciding whether or not your mirror surface is protected: if you do NOT know, assume that it is NOT coated.

Most manufacturers of Newtonian mirrors supply the finished product with a coating of silicon dioxide over the final aluminized or silvered coatings; *ASK* whether your mirror is coated....if you cannot get an answer, then assume that it is not.

On the other hand, most primary and secondary mirrors of popular **catadioptic** (Schmidt-Cassegrain and Maksutov) may **NOT** be protective-coated unless otherwise specified.
There is a good reason that many manufacturers do not put protective coatings on telescope mirrors: they can reduce performance, both in terms of optical figure (irregularities in deposited protective coatings can change the wave front of your mirror) and in terms of reflectivity (many new mirror systems have "enhanced" coatings which contain highly reflective alloys in addition to aluminum. However, since most enhanced coatings also contain the element SILVER, and since silver tarnished instantly with

exposure to oxygen, the chances of enhanced optics being overcoated are pretty good in your favor.

Attempts to clean uncoated optics can result very quickly in permanent damage: sleeking (leaving streaks within the coatings themselves) or spotting is very common, even if the utmost care has been used. Never, should any cleaning agent whatsoever be used on unprotected mirror surfaces or damage will occur 100 percent of the time. Simply do not take the chance.

CLEANING IN RELATION TO MIRROR ACCESS

In some modern telescopes, it may be undesirable OR even impossible to totally remove the primary mirror for the typical consumer and end-user; thus cleaning will likely take place less often than it would if the mirror were smaller or easy to remove from the optical tube assembly.

Remember my *Number One Rule on Optics Cleaning*: "**Don't**....unless you absolutely have to."

Number Two Rule is: "Brush first and then determine if cleaning is still necessary." Brushing optics and carefully using compressed air to blow off particulates such as pollen and dust can usually get the mirror or optics back into top shape, and cleaning should be done only if there are stains or excessive spotting beginning to build up on the mirror's surface.

With catadioptic commercial telescopes, cleaning the primary should be avoided at all costs....prevention is the best care you can give the optics of these telescopes: keep the back opening plugged at all times, even when briefly removing accessories....plug it up until you are ready to insert a new gadget. This keeps both dust and insects from floating in AND it prevent humidity and damaging environmental pollutants from entering the inside of the OTA.

In one of the following procedures, not that I discuss cleaning (essential cleaning only....) the mirror of a commercially-built catadioptic by leaving the mirror IN PLACE. Never attempt to remove the mirror of these telescopes unless you have experienced and competent assistance.

CLEANING SOLUTIONS FOR FRONT SURFACE MIRRORS

To preface any discussion about what is needed for cleaning mirrors, it is important to note that complete immersion cleaning of most large (i.e, Newtonian) mirrors is recommended, and thus the "solution" quantity is much greater. There is no need to make up batches of cleaning solutions and store....just make it when you get ready to clean your mirror. Essentially all that you are going to need are two pairs of surgical cotton gloves, a clean terrycloth towel, a small amount of ***IVORY*** dishwashing liquid, a jug of distilled water and a few white *Kleenex* tissues.....oh, a large sink, bathtub, or basin.

Conversely, to clean unprotected mirrors requires ONLY a very small amount of high pure alcohol content solution and nothing more, since ONLY spot cleaning should ever be attempted; if any unprotected mirror surface becomes so pitted or stained that whole-mirror cleaning is needed.....it is time to send the mirror and/or OTA in for a complete re-coating job. No exceptions.

CLEANING MIRRORS WITH PROTECTIVE OVERCOATING

Again, if there is any doubt whatsoever that your mirror has protective overcoatings, assume that it does NOT have protection or be prepared to face the consequences.
IMPORTANT NOTE: rarely do secondary mirrors and diagonal mirrors have any protective coating; always assume that they have front surface exposed enhanced coatings and never clean except as described later.

1) In a basin large enough to hold your mirror and still have adequate room for your hands to grasp around the edges, prepare a solution of the following:
(based on one-half full kitchen sink quantity....this does not have to be exact! For larger basins, such as a bath tub or wash tub, use proportionately similar detergent-to-water ratio)
a) warm, not hot tap water in which you have added ONE TEASPOON of *Ivory Liquid* dishwashing solution....do not be tempted to use more.
b) a thick folded towel placed on the floor of the basin;
c) turn off all fans, vents and central heating/air during this process!

2) Remove all jewelry, including wrist watch and put one pair of the TWO pairs of new surgical cotton gloves on your hands

3) For Newtonian and similar mirrors, first remove the mirror and its cell from the telescope OTA; then remove the mirror from the cell...remember, NEVER touch the front surface of your mirrors...your fingerprint contains acid and oils and can be the most damaging element to your mirror!

NOTE: as with all glass, telescope mirrors become incredibly slippery and hard to handle when wet. Make every precaution to protect the mirror and you will be safest is you "assume the worst" and prepare for the mirror to slip. This means putting a large folded clean towel in the floor of the basin in which the cleaning will be done; having another clean towel folded against a wall and resting on the floor where the mirror will dry.

4) Place the mirror FACE UP carefully down in the basin, resting on the towel, making sure that you have enough solution to completely cover the entire top surface completely.

5) Allow mirror to **soak** for at least 5 minutes but *NO LONGER* than 15 minutes. Do not touch the surface of the mirror at this time.

6) While soaking mirror, **remove** the cotton gloves and place them in the solution with the mirror to prevent contamination.

7) After about 5 minutes refit gloves but do not touch anything outside of the basin; at this time you are going to very, very gently - with NO pressure - massage against the front surface of the mirror with the tips of your fingers....do NOT rub and do not use any type of cloth or tissue at this point, only fingertips [[note that you MAY use *Kim-Wipes* or *Intrinsic* type pads for this process]]. It is fine to "lay down" your fingers and cover more surface.....your are essentially "buffing" the entire surface with the dishwashing liquid using only fingertips.

8) Once done, rotated the mirror 90 degrees and once again massage the entire surface.

9) Occasionally tilt the mirror out of the water for only 30 seconds maximum and examine it.....if there are places that you missed, it will be obvious; if need be, run a very gentle stream of water out of the tap or pour from a pitcher across the mirror and examine while wet; return to basin and massage needed areas until entire mirror is uniformly clean and free of streaks.

10) Leaving the mirror flat in the basis, remove all soapy water from the tub but LEAVE the towel beneath the glass for safety; as the water recedes, begin flushing the surface of the mirror immediately with cold tap water...**NEVER ALLOW THE MIRROR SURFACE TO DRY!**

11) [note: an assistant is quite helpful at this point!] - Once the mirror has been flushed adequately with tap water, begin tilting the mirror upward at about a 45-degree angle; placing an adequate mass of towels behind it is helpful, but careful to not let the mirror slip in the basin! CONTINUE flushing with tap water while doing this...do not let the mirror go dry......have a pitcher of distilled water within reach and shut off the tap water, and immediately flush with distilled water; allow the flush to drip off the mirror and do it again, using only distilled water.

12) REMEMBER - your gloves are soapy....once you have reverted to the distilled water rinse REMOVE the cotton gloves and work with your bare hands only, being careful to only touch the edge of the glass and never touch the optical surface.

13) Lift the mirror out, keeping the surface vertical to the floor and immediately place on the waiting towel on the floor and lean the mirror carefully at a sharp angle against a wall....use extra towels to assure that the mirror will not roll nor tumble. The angle allows the liquid to roll off the surface, thereby reducing substantially the amount of dry water spotting that can occur. NOW, put the second pair of cotton gloves on your hands for safe handling of the mirror from this point forward.

14) After only TWO MINUTES maximum in the drying position (#12), identify any beads of water that are NOT rolling off the surface; these can be easily removed by "wicking", a process in which you roll up a white Kleenex tissue into a "pencil" and touch to the drop...NEVER RUB.....the tissue will wick the water up off the glass and safely away.

15) *Allow to air dry, (*with ALL VENTS from air conditioning/heating closed!) for one hour.

16) Some dust might accumulate during the drying process....use a quality soft artists square tip (see Cleaning Refractive Optics, Part One) brush to remove such lint, but ONLY after one hour of drying time!

CLEANING PROTECTED MIRRORS WHICH CANNOT BE REMOVED:

The above procedure allows for cleaning a primary mirror which can be removed and immersed in a basin; some protected mirrors (i.e., larger Newtonians, some newer SCT and RC catadioptic telescopes) are made in such a way that mirror removal is very difficult or should NOT be attempted. You clean this in three steps.

[**DISCLAIMER:** From experience, I will state that such mirrors should NOT be cleaned, only brushed and blown off with compressed air; I do not, nor does ASO, recommend the following cleaning procedure; the following procedure can be used by skilled and experienced persons in telescope maintenance, but is not recommended for the normal owner/operator of telescope systems. This is the procedure and technique used in the ASO Supercharge and only used when absolutely necessary, and we do not assume any liability from product damage from any attempts at such cleaning.]

CLEANING LIQUID: For such surfaces, use essentially the same process, but instead of immersion, we are going to give your mirror a "sponge bath" applying the soapy liquid (about one gallon water to each one-half teaspoon of Ivory dishwashing liquid). For this you will NEVER USE TAP WATER, only distilled water for both cleaning and rinsing.

RINSING LIQUID: You will also need ONE OUNCE of pure (**91%** or higher) isopropyl alcohol and one capful of *Kodak PhotoFlo* per gallon of distilled water for RINSE (not wash). The application of both solution AND rinse MUST be done using either *Kim-Wipes* (Kimberly-Clark) or *Intrinsic Pads* (Barnhardt Industries of South Carolina)...never, ever use any cloth, tissue or "lens cleaning cloth" for this cleaning or damage will occur.

FINAL RINSE: A final rinse of pure distilled water is absolutely imperative...you must do this final step.
1) Place first pair of cotton gloves on hands
2) Have your one gallon of cleaning solution (distilled water with ½ teaspoon of *Ivory* Liquid) handy with a pad soaking in it; likewise you must have your gallon of RINSE solution (gallon of distilled water with one ounce of pure **isopropyl** alcohol and one

small capful of *Kodak PhotoFlo*) ready with pad soaking it that as well! Your final rinse with pure distilled water needs to be made immediately, so have that ready as well.

3) Put the telescope so that the mirror is angled sharply, i.e., nearly vertical to the ground....your access to the mirror will limit what angles you might be able to achieve here.

4) Making sure that the wipe or pad is ALWAYS completely soaked, but not dripping all over the inside of the OTA, gently begin wiping (Never rubbing!!) across the top ½ of the mirror surface; immediately ...during this process, it is absolutely imperative that you continue to re-soak and freshen the cleaning pad...***never let it dry out*** so much that surface tension increases against the glass!!

5) Even though you have only done ½ of the mirror (always start at the top), you must now quickly RINSE what you have cleaned, using the fresh pad in the rinse solution; keep an abundant (but never dripping) amount of rinse liquid always against the glass! Once the rinse is made, cover the mirror surface that you just cleaned with adequate distilled water final rinse and proceed to clean the lower one-half of the mirror.

6) Once both halves have been cleaned and initial rinse completed, return to the entire mirror and wipe down with copious amounts of distilled water final rinse (no alcohol); repeat twice. Never use so much that your pad is dripping into the telescope tube assembly.

7) Check for water drops that are not quickly evaporating....use a "*Kleenex* pencil" as a "*wick*" to soak up those drops...never rub!

8) Allow to dry in vertical position, with OTA end cap open but with soft cotton sheet over front, for about one hour.

9) Remove any lint or dust with brush or blow off, but do not attempt until after one hour.

USE EXTREME CARE IF ATTEMPTING THE AFOREMENTIONED PROCESS and never attempt unless it is absolutely imperative.

CLEANING MIRRORS WITHOUT PROTECTIVE OVERCOATING:

Very much unlike the previous discussion, cleaning unprotected mirror surfaces should be a "last resort" and is NOT recommended to the normal telescope user. Only if fingerprints, bug droppings, pollen sap, etc. collects on the unprotected mirror surface should any attempt be made to clean. Brushing is encouraged, but cleaning is discouraged.

For this cleaning, you need ONLY a quart bottle of pure isopropyl alcohol (94% is the minimum....99% is far better) and either Intrinsic wipes (*Barnhardt Industries*) or pure white *Kleenex* with no additives....**NEVER** use cotton or cotton balls to clean. Never use Q-tips for cleaning small surfaces, only the pads or tissues as specified.

NEVER attempt to clean the entire surface of ANY unprotected mirror, whether it be a primary mirror or a small flat mirror in your diagonal assembly. Clean ONLY spots and areas needing cleaning.

IMPORTANT: never attempt to clean any spot larger than one inch using this procedure. Use only the following procedure:

1) Place a bright light so that it shines directly onto the surface to be cleaned; you need to be able to see the reflection of the light as well as move your line of sight to inspect so that the light does not shine directly back at you; viewing both ways allows you to examine for streaks and also can assist in preventing you from "over-rubbing" any cleaned area;

2) Put on cotton surgical gloves and locate your area to be cleaned.

3) Put ample alcohol onto your pad or tissue, making it soaked, but not dripping.

4) Very gently wipe the solution across the stain...do NOT rub at all...not one bit. ***Rubbing will remove your coatings!***

5) Follow that wipe with a second one using a totally different wipe or tissue, also soaked with alcohol.

6) Finish by wiping off excess with a fresh dry wipe.....no rubbing, only a light swipe across the surface!

This method can be used on secondary mirrors, unprotected primary mirrors and enhanced coated diagonal mirrors.

HOWEVER, such cleaning is a last resort....never clean unnecessarily and never clean unless it must be done.

*   *   *

Remember what we have always preached at ASO:
the single most damaging thing that you can do to your precious telescope optics is to *CLEAN THEM.*

While it is perfectly safe to clean the protected primary mirror OR the front corrector plate of a catadioptic (*ASO Part One*), it is an entirely different undertaking to clean unprotected mirror surfaces.

My utmost recommendations concerning cleaning of unprotected mirrors?
DO NOT unless absolutely necessary. Let those with experience do it for you or live with the small imperfections....when they get too big, it is time for new coatings.

Best of luck and enjoy your telescopes...may the stars always shine their brightest through them.

Chapter 18

PRECISE POLAR ALIGNMENT
"Doc Clay's Kochab Clock"
Precise Portable Polar Alignment EVERY Time!

There are many published methods by which extremely accurate celestial alignment can be attained for Polar mounted (not Alt-Azimuth mounted) telescopes. It is my feeling that the reason that telescope users do NOT use Polar mode is because what is perceived to be a time-consuming and many times frustrating alignment procedure.

However, using "Clay's Kochab Clock" you can achieve near-perfect celestial alignment of any fork-mounted telescope in a matter of minutes instead of the "hours" it takes when using the DRIFT METHOD. At my observatory, both permanent telescopes were aligned precisely using this method and will track dead-center all night long except for minor periodic error; even my portable telescopes - when aligned by the "Kochab Clock" method - will routinely track for up to four (4) hours without any perceptible deviation in Declination.

Many of you know me as dead-set IN FAVOR OF Polar mode tracking for four (4) main reasons:

1) sidereal tracking is much more precise, as it only requires ONE curved arc motion (right ascension) to compensate for the Earth's rotation;
2) for GO TO and rapid slews, the operation is much more precise if alignment is accurately achieved;
3) long-exposure photography (even piggyback) is better rendered in Polar mode; and,
4) I AM OLD AND SET IN MY WAYS and that's the way "real" telescopes are "supposed" to operate!

For your portable telescope to track like an "observatory instrument" all you need to know are your exact latitude from which you are observing and a few "sky basics" as I taught them to college classes in Field Astronomy.

Before we get started, a warning: the following procedure merely SEEMS like a lot of information and a LOT of time required; I guarantee you that READING this will take far longer in the ultimate end than actually becoming proficient at USING this technique to perfection. It's like the proverbial bicycle...you may have to fall off a couple of times and balancing may be something that just doesn't seem "natural," but suddenly - once you're up and running - you realize just how sensible and easy this procedure really is!!

THE LOCATION OF CELESTIAL NORTH

Fortunately for those in the Northern Hemisphere there is a bright star very close to the point in the sky toward which the Earth's "axis of rotation" or POLE extends. This position is the *NORTH CELESTIAL POLE* (the "NCP") and is within only one (1) degree of the 2nd magnitude star "Polaris," the brightest star of the constellation Ursa Minor; it is the star marking the "end of the handle" of the little dipper.

Alkaid, the tail-star of the Big Dipper is in line (nearly) with Kochab and Polaris. It can help assure that its really Kochab that you are using. The Big Dipper is much easier to recognize than the little one. Now we need a finder 'scope with a ¾ degree radius circular reticule and mechanically rotatable cross-hairs. Then merely observe the angle to the horizon of Kochab, match it in the finder, then adjust azimuth and the wedge as needed.

Although it is very close to the true NCP, merely aiming a telescope's axis of rotation (the "right ascension," "Polar" or RA axis) precisely at Polaris is not adequate for high magnification tracking, even for short periods of time. Rather, the mounting must be offset almost a FULL DEGREE - about 43 arc minutes (43') - FROM Polaris to actually be aligned to the NCP.

For a reference, note that there are 60' arc in one degree....the full moon presents a disk that is approximately one-half degree or 30' arc.

So, using the moon as a reference - and the fact that nearly TWO lunar disks fill the space between Polaris and the true NCP - you can quickly realize that "Polaris-close" is simply NOT close enough! but WHICH DIRECTION do you offset? And once you know the direction, how do you determine HOW MUCH the Polar axis must be offset?

Is it necessary to merely "guess" at how far ONE degree really is?

Here are two quick answers......we'll get down to putting these answers into action later in this guide.
DIRECTION? True Celestial North is found by drawing an imaginary line FROM *POLARIS* to the second-brightest star in Ursa Minor, *KOCHAB*. The NCP is located 43' arc FROM POLARIS to KOCHAB along that line. Kochab is shown in Figure 1 and in Figure 1A through 1D, and comprises the "top outer" point of the "bowl" of the little dipper ladle as shown. Both Polaris and Kochab are about the same brightness, so the two stars are unmistakable.

HOW MUCH? You will see that, thankfully, we don't have to guess at our distance of about 1 degree. Without the moon up there to compare with, how do we decide how much to offset our Polar axis (RA) once we have "imagined" our line from Polaris to Kochab? If we simply had "two full moons" side-by-side up there next to Polaris it would be a piece of cake and easily close enough for very accurate tracking! There are many finderscopes on the market today that contain reticle eyepieces (some illuminated) that clearly mark the distance for THAT FINDER that the entire telescope (optical tube assembly and mount together) must be moved away from Polaris to zero in on the NCP. However, most of today's modern GO TO computerized telescopes do NOT have such finders since most manufacturers suspect (unfortunately) that the user will be opting for the more convenient and easier "Alt-Azimuth" mode which does not require precise identification of the NCP. Nonetheless, you will see how easy it is to use your own finderscope - no matter what size nor type it may be - to quickly acquire the NCP. You need only to "get friendly with it" to understand the orientation of north, south, east and west and to know how much sky you are covering when viewing through it!

But first things first: It's time you learned "Clay's Kochab Clock" and its convenience and speed in providing you with precise polar alignment every time....on time!

"CLAY'S KOCHAB CLOCK"

Figure 1 demonstrates four major times of the year which I call the "Cardinal Points" of the celestial pole. These occur roughly at about 9 p.m. local time on: MARCH 15, JUNE 15, SEPTEMBER 15 and DECEMBER 15. Looking at the drawing of Ursa Minor and its stars Polaris and Kochab in relation to the NCP reveal why these dates are so very important for "getting a handle on" Polar alignment. These four "points" are the basis to understanding *Clay's Kochab Clock* because you are dealing with the NCP only in terms of either "up from...", "down from....", "left of....", or "right of....." Polaris at these exact dates at 9 p.m.!

Once you have learned the locations and times of year/night of these cardinal points in relation to the NCP, you will be able to quickly extrapolate the NCP from any given position of Polaris and Kochab on any given date or time to within very precise accuracy for most portable observing situations.

Consider the positions at about 9 p.m. of the three features shown (Polaris, Kochab and the true NCP) in this drawing:

MARCH 15 (spring) - the NCP is due "RIGHT OF...." (or EAST) Polaris toward Kochab;

JUNE 15 (summer) - the NCP is straight "UP FROM...." (or actually SOUTH in the sky) Polaris toward Kochab;

SEPTEMBER 15 (autumn) - the NCP is due "LEFT OF...." (or WEST) Polaris toward Kochab; and

DECEMBER 15 (winter) - the NCP is straight "DOWN FROM...." (or NORTH in the sky) Polaris toward Kochab.

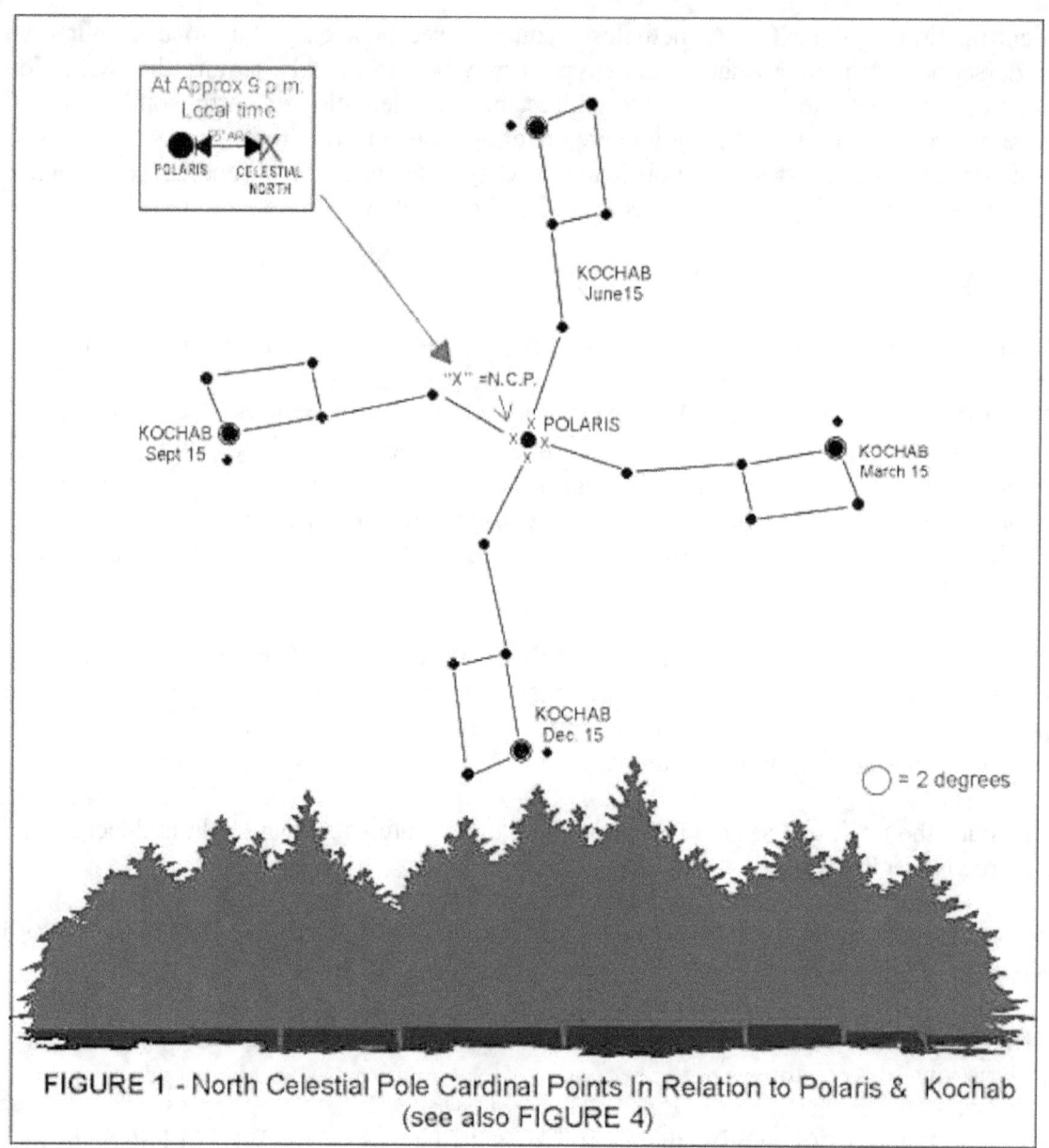

FIGURE 1 - North Celestial Pole Cardinal Points In Relation to Polaris & Kochab
(see also FIGURE 4)

So let's look at one best-case scenario to get familiar with how all this works. Let's say that you want to go out into the dark woods to observer with your telescope on June 15 or thereabouts. In theory, you can quickly se that if you "aimed" your fork-mounted telescope RIGHT AT POLARIS and then - at about 9 p.m. -simply moved the entire scope, mount and all STRAIGHT UP by about one (1) degree and locked it into place, you'd be as close to true celestial north as expected of even observatory telescopes! It really IS that simple....like riding the bicycle after many failed attempts.

Another "best case" scenario falls on September 15, where - after centering Polaris "up and down" in the telescope and clamping the elevation at that position as it is aimed due north - you merely have to move the assembly "TO THE LEFT..." one (1) degree (west) and you are aimed right at the celestial pole! The same simplicity applies to December 15 and March 15 as well....now examine Figures 1A and 1D and determine mentally what direction you must move the telescope assembly to hone in on the NCP for those two dates.

....but those are not the only times we want to observe, right? What about all those days in between? And what if you wish to align, say at midnight and not 9 p.m.? You can STILL use the "Celestial Pole Cardinal Points" (up, down, right, left) throughout any given day of the year and any time of night knowing the REVOLUTION of the constellation Ursa Minor around the celestial pole through the course of one NIGHT. Since it is "circumpolar," or never rising nor setting below the horizon for mid-northern and above latitudes, you can use Ursa Minor's stars the "Kochab Clock" as shown in Figures 1A (an example for March), 1B (an example for June), 1C (an example for September) and 1D (for December) which clearly demonstrate that your offset can be determined through the course of any particular evening as the Earth turns and makes Ursa Minor appear to revolve around celestial north.

Remember, there is nothing complicated about this, even if you wish to align on some date outside of these "cardinal points." Such times will merely be COMBINATIONS of the one-degree offset in "up-down" and "right-left" movements. The distance will ALWAYS be one degree.

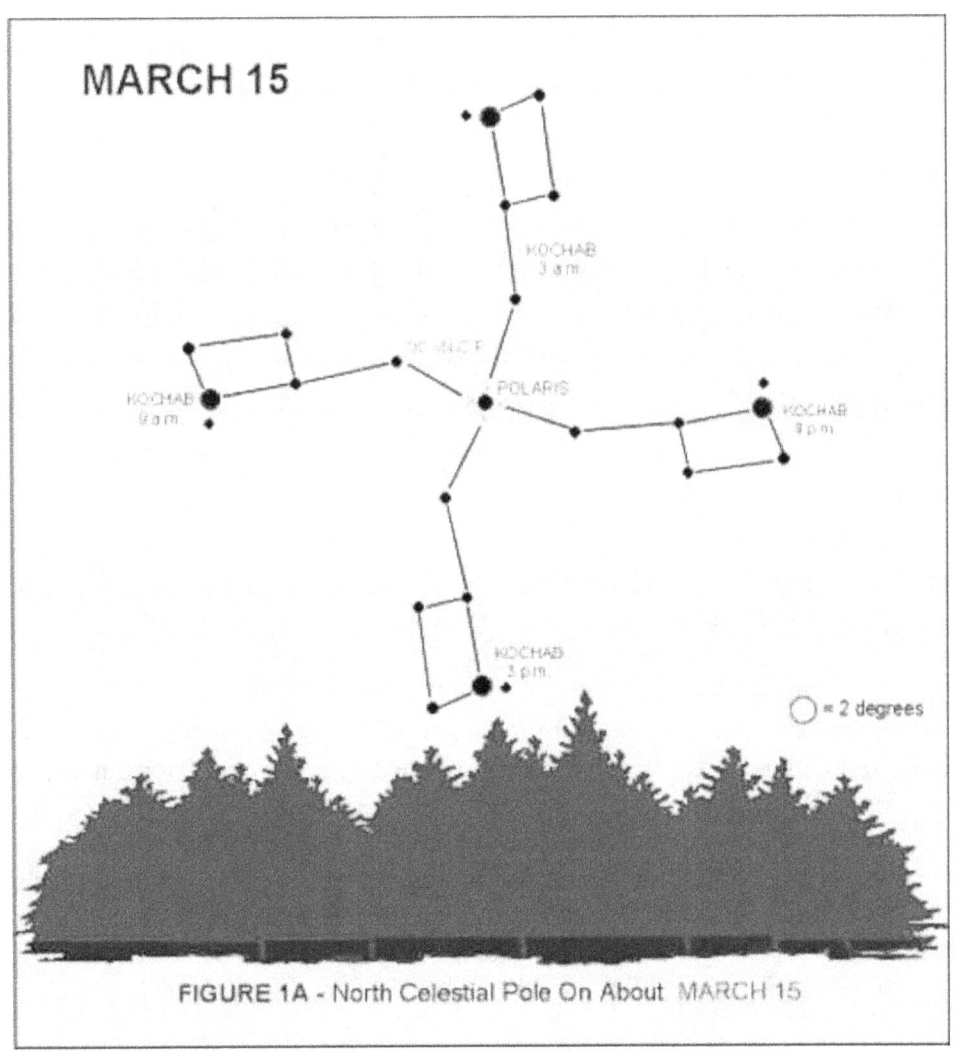

FIGURE 1A - North Celestial Pole On About MARCH 15

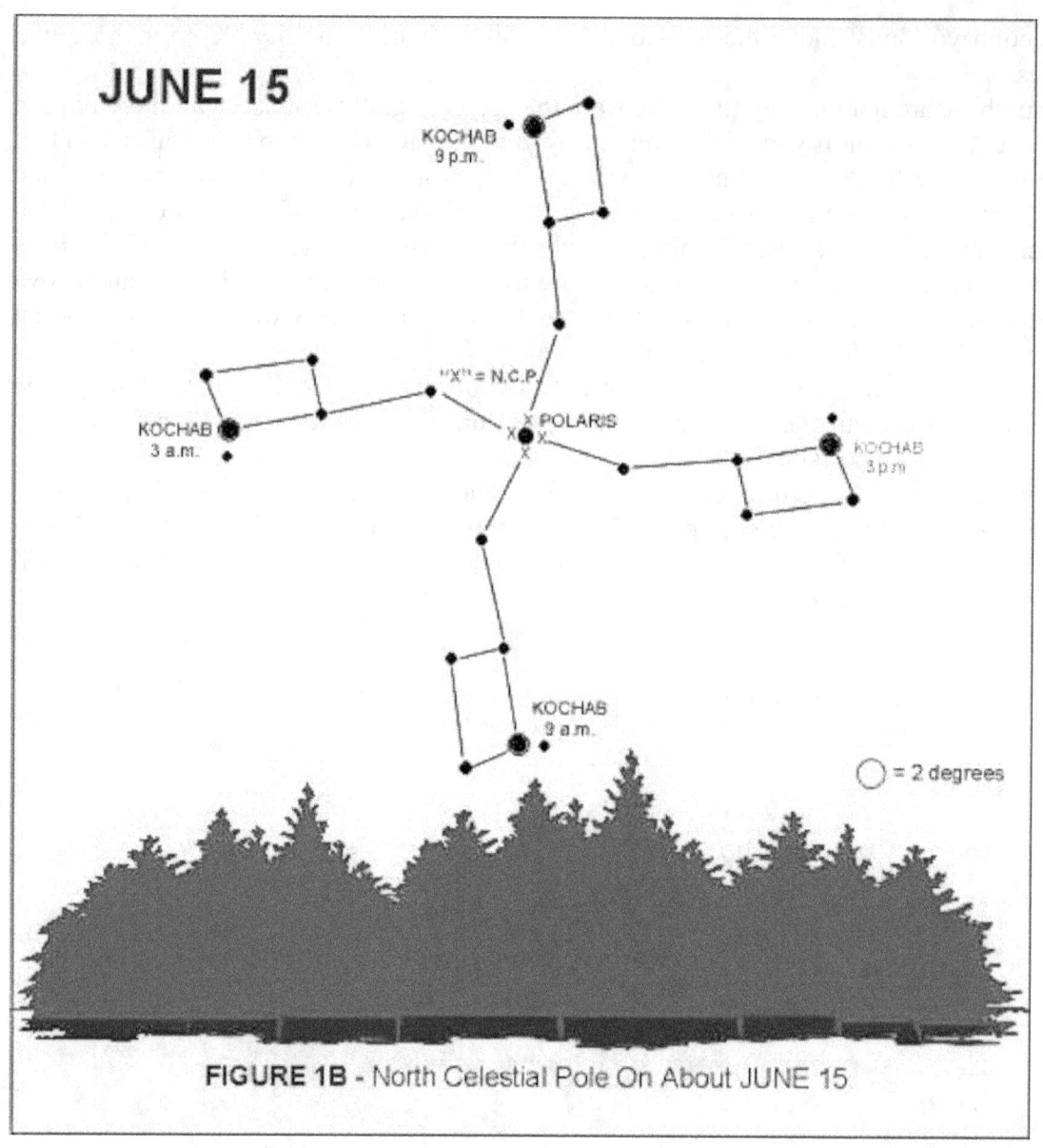

FIGURE 1B - North Celestial Pole On About JUNE 15

IMPORTANT: Although I have provided some quick references for the four cardinal dates in Figures 1A through 1D, a better reference to determine the position of Ursa Minor and Kochab would be through one of the many wonderful "planispheres" or "star dials" that are commercially available at bookstores or through mail-order or e-mail astronomy shops. With these, you merely "dial up" the date and time that you observe, normally hold the planisphere over your head aligned to north and determine the position of Kochab for ANY night and time of the year!

There are two ways to utilize the charts in Figures 1A, !B, 1C and 1D (or better still, the "planispheres" mentioned above). First, you can extrapolate for the positions of Ursa Minor (and thus Kochab, Polaris and the NCP) for months between the "Cardinal Points;" in other words, using the Figure 1A (March 15) and Figure 1B (June 15) you can see that Kochab will be located midway between the two 9 p.m. sky positions about 45 days from March 15 (or, halfway to June 15) which would put you about on May 1.

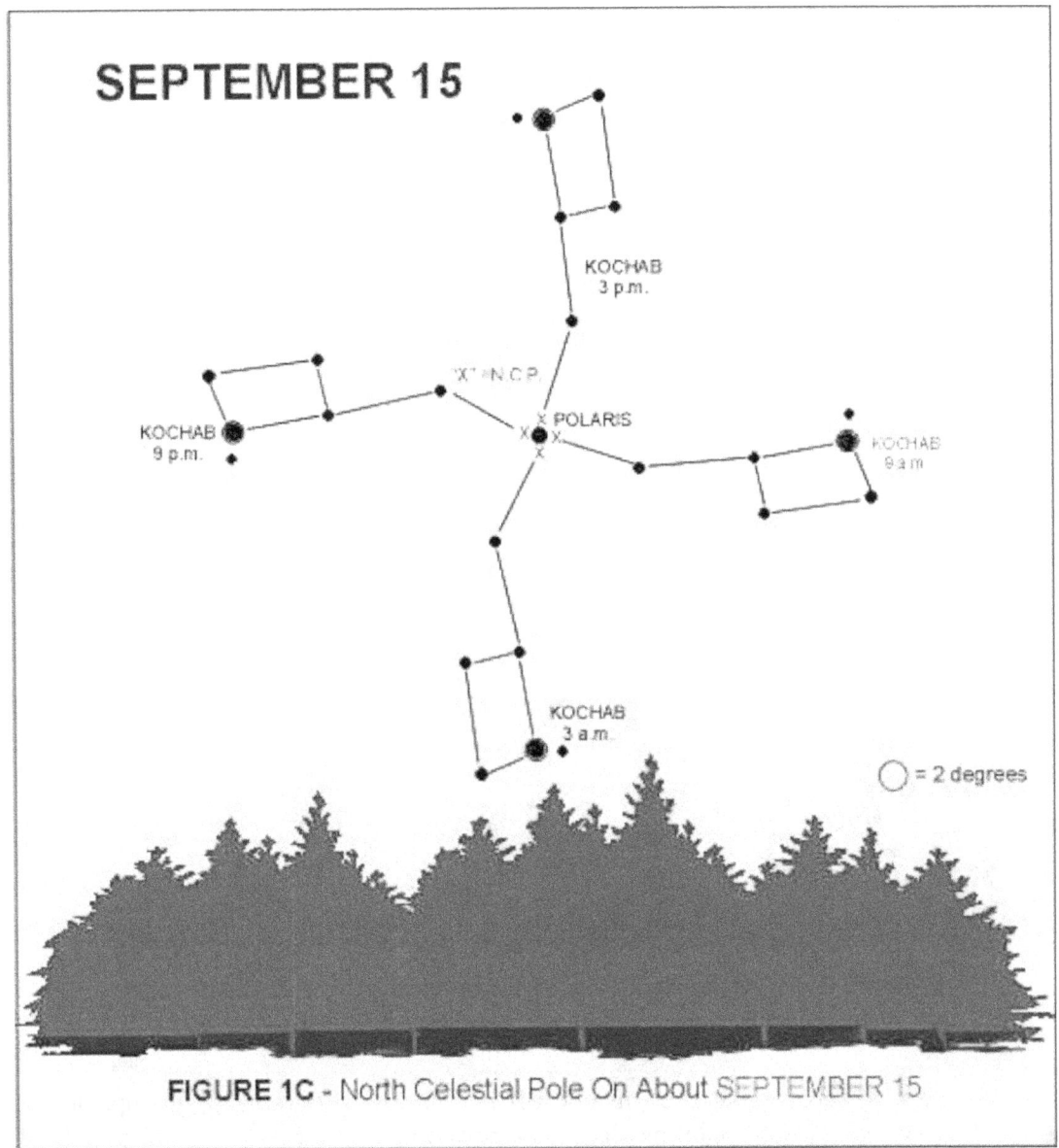

FIGURE 1C - North Celestial Pole On About SEPTEMBER 15

Second, you can use these charts to extrapolate the position during EACH NIGHT; for example, in December we see the position for Kochab at 9 p.m. and again still in darkness at 3 a.m.. If you are wishing to observe about MIDNIGHT, the star will have revolved about halfway from the 9:00 position to the 3:00 a.m. position, and that 45 degree angle can be used as your reference to the position of the NCP.

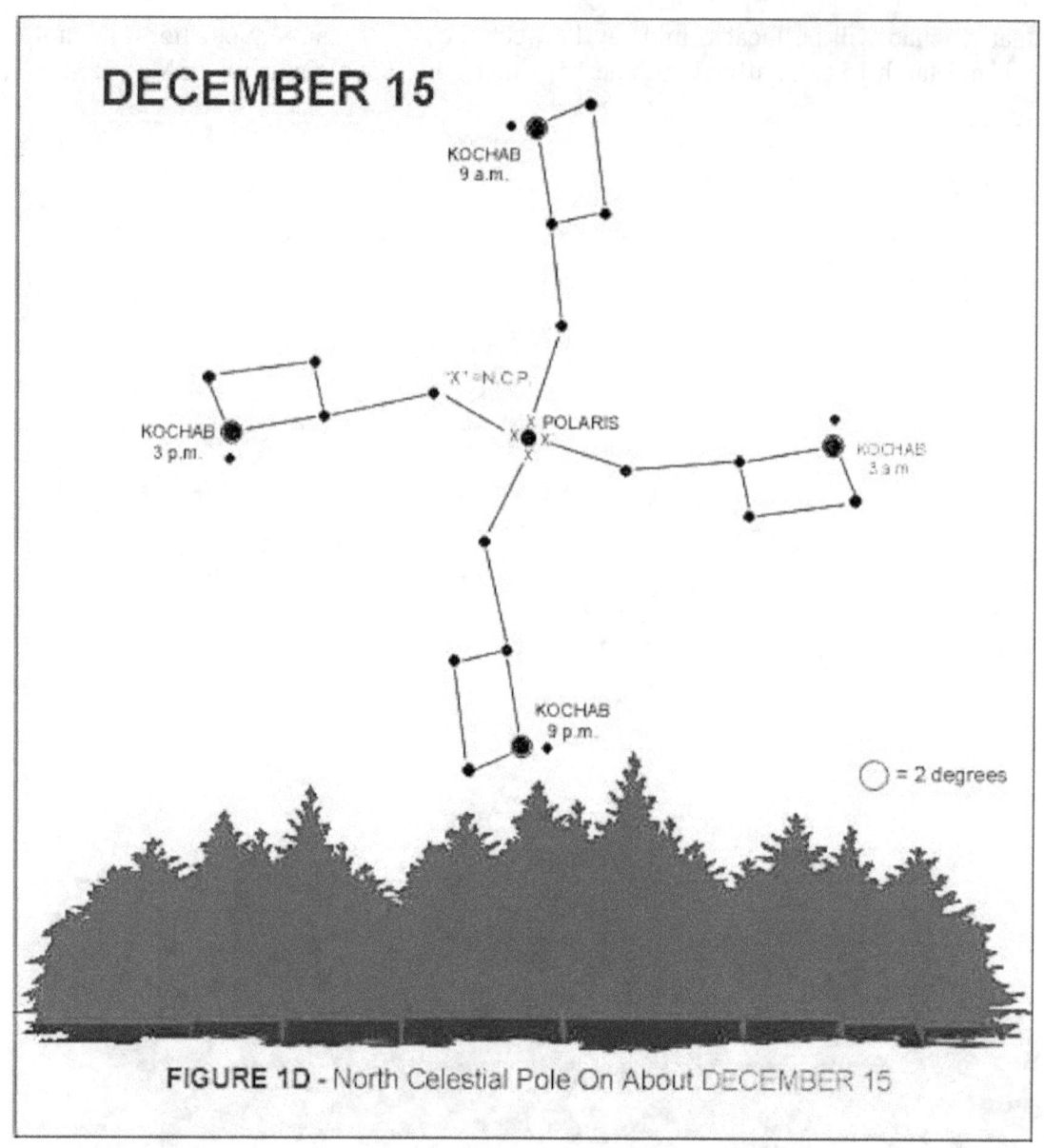

FIGURE 1D - North Celestial Pole On About DECEMBER 15

Ideally, you want Kochab positioned in one of the Cardinal Points (straight up, down, right or left from the celestial pole) as closely as possible; an hour or so on either side of the times given will get you VERY close to true NCP using these stars. The reason is quite simple: all it takes to acquire true north is ONE adjustment either in altitude or azimuth and no guessing in between. As a quick and easy example, look at the chart for March (Figure 1A). If you are setting up your telescope at or around 9 p.m., you merely must move the assembly to point EAST of Polaris by about one degree since Kochab is due east at that time; the NEXT cardinal point for Kochab on the same night will be a 3 a.m., when the telescope mount must be moved UP from Polaris to similarly acquire NCP, again using only ONE adjustment direction to center.

SETTING UP THE POLAR TELESCOPE TO USE THE "KOCHAB CLOCK"

There are many ways to set up a telescope in equatorial (polar) configuration quickly; even simply sighting Polaris in the center of a low power eyepiece when aimed at north and level is suitable for quick observing.

However, more exacting projects such as star parties, astrophotography and research projects requiring accurate GO TO operation, will require more refinement of polar alignment. For use of the "Kochab Clock" for accurate and quick alignment, you merely must know the position of that star in relation to the time and date you are observing (from Figures 1 through 1D), your latitude of the observing site (to within ½ degree) and the FIELD OF VIEW (FOV) of your finderscope equipped on your telescope.

Although the diagrams which follow show a fork mounted telescope, this easy process can be done with any polar mounted German equatorial telescope as well.

Let's go through a quick and easy "Kochab Clock" Polar alignment for the night of June 15:

1) Determine your observing site latitude (using USGS survey maps, a GPS device or any good map will provide the latitude to about ½ degree); remember, your declination setting circle WILL BE USED, but it is accurate ONLY to within about ½ degree for this purpose so any more refinement is overkill. At the same time, look up or determine (see below under "Using Your Finder....") the exact field of view of your finderscope.

2) Whatever base you are using (tripod head, flat plate of the wedge, pier or table top) make sure at this time that it is very level both north-south and east-west. NOW, make sure your finderscope is aligned to the main telescope optical path (you may use a terrestrial object to tweak the finder into position). IT IS VERY IMPORTANT that your crosshairs to the finder are IN LINE with both axes of the mounting, in other words, one crosshair running E-W and the other exactly N-S!

3) First orient your telescope as shown in FIGURE 3 (Polar Home Position) as this will allow you to roughly set the telescope and its mount (on a wedge or tripod) to the correct latitude of your observing site. With the telescope aimed "outward" (north) DIRECTLY IN LINE with the two fork arms as shown in Fig. 3, move ONLY the wedge/tripod or combination (DO NOT USE THE TELESCOPE BUTTONS OR CLAMPS....you must have the scope fixed as shown in Fig. 3!) to center the assembly toward POLARIS. This is merely to orient your equipment reckoned close to true north.

4) Once the main telescope is visually aimed fairly close to Polaris, tighten down all the wedge/tripod adjustments for now; unclamp both axes and rotate the telescope in declination and RA to the position shown in Figure 2, with the optical tube assembly (OTA) pointing straight UP; the setting circle is shown on the "right" fork only for illustrative purposes (it will actually be on the opposite fork arm).

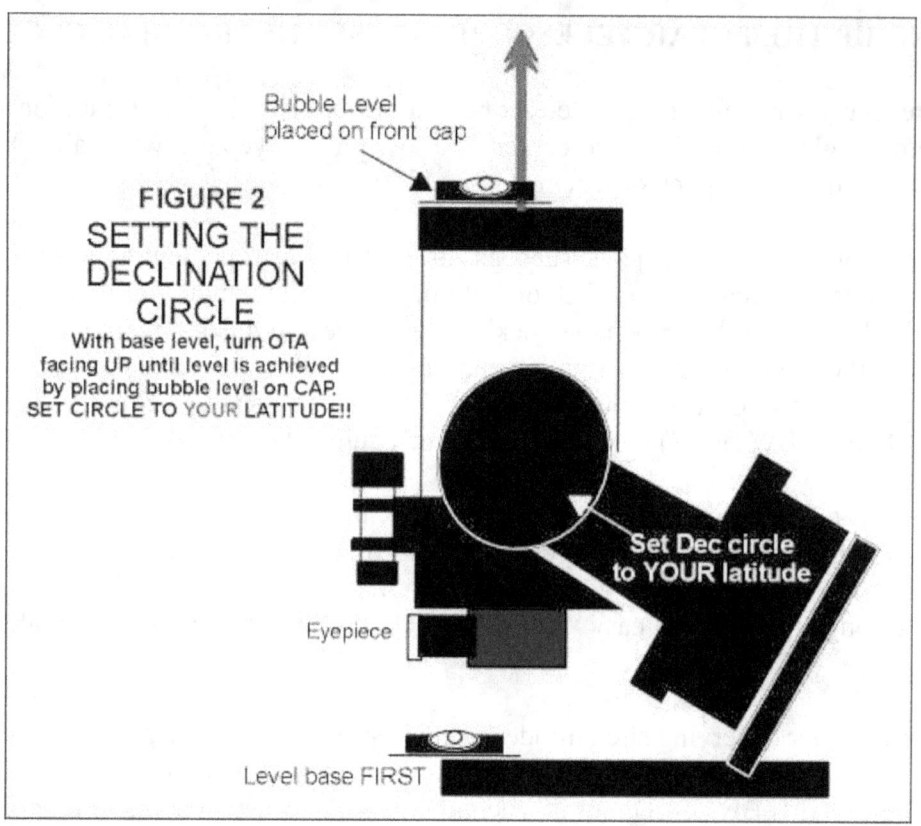

**FIGURE 2
SETTING THE DECLINATION CIRCLE**
With base level, turn OTA facing UP until level is achieved by placing bubble level on CAP.
SET CIRCLE TO YOUR LATITUDE!!

5) With the OTA roughly aimed upward, unlock the RA axis; using the FLAT SURFACES of the lower parts of each FORK ARM (they are very perpendicular to the OTA and can be used for this purpose) place a bubble level that can read perpendicular-level against the fork and rotate the forks in RA until they are straight-up-and-down as indicated by the level; this will assure that the two fork arms are parallel to the ground.

6) Clamp the RA axis again and - WITH THE LENS CAP ON THE OTA - unclamp the Declination axis once more and place the bubble level as shown in Figure 2 across the NORTH-SOUTH direction of the lens cap which provides a nice flat and true surface. Rotate the OTA back and forth slowly in Declination ONLY until level is achieved here and then lock firmly down.

7) When the end of the OTA is level (assuming that your BASE [the wedge or tripod head] is also level) then your telescope will be pointing ASTRONOMICALLY AT THE ZENITH, or perfectly straight up. This angle at which the telescope is pointing in Figure 2 is ALSO the angle of your exact latitude of the observing site! Thus, check the declination setting circle; if it DOES NOT READ exactly the latitude that you pre-determined in Step 1) above, then loosen and carefully adjust until it reads that latitude precisely.

8) All of this trouble so far is merely to get into "home position" and so that you can be assured that your Declination setting circle is reading PRECISE declination.....because you need it to for accurate polar alignment using the "Kochab Clock." Once set, very slowly as to not upset the circle adjustment, unclamp and move the OTA in Declination

only toward due north until the setting circle reads "90" degrees. Refer to Figure 3 to see this configuration [NOTE: it is highly unlikely at this point that Polaris will be in the field of view of your main telescope, but most likely WILL be seen somewhere in the finderscope.]

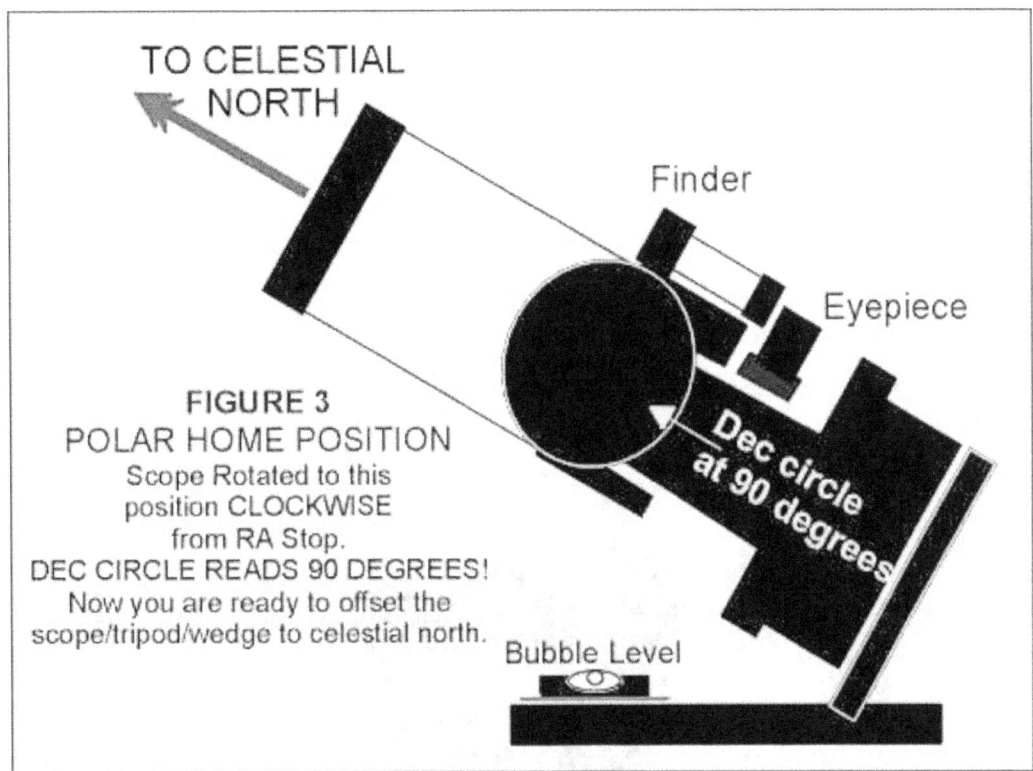

9) Once there, make sure the telescope's RA and DEC clamps are firmly engaged and NOW MOVE ONLY the wedge/tripod adjustments... NOT the telescope control keys or clamps....and center Polaris in the field of view of the finderscope only, making sure that your finder is adjusted to sight exactly with your telescope. (see Figure 4 for details). Do NOT lock down the adjustment clamps for the wedge, tripod or base at this point....you will need to keep them loose to offset the assembly to true N.C.P.

10) Your telescope OTA and the fork mounting are now optically and mechanically aligned and ready for locking onto true Celestial North using the "Kochab Clock" offsets as previously discussed. If you have extrapolated correctly the exact position (up, down, right, left or ANY COMBINATION of two of these directions of offset) then you merely will move the telescope in that specified direction; perhaps you will be lucky enough to be observing on one of those nights or one of those times that fall precisely on one of the "Cardinal Points" for the NCP.

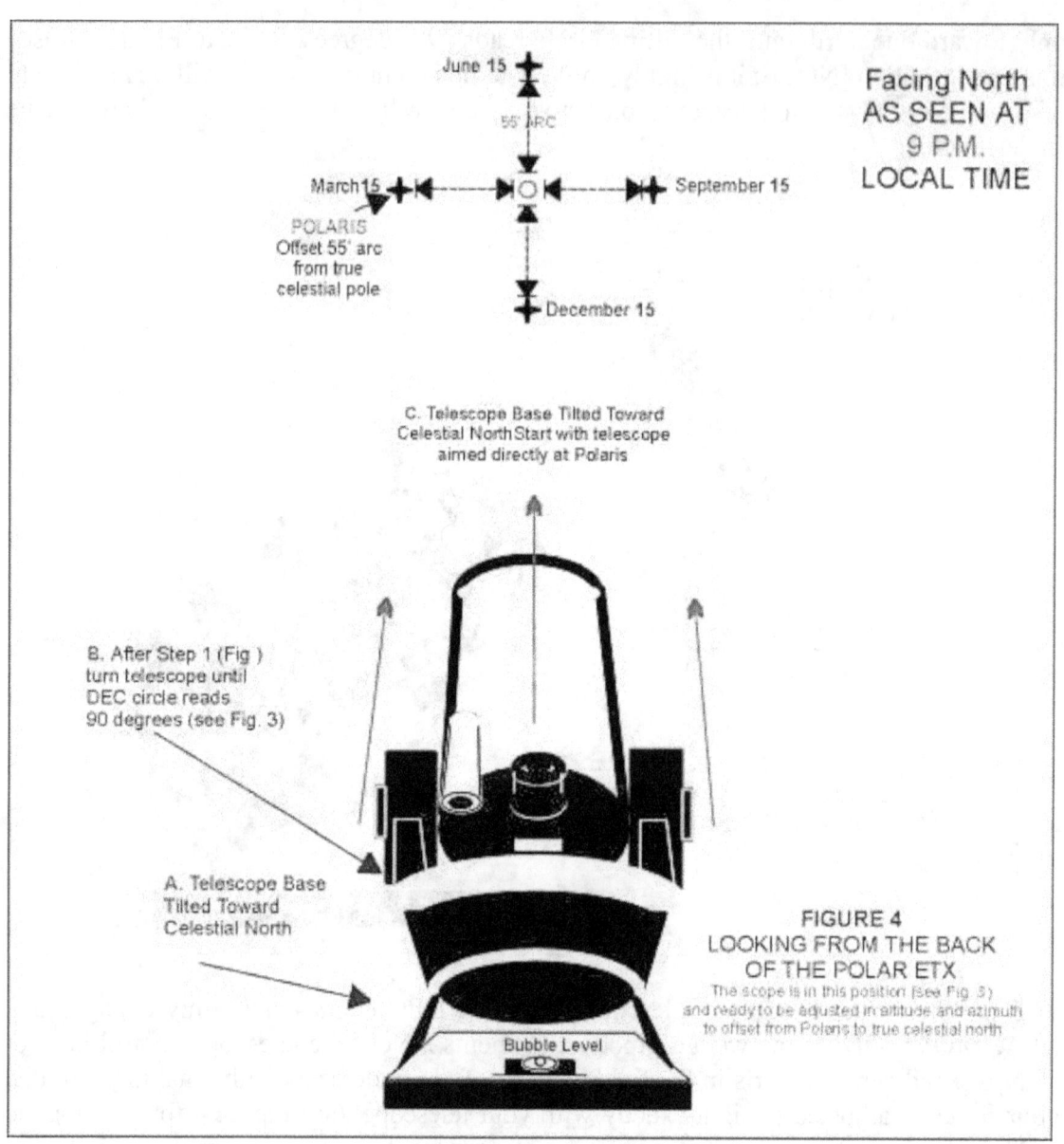

FIGURE 4
LOOKING FROM THE BACK
OF THE POLAR ETX
The scope is in this position (see Fig. 3) and ready to be adjusted in altitude and azimuth to offset from Polaris to true celestial north

To this point, we have identified how easy it is to reckon what DIRECTION to move the assembly toward the NCP, but now we must determine HOW much. Fortunately, the amount of offset from Polaris toward the NCP never changes...it will always be about 43' arc minutes.

IMPORTANT TIP - HERE IS THE KEY: If you can find a time when Kochab, Polaris and the NCP will be at any of the "Cardinal Points" all the easier. REMEMBER, in this example it is at 9 p.m. on June 15, so you first simply CENTER POLARIS in the crosshair and lock down your azimuth adjustment. At the right time, when Kochab is DIRECTLY UP (see Figure 1B), then simply RAISE the elevation (altitude) of your wedge or tripod by about 1 degree....it is the ONLY motion you need to make for alignment! If using the right angle finder to acquire the NCP, then Polaris would be seen to drop DOWN as you raise the elevation of the scope assembly.

On the other hand, if observing on September 15 at 9 p.m. or thereabouts (Figure 1C) you can see that you would CENTER POLARIS in elevation (altitude) and lock THAT adjustment down and - again as soon as Kochab is seen (Fig. 1C) just LEFT of Polaris and the NCP, merely PUSH the mounting assembly to the LEFT one degree to acquire exact NCP accuracy. If using the right angle finder (like supplied on my ETX 125) this would make Polaris appear to shift to the LEFT in the field of view.

USING YOUR FINDER TO OFFSET TO THE N.C.P.

Before discussing the distance needed to offset Polaris so toward the NCP, BE SURE to understand "sky directions" in your finder. The following orientations will apply to all finders for our automated telescopes, no matter what size nor brand:

1) straight-through finders - SOUTH at top; NORTH at bottom; EAST to the left; WEST to right.

2) right-angle finders - NORTH at top; SOUTH at bottom; EAST to right; WEST to left. This is important to know when viewing through the finder and knowing the direction in which the NCP is OFFSET from Polaris toward Kochab (see Figure 4 for apparent actual SKY DIRECTIONS for the NCP relative to Polaris at seasonal times of the year)!

FINDER FIELDS OF VIEW: Since many finderscopes may be used with our GO TO telescopes, I have provided values that give a "close" angular field of view for the common finders that will be utilized. Note that, even from finder-to-finder of the same brand and size, there are variations but these numbers are close enough to provide angular measurements accurate for polar alignment. For each of the finders, note that I have subdivided the finder's field into ½ and ¼ increments; see Figure 5 for an illustration showing these increments.

FINDER: 8 x 21 straight-through.....FIELD: 6.8 degrees / = 3.4 degrees for 1/2 / = 102' arc for ¼ field;
FINDER: 8 x 21 right-angle.............FIELD: 6.4 degrees / = 3.2 degrees for 1/2 / = 96' arc for ¼ field;
FINDER: 6 x 30 straight-through.....FIELD: 5.2 degrees / = 2.6 degrees for 1/2 / = 78' arc for ¼ field;
FINDER: 8 X 50 straight-through.....FIELD: 4.8 degrees / = 2.4 degrees for 1/2 / = 72' arc for ¼ field.

FIGURE 5
Standard Finderscope Fields of View

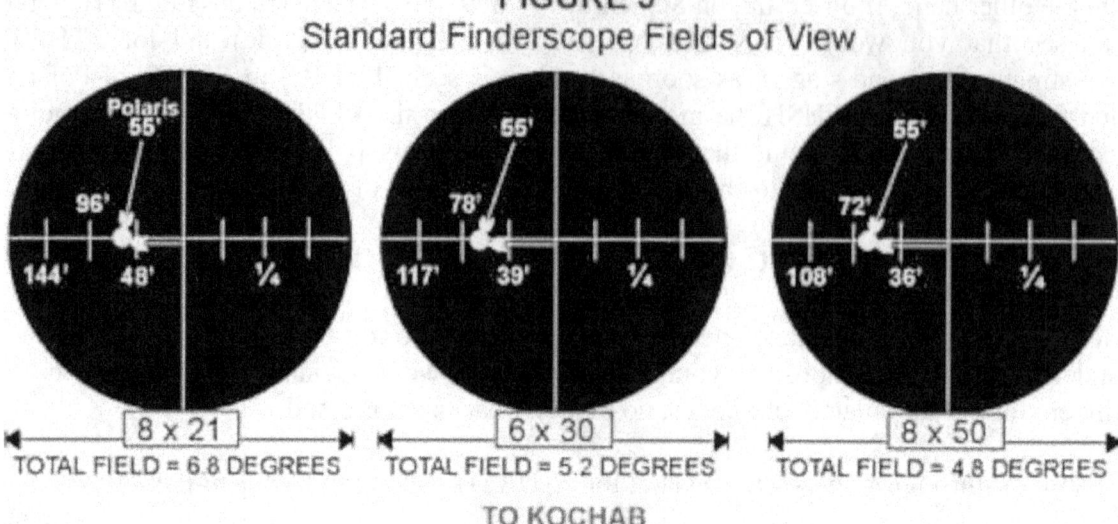

In Figure 5, we are not really concerned with "direction" and thus it does not matter whether this is a straight-through or right angle finder, but only the size or FIELD the finder might show; this is to demonstrate how to SUBDIVIDE your finder to determine the correct distance to move along any one (or combination of two if not offsetting to "Cardinal Points") crosshair to the proper angular distance from Polaris.

Note that a ¼ field for ALL finders in Figure 5 is too great an offset for the NCP from Polaris; rather, determining where slightly more than 1/8th the field of most finders would give us a very close approximation of the needed 43' arc spacing. REMEMBER: In Figure 5, the "direction" of offset shown is for EXAMPLE ONLY. This merely shows how far to move Polaris off of center. Each arrow points to the position relative to center that Polaris would need to be moved in each finder to PUT THE NCP IN THE CENTER of the finder (the offset of course MIGHT need to be up or down, or to the right, rather than to the "left" as shown).

The 1/8th increment is EASIER to visualize that it might at first seem. Since you have a "crosshair" the field of the finder is actually already divided into two halves. So 1/8 the TOTAL field of your finder is merely a fourth the distance from the center of the crosshairs to any edge! We can "think and visualize" in quarter increments a LOT better than in 1/8ths!

The two 8 x 21 finders are ideal for just this measurement, meaning that YOUR offset in such a finder would be slightly more than ONE-QUARTER of the way FROM Polaris (now centered in the eyepiece) toward the edge of the finder field that you have determined is the proper direction to move (Figure 4).

Remember at this point that the field orientations (N.S,E,W) are different as described above for right angle and straight finders; the best thing to do is to determine for yourself by nudging the end of the telescope OTA in a certain direction and watching the motion of stars in the finder as you do. THAT is the only way to be certain you are offsetting properly. Just remember where KOCHAB is....THAT is the direction you want to move the assembly by just about one (1) degree, that imaginary line running from Polaris, across the NCP through Kochab.

A TESTIMONIAL ABOUT ACCURACY

This final thought about using the apparent motions of Ursa Minor and the "Kochab Clock" method:

A few years ago I finished installing a permanent observing station for a college in Texas, which was expand its observing and astronomy program, adding new telescopes and one permanent facility. As soon as the last bolt was installed, I set about to permanently align the pier so that the students could easily initialize the telescope for each night's observing with no fuss about polar alignment.

So....as I have done so many times with permanently mounted telescopes....I began working on the adjustments to the wedge and pier through the ever-popular "drift method," which has taken as much as a WEEK to accomplish to my satisfaction in the past.

The irony of this choice of celestial alignment was that - as I was overseeing the construction - I continued to observe nightly from January from the same pier which I had aligned using my "Kochab Clock." I aligned this way as soon as the concrete pedestal cured, when Kochab was DUE EAST of Polaris about 11:30 p.m. one evening (one of my "best-case scenarios"). I merely waited until Kochab was EXACTLY to the "right" of Polaris, and - with Polaris first centered in the crosshairs and locked tight in elevation - I simply pushed the huge wedge at the right time exactly ¼ of the way FROM CENTER to toward the RIGHT (see Figure 1 and Figure 1A). This allowed Polaris to "scoot along" the E-W crosshair until I decided I had moved it just the ¼ distance necessary (with the right-angle 8 x 21 finder Polaris appears to move to the RIGHT) when pushing the wedge and scope to the RIGHT).

So....for well over a MONTH, the school continued to report perfect GO TO's, perfect slews, "....4-hour tracking on Saturn with no correction," all from a celestial alignment that took less than 10 minutes using *"Clay's Kochab Clock"* method. It was better performance than my own observatory-mounted equipment that I had laboriously aligned using the "drift method." (see following chapters)

THEN, last week I decided to re-align my own observatory telescope using the "drift method" and NOTHING WORKED RIGHT. My GO TO's were marginal at best, I had declination drift....and on and on. I was so unhappy that I immediately "undid" my tedious three-day drift alignment and went back to *"Kochab's Clock"* on March 7 when the moon was near-full. At about 9:30 p.m. Kochab was once again immediately RIGHT of Polaris (after I had adjusted the altitude to put Polaris in the center) so it was in "perfect" position. I moved toward it and locked down in about 15 seconds, tops, and did a test GO TO at 146x on the star Castor.

Dead center. Then I went all the way across the sky to Capella at the same magnification....dead center. Later, about 11:30 p.m. I put the scope on Regulus and did a trial run on tracking at 410X. I went off and had coffee since the full moon was obliterating views of everything else in the sky. The results? At 3:47 a.m. sharp the next morning when I went out to shut down the power to the telescope ....

Regulus, now far over in western skies from earlier that even, was still shining a brilliant yellow in the eyepiece of my telescope.....DEAD CENTER.

*   *   *

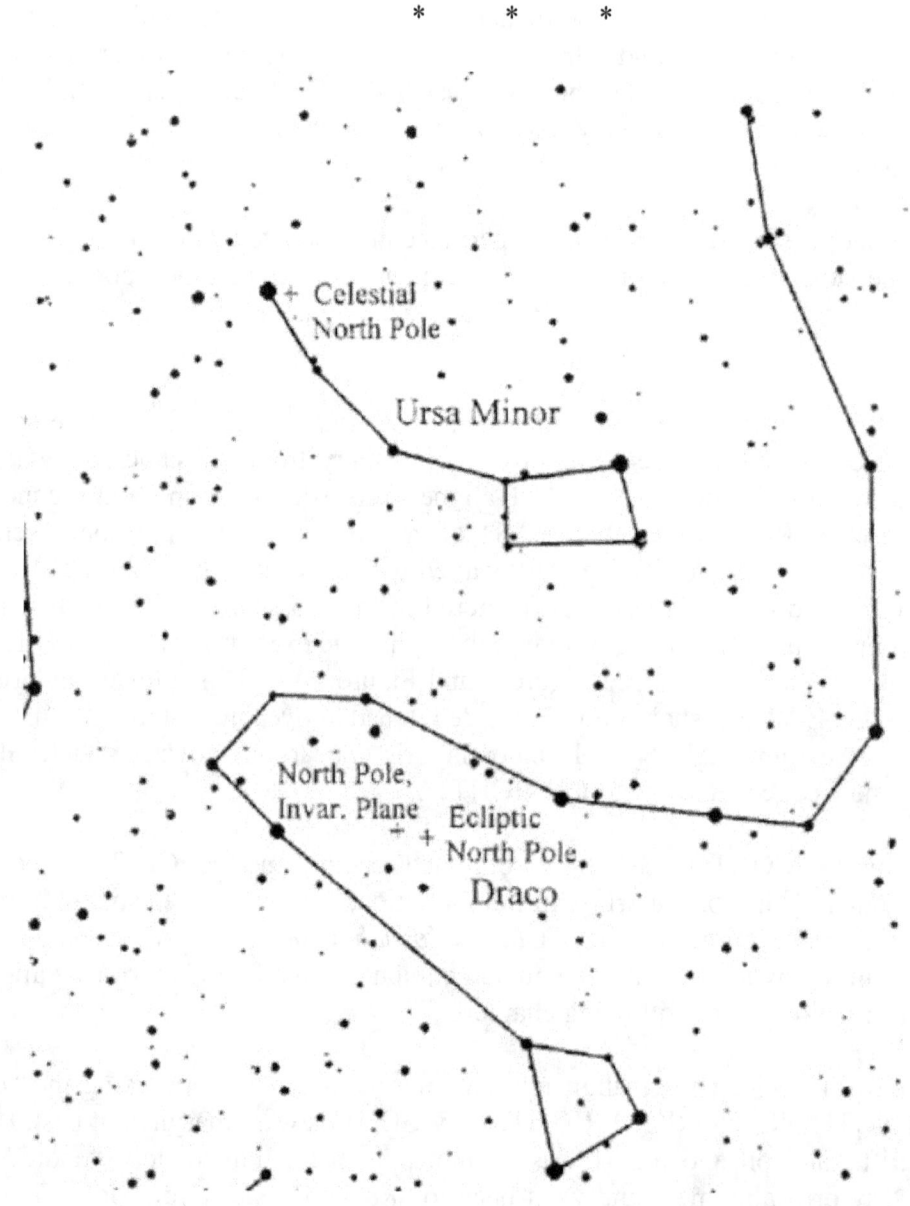

Chapter 19

Dr. Clay's Permanent Polar Alignment Without Polaris

This method allows anyone to get a semi-permanent OR permanent alignment of a telescope which cannot have access to a clear shot at Polaris nor the North Celestial Pole (NCP). This can be done with either a German Equatorial or Fork mounting.

First, hit celestial north as best you can without actually seeing it. When doing the two-star alignment, you can expect to see at least one if not both stars pretty far off; your setting them to center is your communication to the telescope and Autostar that it is not set exactly right and this will help to equalize this.

When you have a moment (during the full moon) a good exercise that can get you closer for permanent installation (you did not say if you were using a permanent setup or bringing it all in every night) is to go ahead and do your initial alignment.

Once done, slew to a star about 30 degrees NE of the meridian with a known declination; look at the DEC circle and note how much you are off from what it should be (be sure and use epoch 2000 coordinates);

Next, unclamp the DEC and turn it until the actual known DEC is entered on the dial properly.....the star will not not be in the field of view. DO NOT move thke telescope via the hand control nor its slow motions, and do NOT unlock its clamps in either axis... adjust the equatorial mount or fork mount wedge so that you can bring that star BACK into center of the telescope and stop once done.

Now, repeat the process by slewing to another star in the NW part of the sky and do the same thing again.....your offset should be about half of what it was before the first procedure.

Once that is done, your final step is to slew to a star ON the meridian and almost overhead (near the zenith). You guessed it: do the same thing to that star. It should be a very small amount by this time, adjusting only the wedge and leaving the two clamps firmly engaged.

It takes some time and maybe a bit of muscle to fine-tune the pointing of your mounting, but the benefits are tremendous.

If you have the tripod and big wedge, why not MARK the feet of the tripod onto the surface that you use most AFTER this procedure is done; them just reset them each time out. Or, you can do like Jim Phillips in Saipan and leave the wedge and tripod outdoors and cover with a good weather proof tarp. A periodic squirt of WD-40 does wonders!

Chapter 20

Precise Polar Alignment Using the Drift Method

There are many ways to Polar Align your equatorial telescope; among the most popular for over two decades is "*Clay's Kochab Clock*", using a slight offset from Polaris toward the bright star Kochab in Ursa Minor (see Kochab Clock in the previous chapter).

But by far the most accurate of all, is the "**drift method**" of polar aligning, which takes about 1-2 hours (can be hastened for portable work with less accuracy). This method today is best done using a CCD camera on "focus" mode or any webcam and monitor where some form of crosshair has been added. This electronic method is not only easier on the observer, but far more accurate and less prone to error.

No doubt the Internet is packed with links to "Drift Align Methods" that take the process to the extreme….I have seen some descriptions on the process that require as many as 10-12 pages of text and intricate diagrams.

Folks, it is not that difficult and not worthy of worry and fear of failure. Drift align is very simple: if you can watch a star on some crosshairs of a telescope eyepiece, or on the computer screen and simply determine if it is "moving UP or moving DOWN" then you can understand this process and carry it out masterfully.

I have been using the drift alignment method for nearly six decades with success from the very first time.

Spend the time doing your drift alignment for your permanently mounted equatorial telescope (portable telescopes need not bother) when the moon is full, or there are partly cloudy conditions. Drink lots of coffee, or iced tea….listen to some music.

All that is required is for you to watch the North-South drift of a star while your telescope tracks. You do not need to know celestial mechanics….no need for logarithmic functions.

Most methods of polar alignment is limited by the accuracy of your telescope's setting circles and how well the telescope is aligned with the mount. The following method of polar alignment is independent of these factors and should only be undertaken if long-exposure, guided photography is your ultimate goal. The declination drift method requires that you monitor the drift of selected stars. The drift of each star tells you how far away the polar axis is pointing from the true celestial pole and in what direction. Although declination drift is simple and straight-forward, it requires a great deal of time and patience to complete when first attempted. The declination drift method should be done after the previously mentioned polar alignment steps have been completed.

To perform the declination drift method, you need to choose two bright stars. One should be near the eastern horizon and one due south near the meridian. Both stars should be near the celestial equator (i.e., 0° declination). You will monitor the drift of each star one at a time and in declination only. While monitoring a star on the meridian, any misalignment in the east-west direction is revealed. While monitoring a star near the east horizon, any misalignment in the north-south direction is revealed. As for hardware, you will need an illuminated reticle ocular to help you recognize any drift. For very close alignment, a Barlow lens is also recommended since it increases the magnification and reveals any drift faster. When looking due south, insert the diagonal so the eyepiece points straight up. Insert the cross hair ocular and rotate the cross hairs so that one is parallel to the declination axis and the other is parallel to the right ascension axis. Move your telescope manually in R.A. and DEC to check parallelism.

First, choose your star near where the celestial equator (i.e. at or about 0° in declination) and the meridian meet. The star should be approximately 1/2 hour of right ascension from the meridian and within five degrees in declination of the celestial equator. Center the star in the field of your telescope and monitor the drift in declination.

If the star drifts south, the polar axis is too far east.
If the star drifts north, the polar axis is too far west.

Using the telescope's azimuth adjustment knobs, make the appropriate adjustments to the polar axis to eliminate any drift. Once you have eliminated all the drift, move to the star near the **eastern horizon**. The star should be 20 degrees above the horizon and within five degrees of the celestial equator.

If the star drifts south, the polar axis is too low.
If the star drifts north, the polar axis is too high.

This time, make the appropriate adjustments to the polar axis in altitude to eliminate any drift. Unfortunately, the latter adjustments interact with the prior adjustments ever so slightly. So, repeat the process again to improve the accuracy, checking both axes for minimal drift. Once the drift has been eliminated, the telescope is very accurately aligned. You can now do prime focus deep-sky astrophotography for long periods.

NOTE:
If the eastern horizon is blocked, you may choose a star near the western horizon, but you must reverse the polar high/low error directions. Also, if using this method in the southern hemisphere, the direction of drift is **reversed** for both R.A. and DEC.

Chapter 21

The Iterative Method of Polar Alignment

One of the most simple of all polar alignment techniques that is applicable to both German equatorial mounts as well as fork mounted telescopes is that of the "Iterative Method" whereby you use the GO TO function of your computerized telescope to hone in ever-so-closer to true celestial north by using a sequence of repetitive steps with the star POLARIS and your choice of bright stars.

This method is also tried-and-true and highly recommended for quick and easy polar alignment for field setup at your dark sky site as well as incredibly precise polar alignment for any observatory telescope.

You will use three stars: Polaris, one star about 45 degrees west of overhead, and a third star 45 degrees east of overhead.

If using a German equatorial mount (GEM), please be careful when slewing to Polaris and by all means watch the telescope assembly as it moves around your tripod or pier. Slewing to Polaris is a very difficult position for the German equatorial, but quite easy for the fork mounted instrument.

The process is quite simple and effective:

1) set up the scope and level it with the fork arms aimed NORTH or with the GEM have the mount's RA axis pointing toward Polaris and set as close to possible with the angle of Polaris.

2) turn the tube so that it is also facing north, or straight out the fork arms and reading 90 deg. on your DEC circle. Lock it down.

3) if using a fork mount, rotate the OTA until you can feel the hard stops to make sure that the scope is positioned on the wedge so that it can rotate through from horizon to horizon

4) then move the scope into **Polar Home Position** (see your telescope model instructions on Home Position), still aimed due north in line with the fork arms BUT with the finderscope DOWN on the underside (fork mount only).

5) if you can see Polaris in the finder, then go ahead and CENTER it using ONLY the adjustments on the GEM or WEDGE to move the mounting...do not unclamp the scope. You are NOT moving the telescope relative to the mount.

6) NOW TURN ON THE SCOPE.....make sure that your Telescope/Mount/ is defaulted to POLAR if your mount can be used in dual Polar/Alt-Az modes....(!!)

7) when you get to ALIGN, select Polar ONE STAR (**never** Two Star)

8) Press ENTER....scope will rotate in RA until it thinks it has Polaris....it will tell you to CENTER Polaris, first in the finder then in the scope, using ONLY the mount adjustments or wedge, not the hand control.

9) Once centered, you press ENTER again and it will slew to a bright star....at this point you USE THE KEYPAD to center that selected star.

10) Now, making sure the star remains centered, simply press ENTER again once centered....then press it AGAIN and HOLD for 3 sec...that will Sync the scope on that star. Note that star syncing (or "Calibration" for some model telescopes) process will vary from model to model, but all computerized mounts can sync or calibrate on any star.

11) Now do a GO TO back to Polaris.....

12) It will likely NOT be centered in the scope but should be close.....

13) Again, use ONLY the MOUNT or WEDGE to center Polaris yet again. **Do not** press any keys.....you are mechanically moving the mount.

14) Once you have centered Polaris via the Wedge or Mount, then do a GO TO to the bright star in the western sky that you are sure you identify correctly.

15) Center that star with the keypad and press ENTER and hold for 3 sec. to again Sync. (or Calibrate, based on your keypad requirements)

16) Now do another GO TO back to Polaris and once again center that star using ONLY the mount or wedge adjustments, but only move it ½ way toward center and stop.

17) Do a GO TO to the star in the EASTERN sky now and center using only the keypad.

16) You can keep doing the GO TO back to Polaris as often as you want and using the wedge or mount to get ever-closer....but you should be quite close at this point.

This process takes less than 10 minutes tops. If you have a computerized GO TO telescope, you will love this alignment method.

Chapter 22
Telescope Setting Circles and How to Use Them

Introduction

To newcomers to astronomy there is a certain "mystique" surrounding those numbered "dials" which come as standard equipment on a quality equatorial telescope. A telescope equipped with an equatorial mount is one which can be correctly "aimed" at celestial north (Figure 1) thereby allowing the easy tracking of celestial objects as the earth turns on its axis. Once so aligned, the telescope - through the motions of the equatorial mounting - is able to "keep up" with the object through just one slight motion compensating for the earth's rotation.

Believe it or not....faint objects in the sky CAN BE located without the use of a microprocessor and a hand paddle!

To allow an observer from anywhere in the world to access (without an AutoStar) an object that might be too faint to see with the naked eye, the sky is mapped much like the Earth's globe. On Earth, the sphere of our planet is marked through latitude and longitude and all permanent objects on its surface are mapped accoringly. To understand the coordinates of the sky - declination and right ascension - you must first understand how they correlate with these similar positional measurements on Earth.

Because they are so distant and do not appear to change, it has allowed astronomers over the centuries to catalog and accurately map the stars, the many clusters, galaxies and nebulae within them into catalogs and star atlas. So, this method of charting the sky into right ascension and declination allows us to map the celestial objects on the celestial sphere just like latitude and longitude on the sphere of our Earth have allowed us to geographically specify the location of each and every city, mountain, lake, and even HOUSE on the surface of our planet!

Latitude and "Declination"

Latitude is the NORTH/SOUTH measurement of the earth, in degrees from the equator (0°) to the north pole (+90°) or to the south pole (-90°). As an example, the city of Arlington, Virginia is located from the equator toward the north pole at about Latitude +37°. By contrast, the southern hemisphere city of Melbourne, Australia is positioned from the equator to the SOUTH pole at about Latitude -37°. No matter east-to-west on earth, the measurement of latitude remains the same.

In the mapping of the sky, the equivalent (north-to-south) measurement of latitude is called "declination" and is measured - like on the earth's sphere - from the celestial equator (0°) north (positive degrees) and south (negative degrees). Thus, "Celestial North" is +90° while "Celestial South" measures -90°. If an imaginary line is drawn through the earth, extending through both the south pole and the north pole, this line would point in either direction to the "Celestial Poles," south and north, respectively.

In the northern hemisphere, this imaginary line extends spaceward toward the relatively bright star POLARIS in the constellation of Ursa Major, and is used as a guide for aligning telescope equatorial mountings as described in "Kochab's Clock," in the preceding chapter.

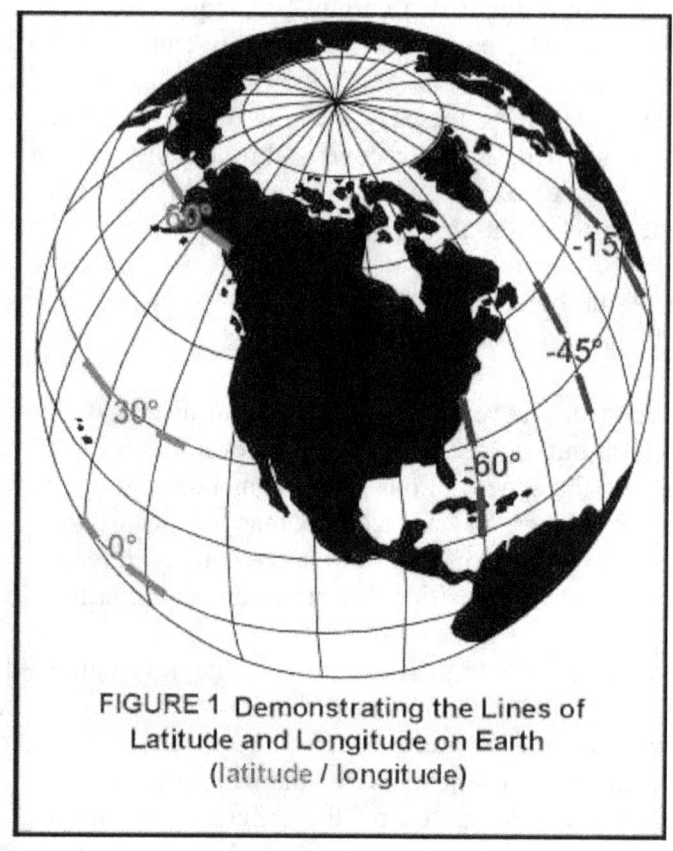

FIGURE 1 Demonstrating the Lines of Latitude and Longitude on Earth
(latitude / longitude)

Unfortunately, no such bright star exists at or near the south celestial pole. No matter where you are on earth, from an east-west standpoint, the measurement of Latitude remains the same. "Figure 2" shows lines of declination - similar to "latitude" - in the sky in relation to the position and angle of the Earth.

Longtitude and "Right Ascension" –

On Earth the mapping of EAST/WEST direction is in degrees of Longitude. Arbitrarily beginning at the Royal Observatory in Greenwich, England (0° Longitude), longitude is measured eastward (positive degrees) and westward (negative degrees). At the International Date Line, both "meet" and equal a common 180°. For example Memphis, the southern city in the United States, is nearly one-quarter the way WESTWARD around the globe from Greenwich, at a latitude of -91°; going the other direction from Greenwich, the Russian city of Yartsavo is almost one-quarter EASTWARD, at +90°. As with east-west direction to measure latitude, it does not matter if you are north or south of the equator to measure longitude.

The measurement of east-west direction of the sky is markedly different than Longitude on Earth. Called Right Ascension in the celestial sphere, it is measured NOT in degrees, but in hours, minutes and seconds, just like a clock for good reason. Each hour of right ascension equals 15° of sky.

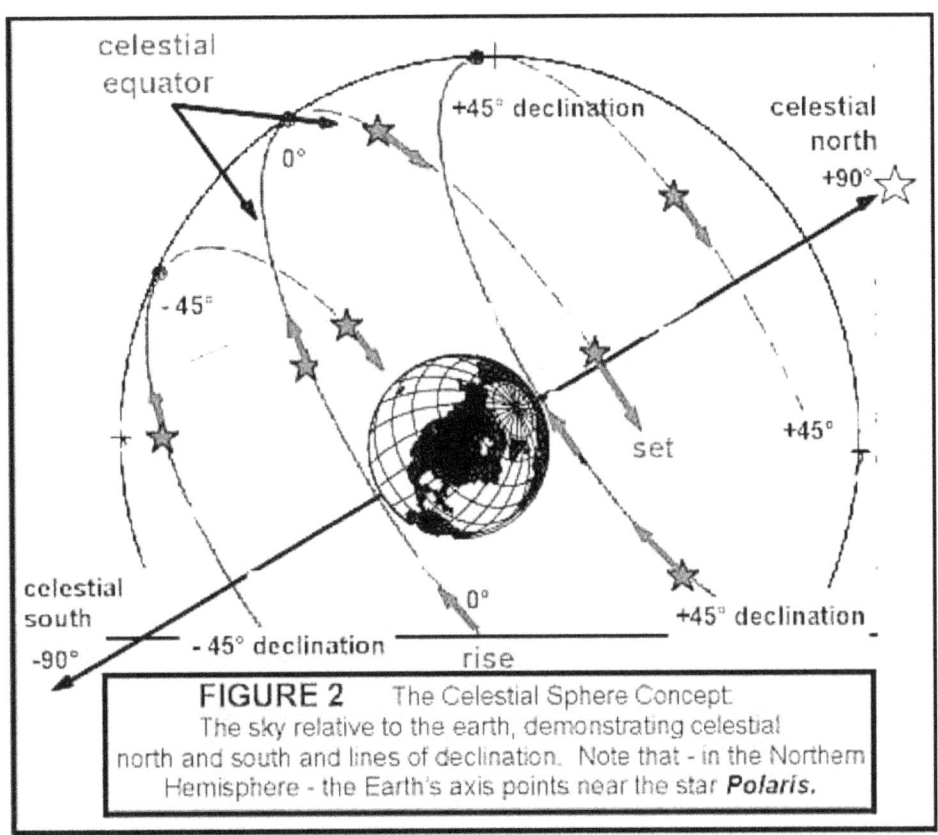

FIGURE 2 The Celestial Sphere Concept: The sky relative to the earth, demonstrating celestial north and south and lines of declination. Note that - in the Northern Hemisphere - the Earth's axis points near the star **Polaris.**

Our 24-hour clock is measured in solar time, the time it takes for the Earth to rotate one complete time on its axis, measured from midnight to midnight. Because it is measured relative to the sun, this is a very accurate and little-changing period of 24 hours. But - if measured relative to the stars, and not the sun, the Earth keeps different time - sidereal time. This is because, in addition to spinning on its axis the Earth is ALSO revolving around the sun, one complete trip every 365 days. This results in a very slight differences between star time (sidereal) and local time (solar) of about four (4) minutes each day.

In simple terms....since this is not a treatise on celestial mechanics.....a star that is EXACTLY OVERHEAD tonight at 10 p.m. local time will pass directly overhead tomorrow night at about 9:56 p.m., or four minutes earlier. In one month, that difference in sidereal vs solar time is a whopping two hours (30 days x 4 min/day = 120 minutes)! This "gain" results in different constellations and stars gradually rising earlier each successive night. Eventually, there will be different constellations seen at different seasons. But in the course of a lifetime....each season will ALWAYS show its own set of constellations!

Apparent Motions of the Stars Throughout the Year - The Earth's Revolution: Our "Year" - Local Mean Time vs. "Star Time"

Because of the difference in sidereal time and local solar time, (resulting in all stars appearing to rise about 3 minutes 57 seconds earlier each successive night) you will note that the reference stars will change monthly as a result. This constitutes the Earth's "Year." Thus, your ideal reference stars are those identified as bright and visible, close to your desired object for that particular night you are observing.

Apparent Motions of the Stars In One Night - The Earth's Rotation: Our "Day"

As the Earth turns on its axis in its 24-hour day, the stars appear to rotate from east-to-west throughout the course of every night. A star rising a 9 p.m. will - at midnight - be on the meridian, an imaginary line extending from the North Star, Polaris, directly overhead and continuing to the due south horizon (this is as high as ANY star of a particular declination will reach from your location); the same star will set at approximately 3 a.m. (See Figure 3)

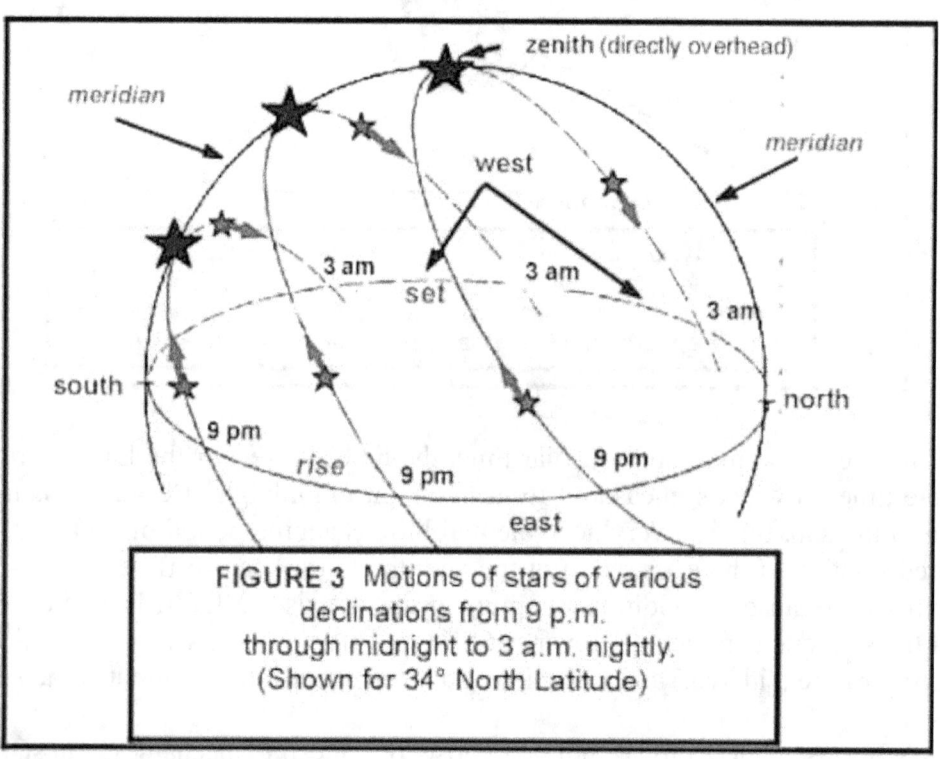

FIGURE 3 Motions of stars of various declinations from 9 p.m. through midnight to 3 a.m. nightly. (Shown for 34° North Latitude)

A Note About Star Motions and Using Setting Circles Through the Course of the Evening - IT IS NOT ADVISABLE TO USE THE SAME REFERENCE STAR FOR SETTING CIRCLE ACCURACY THROUGHOUT THE NIGHT; as described following, always choose a reference star (particularly for setting circle and computer navigation) CLOSE TO THE OBJECT/AREA you wish to explore. This will increase your accuracy for telescope pointing via circles or computer (as well as assist you in quickly learning the many fascinating stars and constellations of the nighttime sky!). OR, you might choose to re-set the RA circle (as described following) and simply move from your present object's coordinates to those of the new object you wish to view!

POLAR ORIENTATION MODE AND THE USE OF SETTING CIRCLES

By now you have undoubtedly figured out that "Alt-Azimuth" celestial alignment and using setting circles DON'T MIX. You must use "Polar Mode" ("SETUP / TELESCOPE / MOUNT / POLAR") and be tracking with the fork arms tilted to your latitude toward the NCP (see my "Enhancement Guide....", Part 3 and Part 5 for celestial Polar alignment and "home position").

FIGURE 4
Setting Circle Use is Only Possible in Polar (left) Mounted Telescopes, Whereas Alt-Azimuth (right) Scopes Must Rely On Computer Control.

Regardless of if you are a dyed-in-the-wool Polar user (as am I) or a devout Alt-Azimuth activist (as most of you), you STILL must understand how about celestial motions (my crash course above) and HOW to use those mysterious setting circles.

What, for example, would you DO if the batteries went dead? If you were at an observing party and the AutoStar fried? If there was a blackout in Los Angeles? Most of you would be dead-in-the-water as far as finding your ways around the sky....that's what! So at least let's study the basics of setting circle use!

The telescope has TWO types of setting circles (well, "three" if you include the telescope's computer LED), one for DECLINATION and the other for RIGHT ASCENSION. Remembering that "declination," like "latitude" NEVER changes from moment to moment, the DEC CIRCLE will always be fixed, reading sky angles that do not change (unless the setting circle has inadvertently been moved or jostled from a correct position [SEE previous chapter....*Kochab's Clock*,"on very accurate DEC CIRCLE adjustment setting]).

On the other hand the RIGHT ASCENSION (RA) CIRCLE moves its reading very slowly as the telescope moves to keep track of the objects it is centered on. It follows the TELESCOPE and not the pointer on the base! Therefore, EVERY TIME you change objects, you MUST reset your RA CIRCLE to find that object, as described following. That one step is the only difficult procedure (and it's not hard, if you remember to do it!) about using setting circles!

Figure 5 shows a Meade ETX 125 that is aligned in Polar position and ready to go; in this setup, motions of the telescope can be maintained and objects found manually without any power whatsoever.

FIGURE 5
When Properly Aligned, the Polar Telescope Requires Only One Motion in Tracking (RA) and this "Hour Circle" Must be Reset Each Time the Scope Moves to Another Object.

THE DECLINATION SETTING CIRCLE

Noting the description of declination from above, and that every reading will always be somewhere between "0" (celestial equator) and "90" (the North Celestial Pole) degrees or between "-0" and "-90" degrees (south of the celestial equator), once the circle is precisely set, it should give good readings throughout the night provided that you have properly leveled the telescope and set it up in a very good Polar alignment.

I routinely check my setting circle each night before observing; to adequately assure good polar alignment on a portable telescope, you MUST check (and reset) the DEC circle at the beginning of each observing session anyway. It seems that no matter HOW firmly you secure the circle to a precise position, it is prone to slip from accidental motion or even against the inner part of the telescope as you move it!

Every object in the sky has a fixed "degree" of declination in the sky; the only variable is where YOU are to observe it (your latitude and longitude on Earth). A polar-aligned that is properly set up will read your latitude exactly when the base is level and the Optical Tube Assembly is pointing "straight up" in the sky.

At this position, the scale on the DEC CIRCLE will read "[your latitude]" in degrees; now turning the telescope southward until you reach the CELESTIAL EQUATOR, you will notice that the numbers decrease until actually reading "0", the actual declination for the Celestial Equator. If you go the opposite direction and turn the telescope northward until the circle reads "90" degrees, you will then be pointing VERY close to the north star, Polaris.

No matter WHAT bright star you point your scope to and center, the DEC CIRCLE will read the exact (if you are aligned well) declination for that star as you can get either from PC program directories or from a good astro reference book. This is true no matter what time of night you access the star, nor what day (night) of the year. More about declination in a moment. Note in Figure 7 that the setting circle is set to the coordinates of the bright star **Vega**, or about **+38 degrees**.

RIGHT ASCENSION SETTING CIRCLE

The right ascension circle is not so easy to keep up with, for the reasons of star vs. earth motion described previously. We do not need to go into celestial mechanics here, but suffice it to say that the RA circle will continually have to be adjusted relative to the pointer EACH time you move the telescope to another object. As an example in this Guide, we are "starting" with a reference star, **VEGA** which is easy to find; we will be moving FROM Vega to a telescope-only visible object - MESSIER 57, the ring nebula nearby. The RA circle in Figure 7 is set to the right ascension of Vega: "**18h 35m.**" If your clock drive is running, the pointer will NOT be pointing to "18 35" in a short while, but Vega will still be centered in your telescope; hence the need to re-set the circle each move. This large circle moves freely (a bit too freely in my opinion) around the

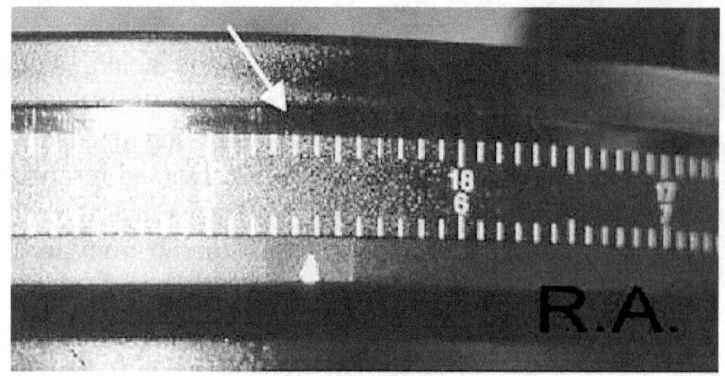

FIGURE 7

"turntable" of the Meade telescopes and care must be exercised when using the circle as it is only fastened into its circle via a small piece of fragile tape at one junction (see Figure 8). Rotate the circle to "dial in" your RA using two hands evenly spaced 180 degrees across the circle; once positioned, it stays very well, even though it may seem too loose to do so.

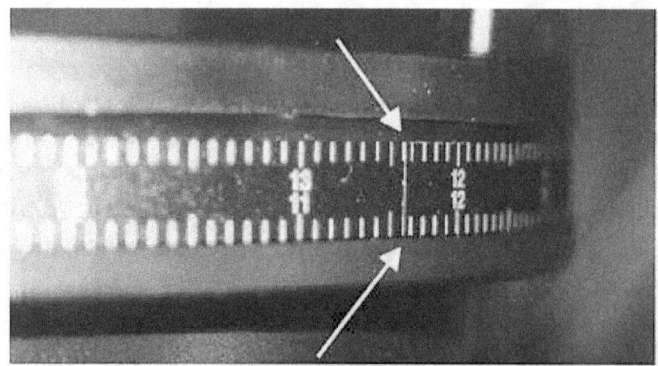

FIGURE 8
Showing Taped Circle Junction

[MAINTENANCE TIP: Figure 8 shows the delicate junction of the RA setting circle "tape." This frequently comes unglued and if so, merely take the two cut ends and match carefully the increment marks to re-attach. ONE DROP of regular "model cement" or similar will re-attach the circle, as will a small piece of permanent "double stick" tape, but: 1) DO NOT apply so much glue or tape that it sticks to the surface of the scope "turntable" and prevents the circle from adjustment; and 2) NEVER get so tight as to prevent easy motion of it as you attempt to rotated the numbered strip.]

NOTE that there are two (2) sets of values on most telescope setting circles; as you are looking from the base of the scope (with the numbers right-side-up) you will be using the TOP row of numbers if observing from the Northern Hemisphere and the BOTTOM set if from the Southern Hemisphere. In the north, the numbers INCREASE in RA the farther east you move your telescope.

The RA CIRCLE is marked in HOURS and increments of hours; just like a clock, RA has an "hour" that is subdivided into 60 "minutes" which in turn is divided into even smaller increments, "seconds", 60 of them. Your RA circle has major marks at each hour, and a lesser mark halfway between them (30' arc). Each very small mark in between represents a distance in angular measure of 5' arc.

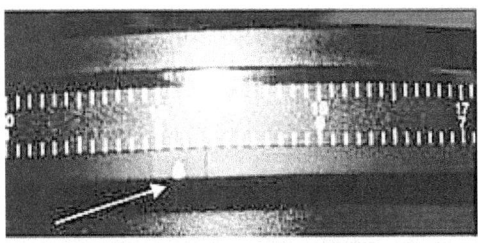

FIGURE 9
Coordinates of M57, Ring Nebula

An HOUR of right ascension is about 15 degree of sky, so it is a large increment; for simplicity, you can assume that if you have Vega centered at 10 p.m. and wait an hour with your motor turned off....there will be an object - not Vega - in your eyepiece ONE HOUR later that has an RA of just about 19h 55m and the exact declination as the bright star!

A TEST SEARCH FOR MESSIER 57

Using the circles is a very simple task and requires only to 1) properly align and level the telescope before starting; 2) remember to move the RA circle each time you proceed to another object after observing the current one.

1) FIND A REFERENCE STAR - A complete list of bright reference stars is included with maps on my "Alignment Star Charts" found on *Arkansas Sky Observatories'* website site under "Observational Guides;" merely select your star (find one close and within the season you are observing) and access the coordinates (through Autostar in this example....your sky program key strokes will vary slightly): "Select / Object / Star / Named / [scroll to desired star and press enter, then scroll to RA and DEC]. For this "test" we use the bright summer star "**Vega**" at R.A. 18h 35m and DEC. + 38 degree 44'.

Always select a bright star NEAR the object you wish to find; look through reference guides and find any star that you can center in your finder/scope with a RA and DEC value very close to the object being sought.

2) With motors tracking, CENTER Vega (or whatever star) in the main telescope and lock on it. Immediately go check your DEC circle to see if it reads CLOSE to +38

degrees; if it is off a bit do not worry, just take whichever direction it may be off into consideration when moving the scope to the desired object.

3) Check in your eyepiece to make sure Vega is still centered and put in a LOW POWER EYEPIECE; now go to your RA circle and carefully (using the "upper numbers" if in the Northern Hemisphere) rotate the circle until your pointer shows the position in Figure 4, "18h 35m". REMEMBER as you look at the circle, the numbers INCREASE from 18h toward the LEFT; thus 18 35 is the first small mark TO THE LEFT past the half-way (30') major mark toward "19h".

4) Immediately after dialing in the coordinates in RA of this reference star, UNLOCK THE SCOPE polar axis and rotate until it turns to the RA reading of the hunted object: RA = 18h 53m (See Figure 9); clamp securely and immediately go to the DEC circle for adjustment. Your telescope will NOW be tracking at the correct RA for your selected object!

5) Unlock the DEC axis and slightly rotate the scope until the DEC circle reads "33" degrees which is close enough to get M-57 into the low-power field of view; lock the DEC axis and look into the eyepiece of the telescope; you should have the Ring Nebula very close to center! (note DEC circle in Figure 9).

You have done well, but the Ring Nebula is VERY CLOSE to Vega; to conclude, try the acid test: use Vega again as your "starting star" and let's use the setting circles to locate the "Whirlpool Galaxy," Messier 51 much farther away in *Canes Venatici*. Use the computer program to find the exact RA and DEC of M-51 and find it via Vega.

REMEMBER! Some time has gone by since you acquired M-57 with your circles, so the RA circle NO LONGER READS the RA of Vega! Use your flashlight and re-set the circle to the coordinate of your reference star each time you prepare to move to the next object! You should, however, NOT have to reset the DEC reading at all.

You will be surprised how accurate you will be....how fun it is to access "by yourself," sans computer program....and just how much fun you can have if "the lights go out!"

* * *

Chapter 23
Balancing a Fork Mounted or GEM Telescope

This is an issue that has been dealt with in many places but never stressed enough as to importance. To properly balance a fork mounting you must use the 3-dimension Losmandy or equivalent counterweight set, have temporary alternate various weight combinations and follow the steps below. The process is the same for a German Equatorial (GEM) mount but may require a bit more creativity.

If you are using your telescope for piggyback imaging, or if you have heavy tail end equipment (robotic focuser, camera, reducers, etc.), there is extra need for the most caution in maintaining balance.

The rule for balance is simple:
"If you unlock either axis and the telescope MOVES....you are out of balance." In such conditions, the telescope will suffer from performance, tracking, GO TOs and even will rapidly wear down over time.

Balancing is simple. BALANCE THE DECLINATION AXIS **FIRST**......once balanced, *then* balance the Right Ascension axis.

There are two major RULES regarding balance:
    a) for balancing the DEC axis (do this first) you ONLY use weights along the Optical Tube Assembly (OTA) and/or on the rear cell; this weight can include accessories added to both top and bottom of optical tube assembly, provided that the weight is offset 180 degrees around the tube assembly by an equal weight....the DEC axis also can be partially balanced by sliding the optical tube assembly forward and backward if that is possible (as with refractors or other telescopes mounted with ring mounts.
    b) for balancing in the RA axis (only AFTER the DEC is balanced) you must use ONLY the fork arms or the counterweight shaft....you **never** balance RA adding or moving weights along the OTA.
    c) no exceptions to this if you want to do it right.

Balancing the DECLINATION axis:

(NOTE that if you have a guide scope or piggyback smaller scope riding on top of the telescope, you MUST have an equal offsetting weight 180 degrees around, on the "belly" of the OTA!)
NEVER use weights under the telescope optical tube if you do not have something offsetting that balance above it! You will see why shortly.

1) aim the telescope due south in the equatorial (polar) mode and tilted about 45 degrees up from the south horizon; if you have weight on both top and bottom of OTA, have the 3-D weight screwed as close to the surface of the OTA as possible;

2) unlock the declination;

3) adjust the tube counterweight in this position by sliding up and down the length of the OTA until the out of balance situation stops;

4) turn the telescope straight up (zenith) and carefully unlock...the telescope will want to tilt either north or south;

5) at this point, sliding tube counterweights and similar weights are useless because they cannot account for the perpendicularity of the torque in this position; here is where the 3-D weight system shines.....the telescope in most cases will want to tilt northward from the zenith...if so, the solution is simple....unscrew the counterweight AWAY from (perpendicular to) the OTA until balance is achieved;

6) if the scope attempts to move southward, then the counterweight is too great and you must go to a smaller size and start over.

NOW....if this does not work, then it is likely because you have heavy equipment on the FRONT end of the scope that cannot be balanced because of the center of gravity so close to the back of the OTA....so:

7) add weight to the rear cell; for most applications other than a very heavy dew shield, the Peterson rear cell balance is an ideal solution....for others it may not be enough;

8) thus, use the REAR portion of the OTA to add weights in any way that you can to achieve front-to-back balance in the south-facing (#1 above) position and repeat the other steps in sequence once that balance is achieved;

Balancing in Right Ascension

Once the Declination axis does not move when unclamped, lock the DEC. and unlock it in RA and turn to the SE sky.....if the scope moves eastward then add some type of temporary weight via a wire tie to the RIGHT fork arm handle or to counterweight shaft (GEM); if to the west, then of course weight on opposite side.

Remember THREE Rules:
1) Always balance DEC first and RA last
2) ONLY use weights along the OTA tube to balance in DEC, never add weights to the mount
3) ONLY use weights on the fork arms, elbows, etc. to balance for perfect RA balance, NEVER use the OTA or weights on that.

Do it in this order and you will be on your way to perfect balance.
Best of luck!

Telescope R*x*

PART TWO

OBSERVING PROJECTS

The Science of Astronomy
And your Telescope

*(adapted from the original works by P. Clay Sherrod:
A Complete Manual of Amateur Astronomy, Prentice-Hall, 1981)*

Chapter 24

The Perseid Meteor Shower and Techniques for Observing Meteors

An Introduction to Meteors via the Perseid Meteor Shower

The famous PERSEID Meteor Shower will pass across the Earth's orbit once again every year in early August, and some years are particularly favorable for MANY meteors to be seen. The Perseids always put on quite a celestial show in various years, yet others are not so optimistic. As Fred Whipple of Harvard once said, "Predicting the outcome of a meteor shower is no easier that picking the winner of a horse race...."

We shall use the Perseid Meteor Shower as a model for our discussion on meteors and observing these exciting celestial interlopers.

The moon for any meteor shower should always be absent from the sky, or in a very small phase, or its light will interfere with sighting of fainter members of the shower. Note that optimum times for any shower to show the peak number of meteors is typically AFTER midnight.

As with the Perseids, do not wait for the exact peak date however....this is a long duration shower and meteors will be easily seen during times when skies are hampered by only a thin crescent moon, or no moon in the sky at all.

It is believed that all meteoroid clouds are the remains of comets which have disintegrated in their orbits about the sun. In 1992 Comet Swift-Tuttle, the parent object that spawns the Perseid meteoroid cloud, shed a great amount of dust in its wake and now every year sets the stage for intense activity as the earth passes through that debris; just what year that the earth has actually gone directly through this debris cloud since it was intensified?

With the Perseids, meteors can be expected to be seen just as soon as the skies darken on the evening of August 11, the peak is scheduled for early in morning of August 12 (about 4 a.m. CDT) and throughout the evening of August 12. NOTE that there is some indication that a SECONDARY PEAK might actually be seen in ideal skies on the MORNING after each peak date for the Perseids, somewhere around dawn or just before on that morning, as the earth will be sweeping through a cloud of material that has been ejected from the parent comet (below) of this dependable meteor shower. In fact, it is possible that this secondary cloud might produce more meteors in any given year than the main swarm.

Note that Comet Swift-Tuttle's (P/1862) one-revolution trail from 1862 will pass inside the Earth's orbit during some years. If there were a closer approach of this comet to the earth, a spectacular meteor storm would be expected...but with these conditions and no prior such close approaches to compare to, it is uncertain what kind of a shower this will give in most years. Because of similar conditions, but with the earth passing directly through the major debris pocket of the comet, perhaps the best meteor shower of history will occur with the Perseids in 2028.

This is a long duration shower, with many (as many as two dozen per hour) being seen from August 9 through the 20th; during the PEAK, expect to see at least 60 or more (perhaps double that number!) around 2 a.m., streaming from the constellation of Perseus, high in the northeastern sky. Best views are afforded by positioning your feet to the EAST and facing directly overhead. A move is underway for observers to actually monitor the MOON during the Perseids via CCD, digital and visual means to look for flashes that may indeed be part of impacts of Perseid meteorites against the lunar surface.

ANNUAL METEOR SHOWERS

There is no month throughout our years that is absent from meteor shower activity. Following is a month-by-month listing of the primary annual meteor showers which are expected to occur every year. NOTE that the actual activity seen will depend on the geometry of the Earth's passage through the debris clouds as well as the amount of moonlight during each of the peaks. The AVERAGE dates of peak activity are given.

January

January 3-4 - QUADRANTID METEORS - The moon will be at a thin waxing crescent phase and absent from the skies for most of this evening for this year's showing of this meteor shower. Always a chance for quite a show...the best that January has to offer each year, so long as the moon does not hamper observation of these meteors. With an incredible short and fast maximum peak of 40 or more meteors possible, it will come and go in a flash (about the time that the sky reaches peak darkness after sunset on the 3rd. In some years under dark skies, observers have seen up to 600 members of this stream per hour, all traveling at a medium speed of about 41 kps. Most are very faint, remember, and distinctly blue in color, so fast film is desired if photographing these meteors. The meteor shower emanates from near and north of the bright star Arcturus in the constellation of Bootes, rising in the northeast about midnight.

January 15-16 - DELTA CANCRID METEORS - Sounding more like a disease than a meteor shower, the Delta Cancrids rise in the east about the same time the sun sets in the west...thus it is nearly directly overhead at midnight each year, in the constellation of Cancer. The shower radiant is actually just slightly west of the bright and well-known naked eye star cluster, Prasepe or the "beehive." Only about four meteors per hour can be seen from this shower under good conditions. I suggest setting up around 7 p.m. local time on Jan 15 for best views. Cold, but fun!

January 18 - COMA BERENICID METEORS - Also coming from very close to a naked eye cluster, the Coma cluster, this meteor shower rises about 10 p.m. (again, a bit of quarter moonlight on this night for this one!) and is directly overhead at pre-dawn. These are among the fastest meteors known....65 kps (compare to the Quadrantids, above)...BUT expect only a couple of these swift interlopers per hour. Perhaps some splendid streaking meteors might be visible for those who brave the typically cold nights of January.

February

February 26 - DELTA LEONID METEORS – Any year when the moon is absent from the sky will prove to be an excellent year for observing the Delta Leonid meteors ......and this shower is the only meteor shower that February has to offer each year. The Earth actually intersects the cloud of cometary or asteroidal (the actual source of the cloud is not known at this time....) as early as Feb. 5 each year and seems to be encountering meteoric material as late as March 19; nonetheless, there IS a definite peak each year that seems to center on the last few days of February. These are moderately slow meteors, traveling at about 24 KPS, and only about five per hour can be expected at most. The radiant, at astronomical coordinates: RA 10h 36m / DEC +19 deg, is found about midway between the moderately bright stars *Zosma* and *Algeiba* (the two that make the long stretch of the Lion's Back in Leo); look for these stars and the meteor shower to be nearly overhead for mid-northern latitudes about 11 p.m. local time, with most meteors seen well after. Since the gibbous moon is in the sky from before midnight to dawn, this should be a great year at best to watch for this mysterious meteor shower.

March

March 16 – CORONA AUSTRALIS METEORS - This brief shower, emanating as its name implies from within the southern constellation of *Corona Australis*, begins typically around March 14 and members can be traced back to that radiant until March 18; from the United States and Europe, this shower never gets above 7 degrees for its radiant, but brighter meteors can be seen streaking from south to north from it; as many as 15-17 meteors can be seen hourly in good conditions.

March 22 – CAMELOPARDALID METEORS - Already high in the sky at dark, this meteor shower really has no definite peak, but a few meteors per hour can be seen coming from this very high northern meteor shower, only 22 degrees from the northern celestial pole; hence it is "up all night" for those braving the cold temperatures of March. Not only are there very few meteors to be seen from this rather dull shower, but the ones that ARE noticed travel the slowest across the sky of all known meteors....only about 7 kilometers per second! We see them as they begin to burn at an altitude of about 80 kms (~50 miles) above the Earth's surface. If you are interested, attempt to spot meteors from dark until about dawn.

March 22 – MARCH GEMINID METEORS - Discovered in 1973 by amateur astronomers, much is still to be learned on this shower, so this is one where you can make a valuable contribution by observing. The radiant is high overhead for northern hemisphere observers at the time the sky truly gets dark, but because of bright moonlight, only the brightest meteors (if any) will be seen. When first discovered in Hungary, nearly 50 meteors per hour in a short-burst stream were seen and this was confirmed again with sightings in 1975. Like the Camelopardalids (above), the meteors in this stream are very slow and there is some possibility that the two showers could be linked to two diffuse clouds of debris from one parent object. Any meteors from this unusual and elusive shower should be reported immediately to the *American Meteor Society* at: kronk@amsmeteors.org .

April

Observe when the moon does not interfere and attempt to observe AFTER midnight for most meteors to be seen! For April, there are no less than NINE meteor showers, some of which provide for wonderful spring sky shows, provided that the light of the moon does not interfere. However, as with a months and times during the year, observers should always be aware that new sporadic meteor showers can occur at anytime from seemingly unknown sources and radiants. *NOTE: one of the most interesting* of all meteor showers is the odd "April Fireballs" (see below) which occur this month.

April 4 – KAPPA SERPENTID METEORS - This is a one-week-duration meteor shower, from April 1 through 7, with somewhat of a mild peak about midway through that period; look for the radiant to rise in the constellation of Serpens about 8 p.m. local time just south of due east and be nearly overhead for observers in southern latitudes of the northern hemisphere at about 2 a.m. Several meteors per hour should be seen from this minor radiant in normal years..

April 7 – DELTA DRACONID METEORS - With no particular peak to speak of, this is one of those "circumpolar" meteor showers for northern hemisphere observers that will be in the sky pretty much all night; it is a very long duration shower from late March until about April 17. Found only in 1971 in the constellation of Draco, the meteors are conspicuously slow and leave very fine trains in their wakes; to view the most meteors from this now-annual shower, set up about 10 p.m. local time and face somewhat northeast; as the night progresses the meteors will be originating more and more from very high northern skies....thus after midnight direct your sights to nearly directly overhead, the ZENITH.

April 10 –VIRGINID METEORS - This is the first of THREE meteor showers which appear to emanate from the constellation of VIRGO during the month of April each year. A two-week display, the meteors can be seen coming from just south of overhead (northern hemisphere) from April 1 through 15 with no definite peak; to differentiate THIS shower from the other two, the radiant is centered at near right ascension 12h 24m / declination 00 degrees.

April 14 - THE APRIL FIREBALLS - Doc's Favorite of All Meteor Showers. Anytime can be an exciting time with this meteor shower, strong moonlight or not….. but then again being bright fireballs, these can be seen in spite of even city lights! As its name suggests, this can sometimes be a pretty spectacular display if the conditions are right and the skies are dark; however, during times of the new moon - as it was in 2010 - , these huge and bright fireballs come streaking clearly across our crisp and clear springtime skies along with countless fainter meteors that are associated with no identified meteor swarm. This unusual display lasts for the last two full weeks of April....there is *no known radiant* or seeming point of origin for this curious group, and they can be seen originating from just about any part of the dark night sky. They likewise are not - or appear to not be - associated with any other known major or minor meteor shower group. The April Fireballs are characterized by tremendously bright meteors, nearly all of which demonstrate beautiful and long-lasting trails through the sky. Even with the bright moon however, with their brightness, the light does interfere for observing these very spectacular meteors. Always look for the April Fireballs late in the night, preferably after midnight. NOTE: several of these renegade meteors have been known to reach the ground as meteorites! Heads UP!

April 17 – SIGMA LEONID METEORS - The Sigma Leonids are no longer "in" Leo....they have migrated it seems into Virgo to become one of our three Virgo showers for April. The radiant is up early, just due south of overhead about 9:30 p.m. local time; this is a minor shower with only a few members seen on dark nights per hour.

April 22 – THE LYRID METEOR SHOWER - Other than some spectacular fireworks from the April Fireballs (see above), this is April's most dependable meteor showers and typically one of the best of each year; this year the morning crescent moon will not interfere with any observing of meteors meteors after midnight, typically the best time to view the greatest number of Lyrids. This shower is comprised of cometary debris from *Comet Thatcher*, a very famous comet last seen in 1861. Although this associated comet was not identified until only 100 or so years ago, this meteor shower from it's demise is one of the oldest known on record, being recorded by the ancient Chinese stargazers first in 687 B.C. As with many meteor showers - and the comets they come from - this one seems to be waning with every encounter with the earth however. It is no longer the sky spectacle as recorded by those earliest sky watchers. Look for the meteors to emanate from a point on the Hercules-Lyra border, very near the brilliant blue-white star **Vega**. The radiant rises about 7:30 p.m. local time, but the best time to see the most meteors each year is always around midnight when the radiant is nearly directly overhead at midnight for northern hemisphere observers.

April 25 - MU VIRGINID METEORS - This is our third of three meteor showers within the constellation of Virgo for the month of April, and is south of overhead about 1 a.m. local time, far in the eastern realms of the large Virgo constellation; it takes dark, moonless skies to see the few - only about 7 per hour - meteors from this annual minor display.

April 23 –GRIGG-SKJELLERUP METEORS- Here is an oddity just by its name...the only annual meteor shower known by the comet from which the meteoroid cloud came! It also is unique in that it is a "localized" meteor shower, visible only in certain parts of the world, but not others, on each pass. For example, there was a brilliant display of these meteors seen in New Zealand in 1977....but not one in the United States. If visible, they will be seen early in the evening, originating south of overhead. At right ascension 07h 48m / declination -45 degrees, these will appear to be coming literally from the south horizon for northern hemisphere observers, perhaps the only way to differentiate them from the other meteors showers in the same direction of sky each April.

April 28 – ALPHA BOOTID METEORS - Coming from a point very near the bright "alpha star" *Capella* in the constellation of Bootes, this radiant is in the sky from dusk until dawn, and nearly overhead at about 1 a.m. Look for these meteors to be few, BUT those that are seen are typically very fine fireballs moving slowly across the sky and leaving beautiful "smoky trails" behind them. Note that the radiant rises about 3-4 hours after sunset.

May

For May, there are three meteor showers, some of which provide for wonderful spring sky shows, provided that the light of the moon does not interfere. However, as with all months and times during the year, observers should always be aware that new sporadic meteor showers can occur at anytime from seemingly unknown sources and radiants. MAY is always an excellent time to go outside and view the heavens and the interloping meteors among them; typically in most locations, whether spring in the northern hemisphere or fall in the southern, the skies are crisp and clear. Most of May's meteor showers occur early in the month.

May 1 – PHI BOOTID METEORS - A really long term meteor shower that actually begins on or about April 16 and persists until May 12, emanating from the constellation of Bootes AND Hercules (the radiant has indeed moved in recent years!), high in the eastern sky at dark, and remain so for most of the night for northern latitudes. The best time to observe the most of these meteors is always about 2 a.m. local time. About 6 meteors can be seen per hour, most medium bright, relative fast and pretty much overhead, all traced back to northwestern Hercules.

May 3 – ALPHA SCORPIID METEORS - The peak of this shower takes place during the week of this year's new moon, so the faintest members of this meteor shower will be seen for several nights before and after the actual peak date. The minor meteor shower is another long duration one, beginning in early April on the Libra-Scorpius border and slowly moving into the constellation of Ophiuchus by May 9! The motion of this radiant is of much interest to astronomers and your detection of meteors from night to night as to where they appear to originate is very important; the radiant will rise in the far southeastern sky about 9 p.m. local time and be overhead at 1 p.m. the following morning.

May 5-6 – ETA AQUARID METEORS - Normally one of the finest meteor showers of each year, the Eta Aquarid meteors were recorded as early as 401 A.D. by the ancient Chinese stargazers.....now we know them to be part of TWO debris clouds left in the wake of famous *HALLEY's COMET* through which the earth passes each year. Meteors can be seen from this shower all the time from April 21 through May 12, but the peak is fairly steep and occurs each year on May 4.....look for brilliant and spectacularly exciting fireballs from May 9 through 11. The radiant for this meteor shower is located very near the star asterism known as "The Water Jar" in Aquarius, but moves a bit northeast each day through the long period the meteoric cloud persists around the earth. Note that this meteor shower for northern latitudes is very low in southern skies...most meteors should be seen coming from the EAST horizon (not overhead like most showers!) about 2 a.m......but by 7 a.m., note that the most frequent meteors appear to originate about halfway from that point to overhead. On most dates with not-so-dark skies up to 10 Eta Aquarids per hour might be expected, most bright and leaving glowing "fireball" trails behind them. Thus, most years are excellent in hopes of seeing these fine meteors, and the very faint as well as the many bright fireballs may be seen.

June

For June, there are no less than 13 (!!) meteor showers, some of which provide for wonderful spring sky shows, provided that the light of the moon does not interfere. However, as with a months and times during the year, observers should always be aware that new sporadic meteor showers can occur at anytime from seemingly unknown sources and radiants.

June 3 – TAU HERCULID METEORS - Beginning in late may and extending through June, this is a month-long minor meteor shower, overhead for mid-northern latitudes at about 10 a.m. The meteor shower is overhead at midnight when most of the 15 meteors per hours might be seen.

June 4 – ALPHA CIRCINID METEORS - This southern hemisphere meteor shower does produce some long-trailed meteors that can be seen low in northern hemisphere skies, traveling from south to north; it was discovered in 1977 by Australian amateur astronomers when 15 very swift meteors were noted per hour; for southern latitudes north of the equator, the meteor shower radiant is actually above the southern horizon at Midnight, so only the brightest meteors can be seen....this shower is in need of observation and continued confirmation.

June 5 – SCORPIID METEORS - A very interesting meteor shower with TWO radiants rather than just one as is typically found with annual meteor showers; both radiants are nearly on the meridian at midnight, so observers are suggested to put their feet to the south and look overhead for these meteors; about 3 a.m. local time; in dark skies observers should normally see at least 20 meteors per hour when the moon is absent. Note that not only are the number of meteors impressive with this shower, but also the sky itself, since the meteors will be coming from near the summer Milky Way star clouds, revealing one of the richest star fields visible to the naked eye and camera. Best to begin observations about 10 p.m. and continue until 3 a.m. local time; radiant average is at R.A. 16h 40m; DEC -17 degrees.

June 7 – ARIETID METEORS - From the constellation of Aries, this is another month-long meteor shower, and can peak on this date with as many as 60 meteors per hour in dark skies. This has been confirmed by radar, but less than that number can be expected visually, perhaps up to 30. Wait until about 3 a.m. local time to assure that the radiant (low on the eastern horizon) is high enough above local haze and moisture to reveal these meteors. These are very fine, slow meteors which leave spectacular trains, and frequently split into Bolides, or "fireballs." The fireballs should be easily seen in all areas of the sky, although the radiant is nearly overhead about the time of peak. This is not a great year to attempt to observe these meteors since the moon will be high in the sky about 2 a.m....... the fireballs can typically be seen in spite of bright moonlight.

June 7 – ZETA PERSEID METEORS - On the same night as the Arietids, this meteor shower is less spectacular, with perhaps 15 per hour visible in earliest pre-dawn skies; radar reveals as many as 40 per hour after sunrise.

June 8 – LIBRID METEORS - A very minor meteor shower from a very large constellation, expect only a few per hour; evidence suggests that this meteor cloud might be dissipating, and no known cometary source is associated with this minor display; observations are badly needed. Coordinates of radiant: R.A. 15h 09m; DEC -28 degrees.

June 11 – SAGITTARIID METEORS - This is a two-week-long meteor shower beginning in early June. The radiant rises in the extreme SE sky about 11 p.m. local time and about a dozen meteors per hour in dark skies might be expected. VERY low in the southern skies for northern observers, at -35 degrees DEC.

June 13 – THETA OPHIUCHID METEORS - Coming from the border of Ophiuchus, Sagittarius, and Scorpius, this radiant rises about 9 p.m., giving a window of good observing ALL NIGHT. However, those that do grace our skies are bright and spectacular, so be alert to these meteors if you are observing and happen upon a fireball from this area.

June 16 – JUNE LYRID METEORS - This is a companion meteor shower to the more-active May Lyrid meteors; The radiant is nearly directly overhead at midnight near the bright star Vega for mid-northern latitudes; since most of these meteors are very faint, observations will be poor this year with strong moonlight interfering. This is but one of many meteor showers that have been discovered by amateur astronomers since 1960....this one has been seen every year since 1966.

June 20 – OPHIUCHID METEORS - The radiant rises highest in the sky at 11:25 p.m. local time. The radiant sets about sun-up, so few meteors should be seen throughout our skies throughout this year's "window"; this is an interesting meteor shower since the number per hour can vary from as few as 8 per hour to over 26 per hour on any given year.

June 26 – CORVID METEORS - This is one of the shortest duration of all meteor showers, lasting only 5 days at most, with perhaps 10 meteors per hour seen to any observer; these originate near the small constellation trapezoid of Corvus, the Crow and the last good showing was in 1937. Astronomers speculate that these meteors are a product of some as-yet undiscovered comet. Since it has been years since a good showing and since the source is unknown, this is a very important meteor shower for a group project. Radiant: R.A. 12h 48m; DEC -19 degrees.

June 29 – BETA TAURID METEORS - Here is a different type of meteor shower....one you CAN'T see~! This is a daylight meteor storm that is of interest to those with ham radios, or those with long-distance shortwave receivers tuned to a distant station toward the direction of the radiant (Taurus. R.A. 05h 44m; DEC +19 degrees); ham operators have recorded a dependable 30+ meteors per hour each year. BUT.....at least the moon can't interfere with THIS one!

June 30 – JUNE DRACONID METEORS - Known in the past as the "Pons-Winnecke Meteors" (from the comet of origin), this can be an incredibly spectacular meteor shower; in 1916 over 100 very bright meteors were seen in fireworks style, but it appears that the numbers may be waning as years progress. Being irregular, observers are cautioned that there may be as few as 10 per hour or well over 100 per hour; with the high declinations (radiant: R.A. 15h 12m, DEC +49 degrees), the shower will rise about the beginning of astronomical darkness and be in the sky all night long, highest just after midnight in high northern skies.

July

July 16 – OMICRON DRACONID METEORS - very high in northern skies. Found in 1971 and few meteors seen since. These are slow-moving meteors; the new moon will be absent from the skies throughout the night this year, so the faintest meteors that possibly are associated with this cloud might be seen during any year with no moonlight during this time. This is a circumpolar meteor shower for the northern hemisphere, circling high in northern polar regions and will be up all night. This is possibly a swarm of debris particles from a long-dead comet that has simply "run out" of material or has been perturbed by the gravity of another object (i.e., Jupiter) and no longer passes through the orbit of the Earth.

July 28 – DELTA AQUARID METEORS (south) - rises about 8 p.m. and overhead about 2 a.m. Normally you should expect perhaps 8-15 per hour; face south and look for meteors overhead and begin your observing about 11 p.m. on the 27th and continue into the dawn of the morning of the 28th

July 23-30 – CAPRICORNID METEORS - From comet *Honda-Mrkos-Padjusakova*, these are bright yellow meteors with many fireballs! The radiant for these meteors is very low in SE sky at dark and south of overhead for mid-northern latitudes by midnight; best chance for the best meteors will be after about 1:30 a.m. local time when the dark side of the earth will be turning directly into the path of the meteor stream. Even in moonlight this can be a spectacular shower, so this year - this year, expect the fainter meteors to be seen.

August

Observe when the moon does not interfere and attempt to observe AFTER midnight for most meteors to be seen! August offers some of the best observing conditions for meteors....the skies are typically quite clear, the cooling night air suggests that fall nights await and fill you with observing inspiration, and August holds five wonderful showers, one of which is the "granddaddy" of all predictable and dependable meteor observing outings.

The famous **PERSEID Meteor Storm** -   Some meteors may be seen during early hours of days other than actually on the date of peak. The sighting of fainter members of this shower will be easy this year on peak day, but expect brighter ones to begin to streak across our skies even in the first week of August....if you can trace their origin back to the constellation of Perseus, then what you are seeing are indeed Perseid meteors. As with most meteor showers, the later you stay up (...yawn...), the more meteors you likely will see, particularly this year with very dark skies if you can venture away from city lights. Begin watching the evening of August 10, and continue until the early morning hours of August 14 for your reward.

However, do not wait for August 12-13....this is a long duration shower and meteors will be easily seen during the first week of the month.

This is a long duration shower, with many (as many as two dozen per hour) being seen from August 9 through the 20th; during the PEAK, expect to see at least 60 or more (perhaps double that number!) around 2 a.m., streaming from the constellation of Perseus, high in the northeastern sky. Best views are afforded by positioning your feet to the EAST and facing directly overhead.

AND YES....there ARE other meteor showers in August!

.**August 1** (and July 31....) – CAPRICORNID METEORS - Wait until after twilight ends (about 1.5 hours after sunset) in the early evening to begin serious sky watching. Remains of comet Honda-Mrkos-Padusakova, about 35 meteors per hour - MANY which are bright fireballs! - can be expected in the morning hours; nearly due south of overhead about midnight.

August 6 – SOUTHERN AQUARID METEORS - look on the meridian, southern skies about 11 p.m. local time for only a few meteors, perhaps 7-8 per hour. This is a curious shower, comprised of two peaks: this one, and another on about August 21-23. Note that meteors from this (these?) showers are not seen yearly and observations are badly needed to fill in the missing gaps about our knowledge of them. Some years no meteors are seen, but since the late 1800's when this double shower was noted and later confirmed, there have been distinct radiants (the "northern" and "southern") seen throughout many years. Observations of this shower are badly needed.

August 20 – KAPPA CYGNID METEORS - Typically many of these meteors are seen along with Perseid meteors, leaving very fine trains of smoke in their wakes! The Cygnid (and the Andromedids, below) will be nearly overhead by 2 a.m.

August 31 – ANDROMEDID METEORS - In 1885, 13,000 Andromedids were seen per hour, all fragments of a now-disintegrated BIELA's Comet. Very unpredictable, this meteor shower needs observations during such excellent times as unexpectedly occurred in 2005. The shower radiant will be nearly directly overhead for mid-northern latitudes about midnight.

September

For September, we can count on at least five (5) meteor showers, some of which provide for wonderful sky shows, provided that the light of the moon does not interfere. However, as with a months and times during the year, observers should always be aware that new sporadic meteor showers can occur at anytime from seemingly unknown sources and radiants.

The advent of crisper skies and cooler temperatures lure many sky watchers outdoors during September to view the impending autumn splendors of the Heavens. While the day of September, worldwide, can be hot and unbearable, the nights can cool remarkably, resulting in some long glances and time spent among the cosmos. Most of the meteor showers for September are modest, minor streams with few meteors; however some are unpredictable and thus the sky is worth monitoring during the times posted below!

September 1 - AURIGID meteors (??) - Note however that meteors from the Aurigid shower have ONLY been seen one year! That was 1935 when about 30-34 meteors per hour were seen, all very bright and quite rapid as they transited across the sky. Observations of this curious "one-time" (?) meteor shower, centered at about RA 05h 38m / DEC +42 degrees (in Auriga) are obviously very desired. Get out after midnight and confirm this meteor shower for us! Plan to observe throughout the night of August 31 and morning of September 1!!

September 6 - LYNCID meteors - from the constellation of LYNX, the meteor shower is about as small and sparce as the constellation from which the radiant of these meteors appears to originate. This was once a magnificent sky show, literally raining meteors through the sky; in 1037 and 1063 it was logged as "the rain of stars" by Korean skywatchers. However, little of this seems to be left. High in the sky during morning hours, look for the meteors coming from RA 06h 40m / DEC +58 degrees; note that the radiant does not rise until well after midnight.

September 7 - EPSILON PERSEID meteors - This is usually a fairly dependable group of meteors producing perhaps 12-15 meteors per hour very low in the NE sky about 10 p.m. when the shower should be its best.....3rd quarter moon tonight, so plan to see what you can of this, and observe this shower pretty much before midnight since it is "circumpolar" and high in northern skies throughout the night for northern hemisphere observers....remember that there will still be some stray PERSEID meteors coming from near the same direction, so it is very easy to get these confused with the Epsilon Perseid shower unless one is very familiar with the sky.

September 14 - SEPT. TAURID METEORS - This is an "iffy" meteor shower, and some years there may be a good showing about mid-month; look for these meteors emanating from Taurus in the early morning hours of Sept. 14, which will be about midway from the eastern horizon to overhead after midnight. About 13 members of this shower were confirmed in 2002, nearly three quarters of which were about 3rd magnitude; there are actually TWO radiants to this unusual meteor shower, both close together in Taurus; This shower rises in the east about midnight.

September 21 - KAPPA AQUARID meteors - this meteor shower is also directly south of overhead (northern hemisphere) about 11 p.m. local time, and continues until the 22nd way past dawn.

September 23 - ALPHA AURIGID meteors – when moonlight is not a factor in late morning skies for the Alpha Aurigid meteors, even the faintest of these meteors are likely to be seen in early morning hours; evening and around midnight are favored. The radiant rises in the NE sky about 8 p.m. local time and reaches nearly overhead about 5 a.m. the following morning after the moon has set and when the most meteors are usually seen. These meteors are very fast and frequently leave fantastic trains of smoke in their wakes....Because the shower is in high northern skies, US observers can plan to view these high northern meteors all night long.

October

The advent of crisper skies and cooler temperatures during fall months lure many sky watchers outdoors during October to view the impending autumn splendors of the Heavens. In most of North America, it has been found that the month of OCTOBER is the "cloud-free-est" and the skies more deep clear and weather-free than any other month of the year; the cool nighttime temperatures and waning mosquito populations (thank goodness this year!) are inviting to all of us to spend more time exploring the dark October night sky.

October 7 - PISCID meteors - A very long duration (Sept 25 through Nov. 2) meteor shower, very low in southern skies; these are very slow moving meteors [only 29 kilometers per second (kps)] on the Pisces-Aries border; about 15 meteors per hour, some very nice with long trains, will be seen after midnight at which time the shower radiant will be located west of overhead. The third quarter moon will be present in the sky after midnight this year, so viewing should be very fair for this shower in 2015, but better for those who observe early in the evening..

October 9 - DRACONID meteors - This is a very short duration meteor shower, unlike the Draconids (above). It lasts only from Oct. 7 through 10th, producing a very erratic number of meteors as the Earth plummets head-on into this stream. For example, in

1933, over 30,000 meteors per hour were estimated....from the Comet Giacobini-Zinner's disintegration, the Earth passed the comet only 15 days prior in 1947, resulting in about 1,000 per hours seen in that year. Note this "circumpolar" northern hemisphere shower to be seen at its best after midnight hours when the Earth is heading directly into the stream....at dawn it will be nearly directly overhead from mid-northern latitudes, The actual number to expect is totally unknown, but you should be prepared for as many as 200-500 per hour during some encounters!

October 19 - EPSILON GEMINID meteors - This is a week-long meteor shower peaking like the Orionids (below) during a time very favorable for many meteors to be seen. The shower actually begins about Oct. 1 with no particular sharp peak, so observing early in the morning on the 14th and 15th might allow for some of the perhaps 5 per hour being seen....these are among the fastest of the meteors striking out atmosphere, traveling at about 70 kps. The radiant rises about 10 p.m. on the 14th-15th, and will be overhead by dawn (northern hemisphere observers). A very favorable opportunity to witness this meteor shower.

October 21 - ORIONID meteors - With its origins in the famous Halley's Comet, this meteor shower is in the high southeastern sky (northern hemisphere) during early morning hours. The radiant, from near the "club" of ORION, the hunter, rises about 9:30 p.m. local time and will be south of overhead about 5 a.m. when morning dawn commences. Normally as many as 30-40 very faint and very fast yellowish meteors per hour can be seen from this group, traveling at actual speeds when they collide with the Earth of nearly 67 kilometers per second. This will be a fine year for early observers for this dependable meteor shower, Both bright and faint meteors should be seen in very dark skies, visible from this spectacular and historic meteor shower. However, there ARE many associated bright meteors from this shower, so heads up!

November

The crisp and cool (sometimes COLD!) night skies of November give way to some of the deepest penetration of earth-based eyes into space, affording thousands of normally not seen stars to glimmer into view; along with this comes a dramatic increase in the visibility of swift and faint meteors that will grace deep Autumn skies; the nights can cool remarkably clear, resulting in some long glances and time spent among the cosmos. Most of the meteor showers for November are modest, minor streams with few meteors but there are a few great showers each year that peak in November, among them the famous LEONID meteors which may put on a moderately good show on the evening of November 18 and into the morning hours of Nov. 19.

In addition, there are MANY other meteor showers which grace our crisp fall nights, some of which are mysterious, some which seem to be vanished from space and others that need observations at every opportunity!

November 5 - TAURID meteors - A very long duration (November 5 -12) meteor shower, that now is defined as having TWO peaks, both seemingly coming from the same radiant at about RA 03h 32m / DEC +22 degrees very close to the Pleiades star cluster; this double clumping is perhaps due to two distinct breakups of the famous *Comet Encke* at two different times and thus one cloud of debris trails the other by a week. Look for the "southern Taurid" meteors to be coming from a point somewhat south and earlier (Nov. 5) than the "northern" Taurids which will peak about one week later, at about Nov. 12. For the peak on Nov. 5, expect about 10 per hour and increasingly slightly after midnight. Note that this shower is well known for producing spectacular fireballs throughout the night and the display can last for many weeks on either side of Nov. 5.

November 9 - CEPHEID meteors - Coming from the constellation of Cepheus, high in northern skies and nearly circumpolar (neither rising nor setting but describing a tight circle around the north celestial pole throughout the night), the Cepheids radiant will be northwest of overhead. This is a new meteor shower, discovered only in 1969 and needing observations badly. The year of its discovery over 50 meteors in a 15-minute period were recorded! So expect to see at least 18 per hour, but only under darkest sky conditions.

November 12 - PEGASID meteors - Like the Cepheids, this radiant is nearly overhead in very early evening for Northern observers in the Americas. A remnant of an otherwise nearly-forgotten Comet Banplain of 1819, this shower still produces perhaps a dozen or so meteors on a good year....the meteors can be seen as early as late October and continuing until early December. Look for the radiant at about RA 22h 54m / DEC +10 in the winged horse Pegasus; normally it is best to observe this meteor shower after about 10 p.m. when the radiant will have moved into western skies.

November 14 - ANDROMEDID meteors - These meteors can be spectacular fireball meteors, leaving very glowing and distinctly reddish trains in their wakes. They are debris left from another famous comet, Comet Biela which split into two separate objects in 1845; shortly later, in 1885 the Andromedids put on a fireworks show with over 13,000 per hour seen, most spectacular fireballs. However shortly after the cloud passed uncomfortably close to mighty Jupiter and since only a very sparse number per hour have been seen. Nonetheless, like most meteor showers, any year can bring a totally different view of the remnant cloud. Many of these meteor are so large that they have reached the ground as meteorites. Shower begins as early as August 31 and lasts until December. Radiant center is at RA 01h 40m / DEC +44 degrees, not too far from the famous Andromeda galaxy. Try to observe this shower throughout the evening from perhaps 10 p.m. until about 1-2 a.m. local time and concentrate on the very beautiful fireballs that this shower is famous for.

November 17-18 - LEONID meteors - Although the Earth is somewhat posed out of the main clumps of cometary material from Comet 55P/Tuttle, the parent object of this debris, there is always a chance of an encounter with a secondary pocket of debris during any year. Most meteor scientists expect the peak each year to be slightly before MIDNIGHT on or near November 17 and perhaps extending into the early morning hours when the radiant will be high in the eastern sky. In many years hundreds or even thousands of meteors might be seen. This year is predicted to be perhaps an excellent showing in terms of recent years as the Earth passes near the thick debris cloud that produced the famous 1466 meteor "rainstorm" that was recorded over all of Europe.

December

December 10 - MONOCEROTID meteors - Observations throughout the the night should reveal several of the brighter members of this elusive meteor shower. Look for these meteors as early as December 1 and lasting through the 17th. They emanate very close to the Gemini-Monoceros border, rising in the SE sky at dark local time and overhead/south about 1:00 a.m., very favorable for both southern and northern hemisphere observers but only when the moon is not in the sky. In some years up to a dozen meteors per hour can be seen from this shower during moonless nights; the point of radiant is: RA 06h 50m; DEC +10d.

December 10 - CHI ORIONID meteors - It is very interesting that the Monocerotid and this shower both peak at nearly the same night....as its name implies, the CHI ORIONID stream has its radiant very near that fairly bright star, and thus the shower members from both showers are hard to differentiate many times; even more interesting is that the Chi Orionid meteors have TWO radiants apparently, one very close to the "horns of the bull" in Taurus and the other further into the constellation of Orion.

December 11 - SIGMA HYDRID meteors - These emanate from the head of HYDRA the mighty water snake, and are among the swiftest of meteors know, most being seen even in morning twilight. They ARE a bit on the faint side because of their speed, but expect about a dozen an hour in dark skies. Have you noticed that THESE meteors too, are peaking on the same night as the Chi Orionid and Monocerotid meteors? However, this radiant (RA 08h 32m ; DEC +02 deg) is far to the east (rising about three hours later) than the other two.

December 13-14 - GEMINID meteors - The faithfully rich Geminid Meteors should present a great show in December if the moon is absent from the sky, giving dark skies for early evening observations; for those who stay up late and into the early morning hour to view some of the brighter fireballs may reward your efforts. The Geminid shower is normally THE meteor shower for December, producing as many as 60 very white meteors in dark skies...only about 3 % of these meteors leave the characteristic "train" or trail, even when appearing as fireballs; this is a very unusual meteor shower in that it does NOT originate from debris of a spent comet, but rather from the MINOR PLANET "Icarus," a very peculiar asteroid that swings by the earth very closely during some passes. The radiant will rise nearly due EAST at dark and will be conveniently located (for northern hemisphere observers) about midnight; wait until about 10 p.m. this year to view this shower. ON THE SAME NIGHT is a very minor and newer meteor shower, the "LEO-MINORIDS", from Leo Minor; it will rise due east also, but about 8 p.m. and be overhead around 2 a.m. This was discovered by casual stargazers in 1971!

December 16 - PISCID meteors - Found in 1973, about 8 meteors per hour were seen coming from the constellation of Pisces near a distinct radiant at 01h 42m, +09 degrees;

few have been seen since, but dark skies provide an good year for sky watchers to "rediscover" this important meteor stream.

December 20 - DELTA ARIETID meteors - If you want one later in the evening, this is IT!; look for about 10 meteors per hour coming from the tiny constellation of Aries. .

December 22 - URSID METEORS - This meteor shower, coming from within the "Little Dipper" will never rise nor set and you can watch it all night; however, best observations would be about 11 p.m. local time and into the early morning hours. The meteoroids in this group have origins with the famous Comet Tuttle, and leave many spectacular wakes and smoky trails in their wakes. Up to 20 meteors per hour under dark skies can be see to any observer looking nearly due north and "up" a bit!

WHAT IS A METEOR?

To refresh on "meteor terminology," let us quickly review the correct word associations for the upcoming event. Of course we are all familiar with the phrase "shooting star," but thank our lucky stars that STARS do NOT randomly fall from the sky. As the point of light seems to suddenly increase in brightness and move across the sky in lightening speeds, it does indeed appear to be a star that is rapidly in motion, leaving a fiery wake behind it. But "shooting star" is merely a misnomer for the phenomenon we know today as "meteor." It is one of three "names" associated with this phenomenon:

METEOROID: an object - usually dust to grapefruit size - that remains as debris from a disintegrating comet in its wake; this meteoroid will follow in essentially the same path as the comet, and thus many times encounter the earth in periodic fashion, just like the debris from Halley's Comet comes in contact with the Earth in October as the Orionid meteor shower and again in spring as the Eta Aquarid meteors. The objects following in the path of Comet Swift-Tuttle are meteoroids until they encounter the Earth's atmosphere.

Compare the two photographs of Halley's comet following, BOTH taken simultaneously on March 14, 1986; the first one I took with a 400mm Nikkor f/3 lens and camera riding piggyback at the observatory for 20 minutes showing the dust and gas tail of the comet when near brightest; it is from this decaying material that largely comprises the tail that many of our meteoroids originate. Now look at the close-up from the European Space Agency spacecraft as it rendezvoused with Halleys on the same night; you can see the solar wind actually eating away at the comet's nucleus, release much dust and debris that can possibly eventually collide with the Earth in October or May of some year in the resulting form of "Meteors."

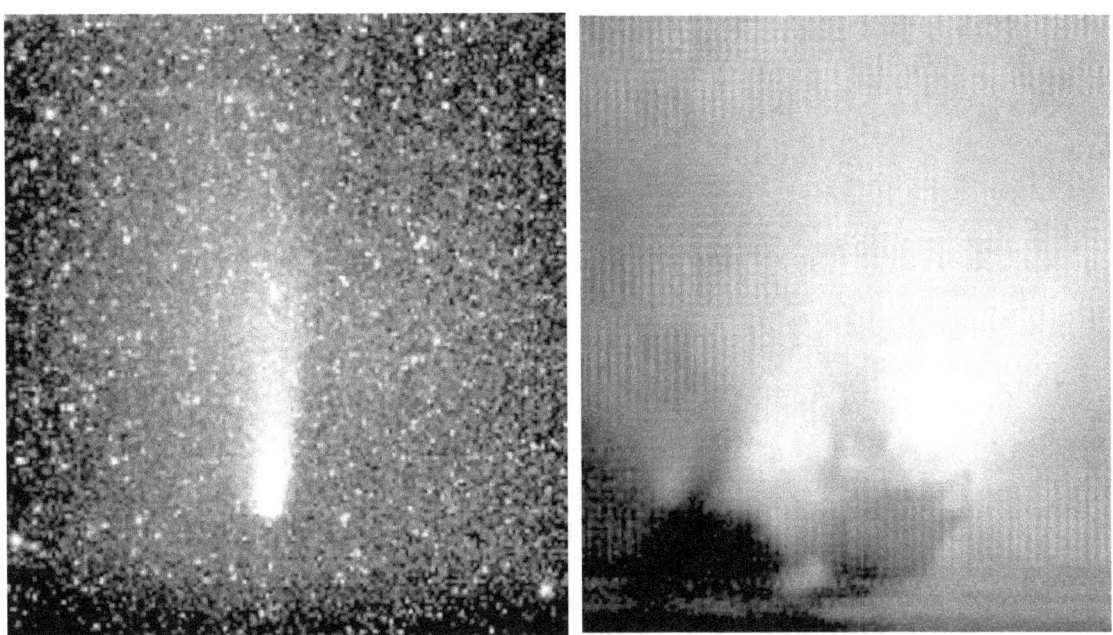

*Halley's Comet 1986 - Simultaneous Images
Left: 400mm Nikor Lens by author; right: ESA from spacecraft*

METEOR: when the meteoroid and Earth intersect, the particles hit the Earth's atmosphere with such velocity (many miles per second) that they ignite from the friction and typically will burn up; only slower-moving meteors and those of larger size can survive the fiery plunge to Earth and remain somewhat intact. The phenomenon of the Swift-Tuttle meteoroids hitting and burning in the Earth's atmosphere is known as the *PERSEID METEORS*. The Perseids usually present the brightest of all known meteors and fly through the sky very swiftly and are bright white and yellow due in large part from this rapid clip through the air. Many break apart slowly and present spectacular fireworks and colors as they pass a fiery and seemingly smoky trail behind.

METEORITE: any meteor which can survive this fiery plunge through the dense air of Earth and actually make it to the ground is known as a METEORITE. Many of the Perseid meteors are so fast and for the most part so small that they rarely will survive the descent, and few if any meteorites found on Earth can actually be associated to the Swift-Tuttle cloud.

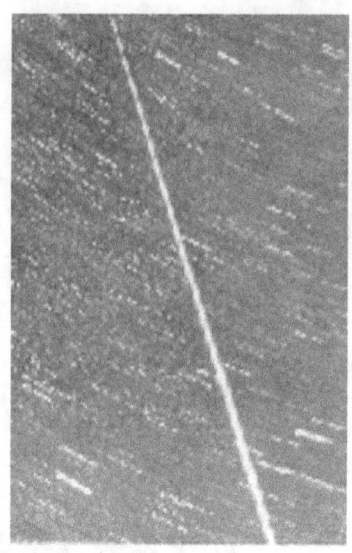

This photograph shows a very bright (magnitude -4, about the brightness of Venus) meteor, or "bolide" as it streaks across Arkansas skies in 1975; this photograph was taken with a regular 50mm camera lens at f.1.7 and focused to infinity; the camera was fixed on a tripod for this 20-minute time exposure. The shot ended as soon as I knew I had captured the meteor in my camera's field of view. Viewers are encourage to use their cameras for such shots; both 35mm and digital cameras are suitable provided that you have a TIME or BULB setting and that the lens is focused to infinity. Mounting on a tripod or piggyback on a telescope is essential. Remember that if your camera is digital or has an electronic shutter, the battery life will drain quickly during long exposures.

METEOR COLORS:

Note that the apparent COLOR of the meteor as it is burned in the atmosphere CAN be a direct indicator scientifically as to some of the chemical components of which it is composed. The nice graphic following (originally in color), courtesy of AccuWeather, shows the different colors that you might witness in a brilliant fireball and the associated KNOWN chemical compounds that can be associated with the burning of the meteor "stuff", much like chemistry lab where certain chemicals can be identified by the "flame test" of a Bunsen Burner.

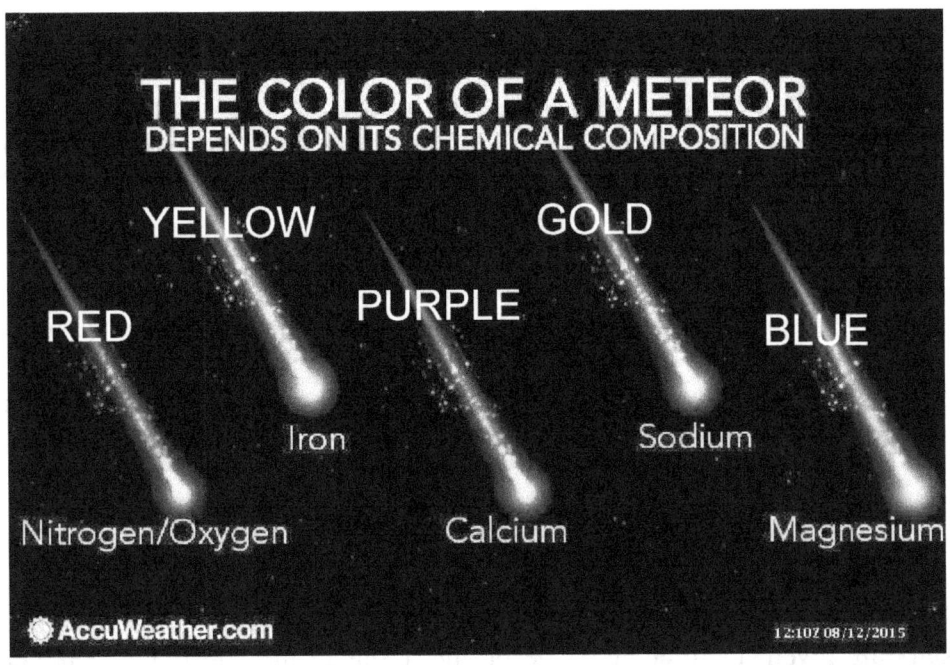

How to Observe Meteor Showers

Like eating a "Reese's Peanut Butter Cup", there are more than one way to watch the Perseids or any meteor shower. The two most popular are:

a) sit back and relax....enjoy the show!; and,
b) sit back and record your observations....enjoy the show!

For the latter, there are a few ways whereby you might actually contribute even casually to the science of "meteor wisdom" and at the same time preserve these memories via your SLR 35mm, video, or basic digital camera. Following is a quick and easy checklist and observing guide for you to enjoy and capture this celestial event .

1. **KNOW THE TIME:** Based on your location and the predicted times given above, don't take any chances; set your clock two hours early (the predictions might be off a bit!) and go outside prepared! Remember, for North American observers, begin your watching about 10:30 p.m. at which time there will be a few early meteors.....we never know for sure exactly when the big hit will occur. So......how about staying UP all night and use this wonderful opportunity to get acquainted with the wonders of the night sky? Note also that many meteor showers are known to produce many meteors as early as two weeks BEFORE and two weeks after the date of the predicted maximum!

2. **KNOW WHAT TO LOOK FOR:** Meteor showers are all named for the CONSTELLATION from which they appear to emanate from; we have the Leonids from Leo, the Capricornids from Capricorn....the Geminids from Gemini. The Perseids appear to enter the Earth's atmosphere toward the direction of the bright star *Algol* in the conspicuous constellation of Perseus , a constellation that appears somewhat as a "running man" (or the "keep on trucking man" to some). This "radiant" is the point at

which the Earth in its eastward motion plunges into the meteoroid cloud. Use the star magnitudes (brightness's - the smaller the number, the brighter the star or meteor) given on the chart to estimate your sightings of as many meteors as possible. The naked eye can see stars as faint as about 6th magnitude on a very dark and moonless night. The constellation and radiant will rise in the eastern sky about midnight local time and will be nearly overhead (for North American viewers) near dawn.

However! It is important to note that the most meteors will NOT be seen in the direction of the radiant, but more often about 30 degrees and more away from it as the meteors are pulled deeper into the Earth's atmosphere the farther they travel. I strongly suggest that observers merely look pretty much directly OVERHEAD to see the maximum number of meteors! This "aiming rule" also applies to your camera's field of view as well.

Let us assume that we are going to observe the famous LEONID meteor shower in November of each year. Note that on the following Leo chart (showing only the brightest stars of the constellation) that the star magnitudes are provided, as well as the approximate location of the meteor shower "radiant." The common Arabic names of the bright stars are given for reference. The brightest star in the constellation is known as "Regulus", the "heart of the lion" while the "tail star" is that marked as "Denebola." The bright star ALGIEBA, above which the Leonids appear to radiate, is a very spectacular "double star" which is known by observers with small telescopes to appear as "...two car headlights far down a distant highway!"

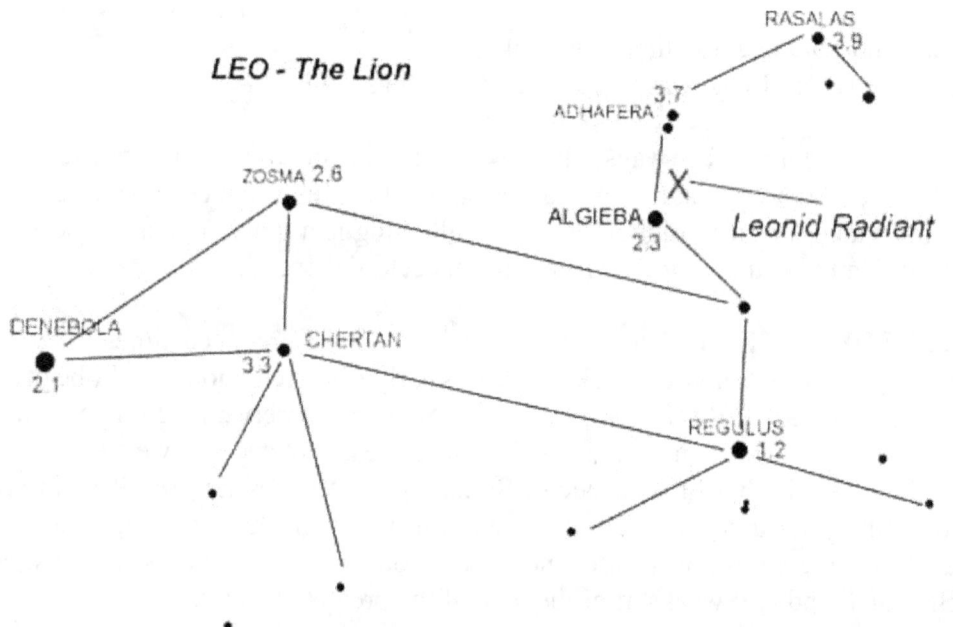

Remember that the MOON hopefully should be absent during the showing of the any meteor shower and you should take full advantage of this opportunity; since all meteor showers present meteors of all brightness levels, the darker the skies, the more meteors per hour (the "Zenith Hourly Rate," or ZHR) will be seen by each observer.

If there are several observers in your group, then orient each observer such that each is looking in a slightly different direction yet some overlap of sky coverage must occur. Each observer will count his or her own meteors beginning at about 15 minute intervals....do not combine the totals for a ZHR.....just add them up and divide by the number of observers to determine a "mean ZHR."

3. **DECIDE WHERE YOU WILL OBSERVE:** Be sure to plan this night carefully! Do NOT plan to observe from the city nor suburbs! Part of that planning MUST be from where you will observe to realize the darkest possible skies to see the faintest meteors.....but don't wait until that night to find a dark site. Plan to go out to your observing location on a night and time prior to the event and make sure that there are no bright "night watcher" lights that will interfere with your observing! If you live in a rural location it is likely that merely going into your own yard to watch the meteors will be fine; however, most "city" observers should attempt to find a very dark sky site to make the most of these relatively faint and fast meteors.

4. **KNOW HOW TO OBSERVE:** this is a really hard part: take out a comfortable reclining lawn chair and a blanket....position yourself with your feet toward the EAST (toward the Perseus radiant), and position yourself comfortably so that you are looking directly overhead! Such conditions are so strenuous that many, many observers have missed entire meteor showers as they nodded off in celestial bliss.

5. **THINGS TO RECORD:** to make the event even more meaningful, there are few minor details that you may want to record:
    a) the time at 15-minute intervals and the number of meteors that EACH person sees during that short interval; this will allow you to compute for yourself and your location about when the actual peak occurs;
    b) the color of the meteors; this is an indicator of both the temperature and the speed of each meteor;
    c) directions of travel; although the radiant position is well known, this is an interesting exercise....merely attempt to trace your brighter meteors backwards in their pathways and you can determine for yourself via the chart above a rough idea of the radiant from your own observations!

6. **PHOTOGRAPHING METEORS** as mentioned the Perseids are fairly bright and fast for the most part; however there ARE members that will be extraordinarily bright and these can be recorded on a standard camera (35mm DSLR with a "bulb" or "time" setting), a digital camera that can allow exposures of up to 16 seconds, and a video camera. For all types, a tripod is required and a cable release is recommended for time exposures for the still cameras. Use a fast ISO digital setting, set your lens aperture "wide open" (f1.8 or similar) and expose for at least five minutes; the stars will trail in these photographs (as you can clearly see in the 15 minute photo taken to record the bright fireball above) from the Earth's eastward rotation.

With your digital camera, merely take sequences of photographs that extend as long as the time exposure setting will allow, and use the widest angle setting and aperture possible. Video camera users can capture the brightest meteors by simply "zooming out" to wide angle and allowing the camera to run continuously. Your best chances of "catching" a meteor with cameras will be through aiming the camera straight up, or alternative between that position and somewhere halfway from overhead to the southern horizon.

7. **WHAT TO DO WITH YOUR OBSERVATIONS:** if you catch a spectacular fireball, by all means submit your photograph to this web site! For ZHR counts and other important information that can add to the body of knowledge of our understanding of meteors and their parent comets, you can submit you records (keep a copy!) via e-mail to: http://www.imo.net/fireball/report which is the International Meteor Organization, a large world-wide group formed and continuing to add to the growing studies of meteors, comets and related phenomenon.

This is an excellent time to take the family and friends on a weekend outing into the dark and cool nights. Find a quiet and restful place far from home, far from the lights and cares of the "big city," and enjoy the wonderful skies filled with the smells of impending winter, the sounds of the cold autumn blasts of artic air and the glorious sights of the winter Milky Way as it pours through constellations of antiquity. Through the bright stars of Canis Major and Orion.....across the blanket of sky that passes behind the stars of Taurus and its visitor Saturn....into the depths of Gemini and the interloping planets, our neighbors.....

Chapter 25

Scientific Studies of the Sun – "The Human Star"

Introduction
The first telescopic views of our sun - hitherto thought to be a "perfect sphere" of light and life - by **Galileo** almost 350 years ago revealed a turbulent and blemished world that would take centuries to fully understand.

Today, when amateur astronomers and those learning the components of our solar system, the first view of the sun is much as Galileo saw - yet today the sight frequently yields signs of disappointment rather than gasps of wonderment. All that is visible of the sun in the spectral sensitivity of our eyes is known as the *PHOTOSPHERE*, a bright yellow-white layer of explosively-hot gases that are occasionally interrupted by small features appearing not unlike craters on the moon - the *SUNSPOTS*.

Even though the sun may look a bit "boring" for dedicated observing, there is much to do on the sun - even on a daily basis - but two things are very needed before setting out:

1) a thorough understanding of the limitations of your telescope equipment when observing the sun; and,

2) special filters and precautions to protect your delicate eyesight from the intense burning rays of sunlight as well as the delicate interior of your precious telescope!

Observing the sun is an ideal activity for many enthusiasts, particularly those who:
* work evening and late night shifts and are unable to use their telescopes during nighttime hours;
* observers who are often frustrated because of bright sky conditions in their primary observing area;
* hosting a weekend cookout or other activity which can include a visit to our nearest star!

PRECAUTIONS TO OBSERVE THE SUN –

Of all objects we observe in space through our telescopes, only the sun poses any real physical threat to us; the very orb that provides the heat and light necessary and responsible for life on Earth is also potentially dangerous to your eyes, your skin and in some cases your general health.

Long ago, when I was a kid telescopes came with "sun filters," small little attachments that you would put on top of your eyepiece that reduced the intensity of the sun so that you could bear to look at it. These did not protect anything.....they merely allowed you to get close enough to the eyepiece without setting your eyeball on fire. Nothing but a small piece of common welder's glass; many times after several minutes of collecting the heat focused from a small telescope, these "filters" suddenly would crack and pass intense light into the unsuspecting viewer's eye.

If you ever have a chance to use one of these jewels....don't. Throw it away.

Today we have specialized filters of all types for our telescopes that are entirely safe for both the observer and the equipment. Regarding equipment, you should NEVER allow sunlight to pass into a closed tube such as a Maksutov, Schmidt-Cass or refractor telescope; the heat will accumulate until eventually something is going to melt or pop. The lens of your telescope is less effective at "starting a fire on a leaf" than is the common magnifying glass focusing the sun's rays to a tiny and fiery point.

Full-aperture, optical glass mirrors are desirable for solar observing; these provide excellent "true" color (yellowish-orangish-yellow-white and all combinations). The "Mylar" foil filters are safe, but they have a false blue coloration and provide very poor contrast when compared the the better (and more expensive) glass types. Good optical glass filters reflect MOST of the light coming from the sun and REJECT the harmful ultraviolet and infrared radiations that most severely burn the eye's retina.

FIGURE 1 - The sun as seen through a quality solar filter
Photo Courtesy Mike Weasner - Ricoh RDC 4200
Afocal Photography, ETX 90 RA, Sept. 20, '00

Although there are many brands available, I have actually tested two - from Orion and from Thousand Oaks - for optical quality, contrast and protective transmission/rejection that I would highly recommend to anyone; they are made in sizes to fit your telescope without any adaptation.

These filters fit like a cap covering the FRONT of your telescope; NOTE: even though these filters are provided with a metal cell and mounted, be sure to shim within the edge of the filter cell to assure a snug fit against the outer perimeter of your telescope tube. Some filters, if not properly snugged, will loosen and fall off, resulting in sudden surges of unwanted and unhealthy light to the unsuspecting observer.

As Figure 1 demonstrates, a good solar filter will show the visible layer of the sun's hot gases, the photosphere. Among the features that you might see are: 1) white-light flares (rare, white streaks usually near sunspot groups); 2) sunspots and sunspot groups (can be seen in Figure 1); and oftentimes "granulation," considerable mottled areas that are tops of turbulent mixings of the sun's gases. **In addition**, there are NOW very high resolution "Solar Telescopes" on the market for non-professional astronomers which will indeed show PROMINENCES, solar flares, and extensive activity not seen with white light filters. Your dealer can demonstrate these remarkable telescopes.

THE SOLAR AND SUNSPOT CYCLE –

The sun, like all other stars and like Galileo pointed out for the first time, is NOT a perfect object. It turns on its axis (a point which eventually resulted in Galileo's house arrest!) as can be attested by observing the sunspots (following in Guide) as they slowly move east-to-west across the visible disk of the sun; the sun makes ONE complete rotation in about 26.7 days at the equator and slightly slower nearer the poles.

Our sun goes through periodic cycles, just like any variable star, every 11 years. The cycles are fairly predictable from one to the other. From MINIMUM sunspot activity it takes only 4.5 years to surge to MAXIMUM sunspot activity, then longer to subside or about 6.5 years. Over 300 sunspots can be observed on a single presentation of the sun during maximum and there are times during minimum that NO spots can be found even with quality larger-aperture telescopes.

In addition to the major 11-year cycle, the sun also exhibits another major cycle of 22 years, and another that consists of minor changes within a 200-year span. Even one estimated at near 10,000 or more has been determined based on climatic changes (tree ring analysis) through the centuries.

OBSERVING AND IDENTIFYING SUNSPOTS AND SUNSPOT GROUPS –

This guide is focused on some interesting and rewarding observing that you might undertake with your filtered telescope to monitor the sunspots, their growth and demise, their rotations with the spinning of the sun and how to determine the *ZURICH DAILY NUMBER*, or a simple "sunspot count."

SUNSPOTS: Every day will bring a new look to the sun; the main objects you will see with good solar filters and your smaller telescope will be "*sunspots*" and *sunspot "groups."* The changes in these spots and groups happens unpredictably and intense solar activity (i.e., when a lot of auroral activity is forecast) usually is a precursor of good observing for sunspot changes.

Figure 2 clearly shows the nice detail that is visible in sunspots; note that in the two excellent photographs in Figure 2 that we are actually looking at a "sunspot group," comprised of many individual sunspot "cells." Typically, a sunspot - or group - will have a very dark (black-appearing) *UMBRA* near the center axis surrounded by a lighter and often-detailed *PENUMBRA*, or lesser dark halo. Both features are clearly visible in the photographs.

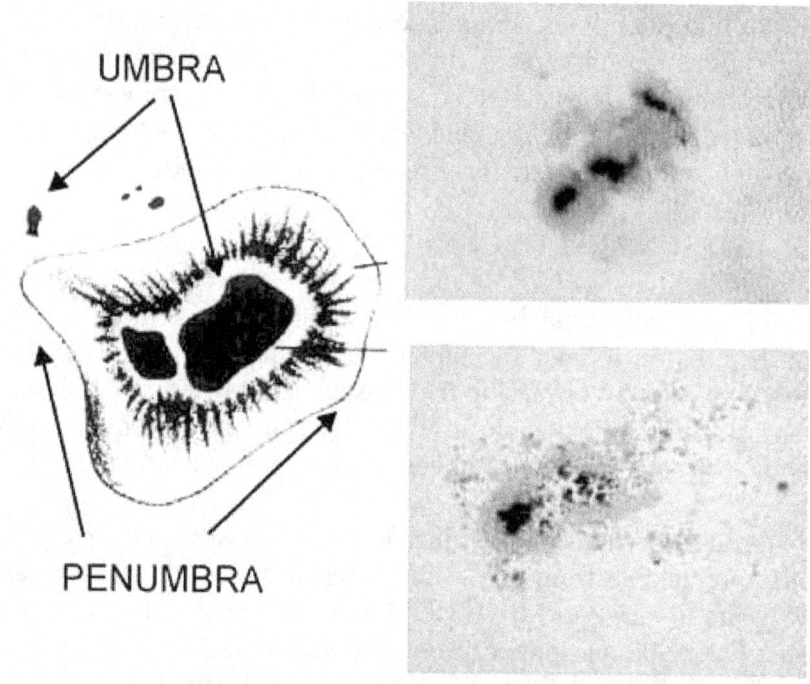

FIGURE 2 - a) above - sunspot features
b) top R - Photo Courtesy Mike Weasner - Ricoh RDC 4200
Afocal method, ETX 90 RA March 11, '00
c) lower R - Photo Courtesy Peter Vasey, 8" Celestron
September 23 '00

Note that *Peter Vasey's* close-up photograph at lower right corresponds to *Mike Weasner's* full-disk image of the sun, both taken on September 23, 2000; the close-up shows the sunspot that Mike captured with his ETX 90 as seen just above the middle on the extreme right edge of the photograph.

Sunspots will perhaps take on a new meaning for you when you realize that they are actually some of the BRIGHTEST objects in our solar system! They only "appear" dark or black because they are slightly cooler (several thousands of degrees) than the surrounding hot photosphere, and thus appear less luminous than the yellow surrounding gases. If you could (thank goodness this is hypothetical) take a single small sunspot and put it in a dark room, it would be blindingly bright and of course very deadly!

Long before Galileo and the advent of the telescope, ancient *Chinese* astronomers were recording major sunspots with the naked eye; still today - particularly during sunset and sunrise when the Earth's air filters most of the light from the intense orb - we can see major sunspots to the unaided eye from times of great solar activity.

For sun observers, it is important to learn both the method to determine the ZURICH DAILY NUMBER (of sunspots and groups - this published regularly in astronomy journals and magazines) as well as "Sunspot Classifications" based on sunspots and their characteristics both as individual spots and in groups.

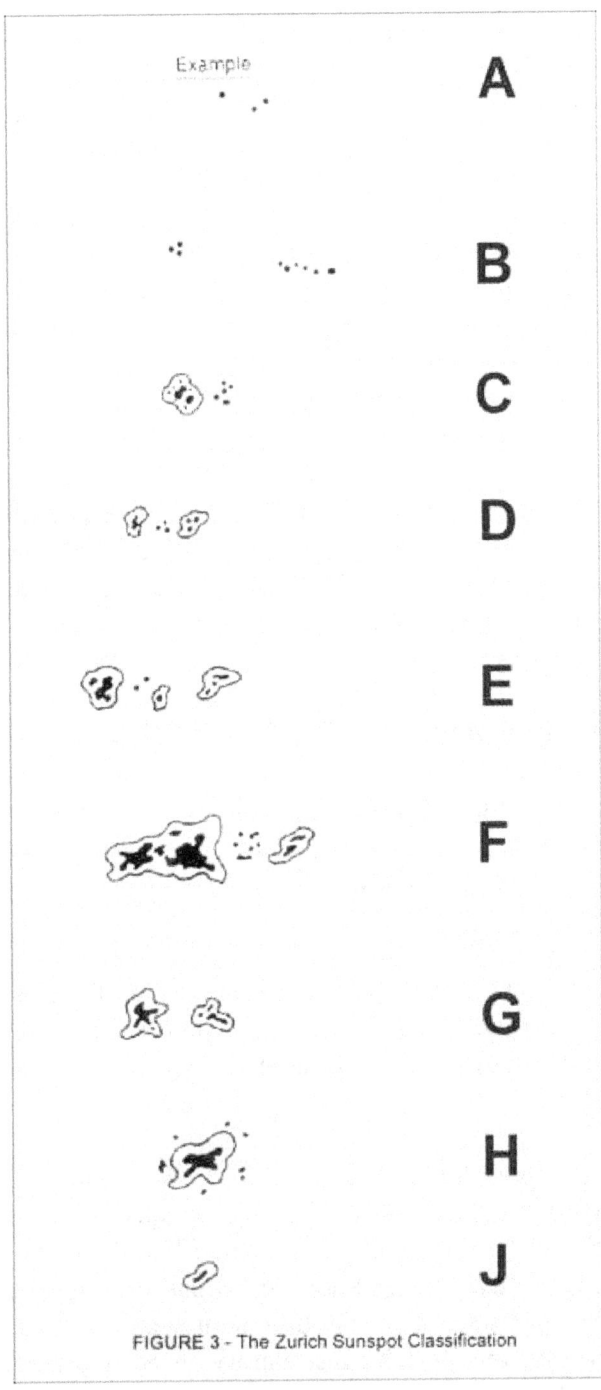

FIGURE 3 - The Zurich Sunspot Classification

Figure 3, (as well as Figure 4) following, is taken from my book "*A Complete Manual of Amateur Astronomy*," [Prentice-Hall, 1981] and demonstrates clearly the major categories of sunspots and groups (no, there is is "no [i]" group!) as based on their distinguishing characteristics.

I will quickly and concisely describe each of these categories:

CLASS A - a small individual spot, or a spot group that shows NO penumbra;
CLASS B - similar to "A" but spots suggest some association to one-another, or they show "symmetry" - (BIPOLAR SPOTS); still NO penumbra is present;
CLASS C - "bipolar" (see above) groups in which only the largest spots have penumbras;
CLASS D - like "C" except ALL major spots show penumbra;
CLASS E - a very large bipolar group (very much like those of Figure 2), larger than 10 degrees across, in which the major spots show lots of detail within the penumbra; there may be many very small spots between the major spots;
CLASS F - largest of all bipolar groups, sometimes as much as 20 degrees, surrounded by very detailed penumbra and many random smaller surrounding spots;
CLASS G - just like "F" above, but NO smaller random spots around;
CLASS H - one large spot with penumbra surrounded by small random umbral spots;
CLASS J - one large single spot (POLAR) with a penumbra....nothing more.

Test your ability to differentiate the spots that you can see. Remember....the image of the sun that you observer will LIKELY exhibit many different categories of sunspots and groups all at the same time. Look around! Using medium magnification (about 30x per inch aperture) will reveal great detail within the spot groups and even some granulation that may surround nearby spot groups.

SUNSPOT COUNTS - THE DAILY ZURICH NUMBER

You would think that "counting sunspots" would be like counting sheep: 1) it seems like a pretty straightforward "1, 2, 3..." process; and, 2) it would put you to sleep.

Neither is the case.

The basis of the "*Zurich Daily Number*" is to use sunspots as a gauge to accurately monitor solar activity on any given day. In this procedure a SUNSPOT GROUP is considered "more active" than is an individual SPOT, so it is factored "higher" than would be an individual spot.

Thus, when counting sunspots for the *Relative Spot Number* (RSN), spots count as "1" and groups count as 10 times the activity of a single spot. IN ADDITION, like so much in today's world, there must be some standard for the SIZE AND TYPE of telescope used to monitor the counts. Obviously we all know that a larger telescope is capable of seeing MORE sunspots because of its greater resolution, so it stands to reason that the "home" of worldwide sun monitoring, the Zurich Observatory in Switzerland, would be using as BIG a telescope as possible to watch the sun....right?

Wrong. The world standard is set from a **2.4" refracting telescope**. Thus, a small scope is an ideal telescope to match the world leader in sunspot observing! Anything larger will reveal TOO MANY sunspots that will skew the actual total, so we must FACTOR those larger instruments, even a small refractor. More on that in a minute.

The RSN Formula - there is a simple formula that takes all this into account and we will apply it to some actual sunspot counts as a test. First, we know the rules:

1) spots are counted as "1" even if they are contained in a "group;"
2) groups (the entire mess of them) count as 10 times the value of a spot - each group is "10" no matter how many spots;
3) a telescope of 60mm is a factor of "1" (multiply the count by this factor); other scopes must factor differently.

Your Telescope Factor (the "k" factor) - if you are using a telescope of 60mm or less, you use a factor of "1" which matches the Zurich world standard. If not, try the values I have determined below:

Scope Size (mm)	Factor (multiply by)
60mm	1
90	0.8
125	0.6
200	0.4

The Formula - don't be too stressed about this...it is easy and we'll put it to the test for you in a moment.

RSN = [(g x 10) + n] x k,
where,

RSN - "Relative sunspot number"
  "g" is the number of groups you can see
  "n" is the total number of sunspots, even those counted INSIDE of groups;
  "k" is your telescope factor as provided above.

It is really very simple to use this formula (always attempt to use low power so that the entire sun's disk (Figure 1) can be positioned in the field of view:

1) count the number of GROUPS ("g") that you can see in a low-power field of view (do not change to higher magnifications at any point during your daily count!) and MULTIPLY that number by "10"...if you see three (3) groups, then you will have that value as "30" (the "g x 10" above in the formula);

2) now simply ADD to that number (in this example, "30") ALL of the individual sunspots ("n") that you can see AT THE SAME MAGNIFICATION; do not raise the

magnification until AFTER you have counted....you can see more sunspots at higher magnifications; even though you have already counted the "groups", you still must COUNT THE SUNSPOTS within the groups as part of the "n" total;

3) if you have counted 70 individual spots, merely add that to the ("g x 10") total of 30 that you have obtained for the groups, and you get a count of "100" for that day;

4) if you have used an 3: scope, for example, you must NOW multiply that "100" by its "k" factor of "0.8" provided above to adjust your telescope to that of Zurich; so, very simply you multiply the "100" by 0.8 and obtain "80" as your RSN number for that day.

That's it....simple as that. Since an 8-inch scope will likely see more individual spots than the 3-inch, the factor number SHOULD result in a final number for both telescope that is fairly equal.

WHAT DO YOU DO WITH ZURICH SUNSPOT NUMBERS?

Okay, you've stood out there and counted black spots on the sun and used remedial math to figure out the RSN....what for? Is it worth anything? Who do you report this to?

Yes, is very important and as they say...."somebody's gotta' do it!" The RSN as mentioned is an indicator of relative solar activity on any given day; the sun's unpredictable nature makes it necessary for us to monitor this activity to better predict and understand the results of solar outbursts to us here on Earth. The sun's activity likely affects our weather and climate, and certainly affects the ionosphere through aurora, radio and microwave disruption, satellite transmission and other adverse affects.

We DO NOT totally understand the nature of the sun no more than we do that of any other star in space; but YOU, with some degree of diligence as you record the RSN can contribute to the body of knowledge of our closest star.

For example, if tomorrow you heard on the news about a huge solar flare that is causing all sorts of communications on Earth, you would immediately take notice. Now, how about if you HAPPENED TO HAVE BEEN OUT the day before that happened and recorded the sun's RSN?? It would really mean something to you then, wouldn't it? As a matter of fact, you may well likely be able to go back over a week's previous sunspot counts and realize that "....something was up!"

Your counts are of no useful purpose if you merely leave them in a notebook in the closet. You can log your RSN as frequently as you record it by contacting the American Association of Variable Star Observers (**AAVSO**) at:
http://www.aavso.org/committees/solar/

where you can obtain important and very detailed information about solar observing programs, forms and the necessary framework to build a good solar observing program and the correct procedures for reporting your valued observations.

LET'S TAKE A QUICK TEST –

Figure 4 shows three rough sketches of "the sun" on different days with three different combinations of imaginary sunspots and groups. Using the method described above, do INDIVIDUAL counts of each Day 1, Day 2 and Day 3. First, count the "groups" and multiply by "10"; then ADD to that number the number of total spots you see (including those in the groups) and then multiply by your telescope factor. What do you get?

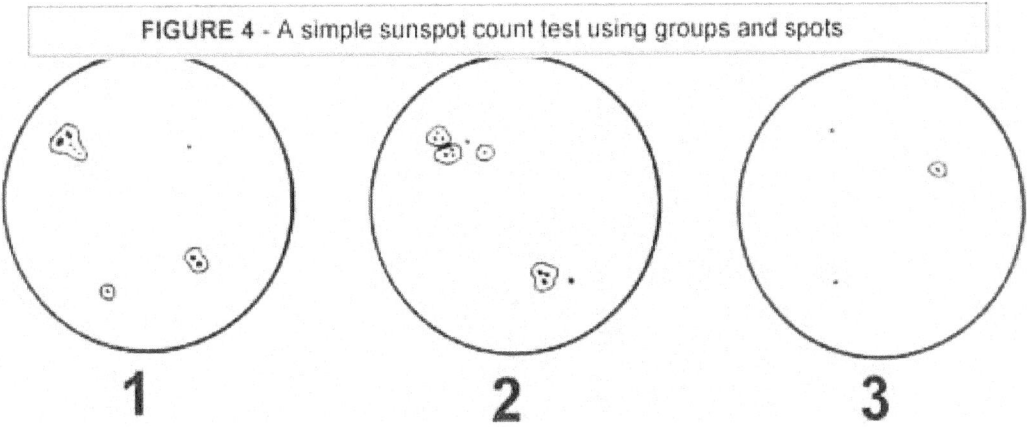

Here are the actual totals (I am using the "k" factor of "1"); go ahead and see what it would be with YOUR "k" factor as well:

Day 1 - four (4) groups ("g"); nine (9) individual spots ("n") / RSN = [(4 x 10) + 9] x "1"
/ **RSN Day 1 = 49**
Day 2 - two (2) groups ("g"); 12 spots ("n") / RSN = [(2 x 10) + 12] x "1" / **RSN Day 2 = 32**
Day 3 - three (3) groups ("g"); three (3) spots ("n") / RSN = [(3 x 10) + 3] x "1" / **RSN Day 3 = 33**

How did YOU do?

You really haven't "stargazed" until you have a chance to "gaze" at your own star...the sun. Too many amateurs neglect the sun because it seems so static and uninteresting....like counting sheep. but payday always comes around when a major solar outburst is recorded and you have the thrill of going back over your OWN PERSONAL records of the sun's recent life.

Only then, can you turn to your family after hearing about solar shock waves on the nightly news and say:

"Yeah....but let me tell you what was going on BEFORE that happened..."

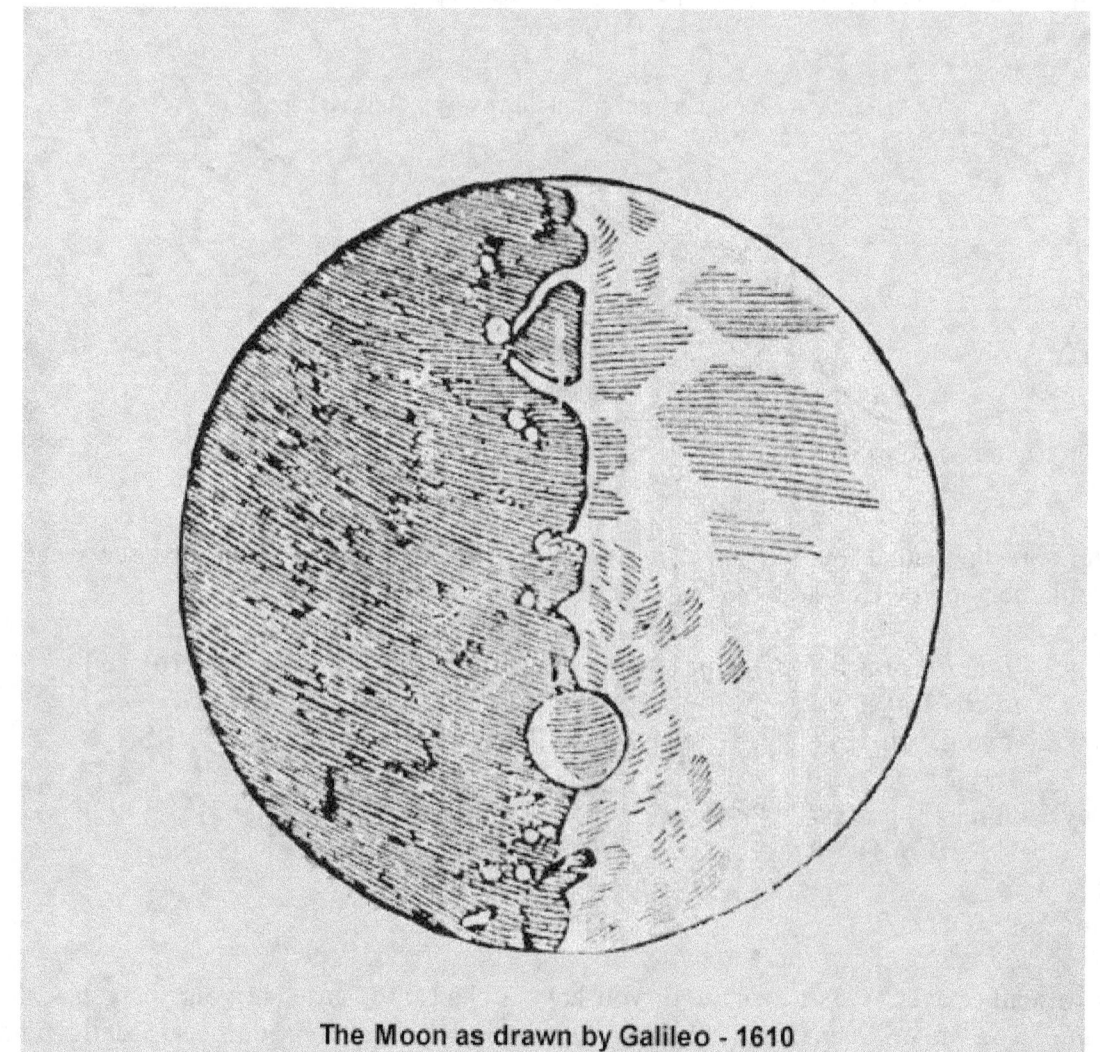
The Moon as drawn by Galileo - 1610

Chapter 26

OUR ROMANCE WITH THE MOON

INTRODUCTION

With all of the latest sophistication in astronomy.....the Hubble Space Telescope, deep space probes that land on potato-shaped asteroids....telescopes that GO TO objects and don't even need an observer to point them...

.....we tend to overlook the fundamental excitement of astronomy and why we are compelled toward it in the first place: DISCOVERY. It is the "unknown" that first lured us to astronomy, the challenge to use our minds and imaginations to unlock mysteries that we simply did not know answers to....and, for the most part, never will. It is still human nature to "explore and discover."

And "discovery" is what led us to the moon. We've been there - and as the television commercial hawking cheese says: "we found out that the moon was NOT made of cheese....and we haven't been back since."

Nonetheless, the moon remains a wonderful object to explore, even in this high tech era of computerize telescopes and satellites filling our skies. It's so close, and yet we tend to ignore it! I certainly do not know EVERYTHING there is to know about the moon....do you? Not one of us can randomly aim a telescope at high power toward the lunar disk and call out the names of even a fraction of the craters seen, their history, their approximate ages.....so WHY are we not looking at this grand neighbor more?

The moon presents as a wealth of exciting craters, mountains, valleys, plains and even perhaps darkened volcanic cones, all of these clearly visible in our modest telescopes; not rarely there are lunar eclipses where the Earth's shadow paints a reddish hue across the lunar face. Also the moon creeps eastward each night, covering up stars in its path making for an exciting night of celestial "hide-and-seek" of distant stars.

So let's explore some wonderful some wonderful facts and exciting events about our moon so that you can explore the lunar surface and lunar events with your telescopes tonight! We CAN bring back our romance with the moon! In case it has been longer than we care to admit, or simply the fact that we never bothered to learn much about our companion, this study contains a few textbook "basics."

One study that is fun for the entire family or serious observers is the observations of "lunar conjunctions" and "occultations." Not only are these visually-exciting events (see "Welcome to My Neighborhood" photo), but in the long run an amateur astronomer and his or her modest telescope can make a substantial contribution to astronomical science as well.

One thing is for sure....you most definitely will not need the GO TO for this one!

PART 1 - THE MOON'S PATH: Conjunctions & Occultations
Daily Motion of the Moon –

AN OCCULTATION WAITING TO HAPPEN
A Lunar Occultation of the Waxing Crescent Moon
of the Planet Venus in 1978
(see the entire event in Figure 4!)
Such events are spectacular to watch both
with the naked eye and with a telescope.
200mm f/3 lens. P. Clay Sherrod

In addition to orbiting around the Earth with respect to the background stars in a period of about 27 days 8 hours, the moon is part of the Earth-Moon duo orbiting the sun almost $1/13^{th}$ of a year; simply put, this results in its revolution around with respect to the sun at a longer 29 days 13h, its synodic period, that on which the modern calendar is based. It is the difference between the two that also allows for the varying lunar phases - from new to quarter, from quarter to gibbous, from gibbous to full, from full to waning gibbous, from waning gibbous to three-quarter, and from three-quarter back to new - all in the course of about what we call a "month." Aspects of the lunar phases and observing are discussed following.

The moon, if observed in your telescope against the background stars at night, moves about 13° eastward in the course of one Earth day. As we see the sun from Earth on the other hand, its motion eastward is only about 1° eastward in the same period.

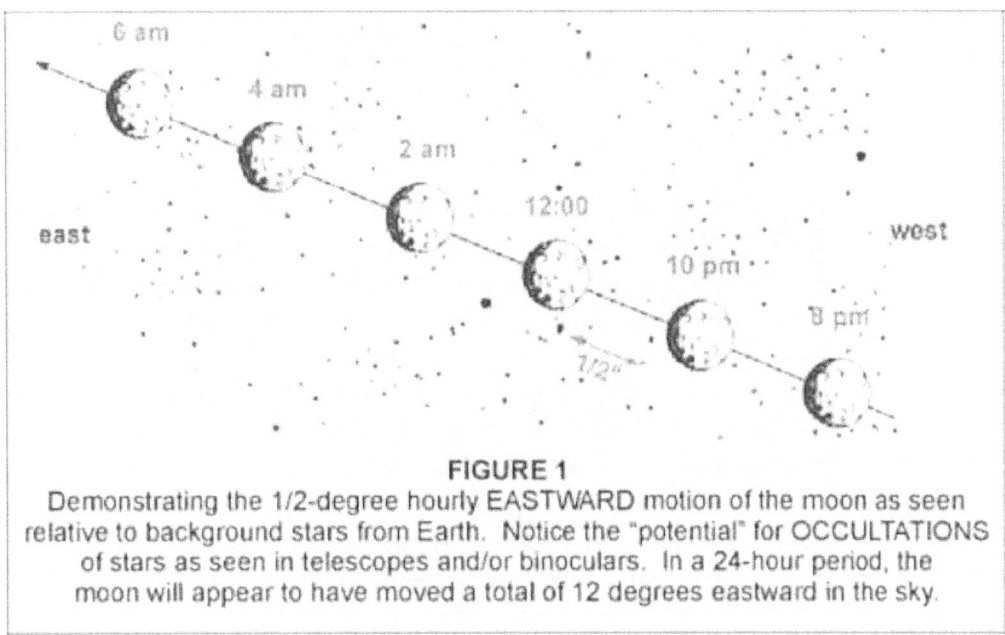

FIGURE 1
Demonstrating the 1/2-degree hourly EASTWARD motion of the moon as seen relative to background stars from Earth. Notice the "potential" for OCCULTATIONS of stars as seen in telescopes and/or binoculars. In a 24-hour period, the moon will appear to have moved a total of 12 degrees eastward in the sky.

The resulting net 12° eastward motion of the moon each day delays the rising of the moon on subsequent nights by about 50 minutes. If the moon transits overhead tonight at 10 p.m. local time, tomorrow night the transit will occur at 10:50 p.m. local time. Thus, if bright light of the full moon (transiting overhead at midnight local time) is blocking your viewing of faint constellations and stars tonight, remember that in only six days the moon will not even rise in the east until slightly after 11:00 p.m.! All the hours of darkness prior to that time can be spent stargazing in a moonless sky!

In addition to this long-term effect of daily motion, we can experience first hand the motion of the moon in an even shorter period. The 12° daily eastward motion of the moon can thought of in smaller terms: each hour this translates in the moon moving its own diameter - one half degree - eastward. (See Figure 1) What does this mean to all of us with small and large telescopes? Using very low magnification - even binoculars - it is possible to watch during that one hour the moon creep slowly eastward against the far-distant stars, the stars appearing motionless. Indeed, there are many, many time in the course of a 24-hour day that at some locations throughout the world that the moon's hourly path actually takes its "disk", or globe, to pass in front of the stars, a phenomenon known to astronomers as lunar occultations. A bit later, I will discuss the occultation phenomenon events that you can observe through your ETX 90, ETX 125 and LX 90 telescopes!

Do not confuse the "occultation" with a more common event: "the lunar conjunction," in which the moon appears very close in the sky to a bright star or planet. The "Welcome to my Neighborhood" photograph shows a beautiful lunar occultation with the planet Venus in the late 1970's. Conjunctions are fairly common events, particularly when Venus is high in the evening western sky during "early" crescent phases (See Figure 6) of the moon, and when Venus is a morning object in the eastern sky. All of the bright planets are frequently seen "in conjunction with" the moon.

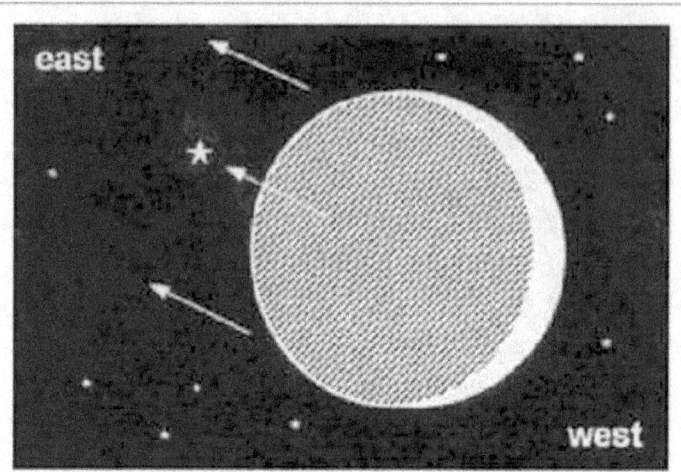

FIGURE 2
A Lunar Occultation of the Waning Crescent Moon
(this would be in EVENING hours)
Note the dark UN ILLUMINATED terminator
of the moon is now EASTWARD,
thereby making first contacton the "dark side"
to instantaneously cover the star.
A star disappearing at approximately this
mid-way point would reappear on the
opposite, illuminated side in about 1 hour.

An "occultation" in astronomy is an event in which one celestial object covers another.....indeed, even a solar eclipse is an occultation in itself, as the moon simply covers the sun. Although not rare, they can be spectacular if a bright star or planet is involved. As well, occultations are of valuable scientific importance. Astronomers utilizing very high power in large telescopes can monitor the exact time (to $1/10^{th}$ second by eye and $1/100^{th}$ second using electronic photoelectric devices) at which the star "blinks out" as it is covered by the dark east edge (or "limb") of the moon. This allows mathematical computation of the exact location and distance of the moon over a long period of time, a calculation that has revealed that the moon is very slowly - very slowly - moving ever-so-farther from the Earth, at a rate of about 3--4 cm/yr.

The tidal bulges on the Earth main toward the middle of the oceans), raised by the Moon's gravity, are rotated forward (ahead of) the Earth-Moon line by the Earth's rotation since it is faster than the Moon's actual motion in its orbit. The gravity from these leading and trailing bulges impels the Moon forward along the direction of its motion in orbit and the force transfers momentum from the rotating Earth to the revolving Moon, simultaneously dragging the Earth and accelerating the Moon.

This ultimately causes the moon to recede from the earth as it angular momentum increases.

For the purpose of this Guide the daily motion of the moon is very simple to keep up with and very interesting as well, since it allows for the moon to actually be observed through your telescope moving against the stars in the sky during an evening's observing. Understanding the not-so-simple movement of our natural satellite also can help you appreciate - and explain to the family - how the moon managed to "....creep away from Venus since last night" and show up far east of where it was only 24 hours ago, and why that thin crescent is destined to become a bright, full moon rising in the east in less than only two weeks!

The Ecliptic and the Motion of the Moon –

If you watch the motion of the moon, the sun and even the planets from day to day, month to month and ultimately from one year to the next, you will soon realize that all these solar system objects move in a well-defined, narrow and repeating path across the sky.

Many solar system objects appear to move very fast against the stars, like the sun and moon. Part of this apparent rapid motion is due to the short distance of the object to the Earth, but a great deal of it is due to the motions of the Earth itself (both from ROTATION in a 24-hour day, and its REVOLUTION around the sun for the 365-day "year"). Still other objects - Jupiter, and Saturn for example - appear to move very slowly because of their great distance from our "observatory": Earth.

So in addition to our moon, all of the objects in the solar system move within a very narrow "belt" of the sky that astronomers today call the ecliptic; even the earliest star gazers of yesteryear realized the motions of "the wanderers" (the seven naked-eye "planetes" including: Mercury, Venus, the Moon, Mars, Jupiter, Saturn and the Moon) in this band and entitled it the zodiac (Greek, for "Zone of the Animals)." Through the twelve constellations within the zodiac the wandering planets, sun and moon all travel, and give rise to the centuries-old attention to the pseudoscience of astrology.

The twelve zodiacal constellations and their common names are (beginning with the First, Aries, and progressive eastward): ARIES (the Ram); TAURUS (the bull); GEMINI (twins); CANCER (the Crab); LEO (a lion); VIRGO (young virgin); LIBRA (balance - the only "non-animal of the group!!); SCORPIUS (the Scorpion); SAGITTARIUS (an Archer); CAPRICORNUS (a "sea Goat"); AQUARIUS (the Water-Bearer); and PISCES (the Fish).

Each of these constellations of the zodiac (a "sign" in astrology) holds special meaning to those born "under the sign," even to astrologers of the twenty-first century.If the moon, as well as all solar system objects, appear to move only within this narrow band of sky, does that mean that occultations of stars take place every month as the moon orbits the Earth??

The answer is "yes".....and, "no."

YES: Occultations of stars and other objects by the moon occur every night as seen from Earth. The moon is moving so rapidly eastward across our sky that not a night goes by at someplace on the planet that multiple occultations do not take place.

NO: The same objects are NOT always occulted each month as they were the previous month, nor are they occulted by the moon necessarily in the course of one year. The actual reason is complex, but the moon, relative to the Earth, sun and stars returns to exactly the same position as we see it only every 18.6 years.

All solar system objects, the moon included, follow a pathway through the ecliptic, an imaginary plane intersecting the sun's equator and containing all of the orbits of the planets and of their natural satellites. Imagine the sun as a huge semicircle, a globe cut in half and laying across a large table. On the table marbles roll around the semicircle like planets around the sun.

While all planets move in individual orbits of their own, their apparent motion is a result of these orbits, and the way we see them moving from Earth.
Part of this combined motion can result in the apparent motion of any celestial object varying as much as a 13° "up and down" (north and south) from the true plane of the ecliptic (the flat "tabletop"). So - although the moon passes close to the same stars at least once monthly - its proximity and ability to intersect and occult those stars will vary, albeit predictably by astronomical calculations.

Simply stated, if the bright star Spica is occulted by the moon tonight, an occultation of the same circumstances (see Figure 1) CAN occur from your same exact location on earth again in about 18.6 years.

Occultations for the Casual Observer - A lunar occultation is a sudden and spectacular event; many times, even with binoculars and unknowing of an upcoming event, an observer might spot a bright star "flash away" as the dark edge of the moon suddenly covers the star (Figure 1). However, the nature of the moon's motion and the seemingly fixed positions of stars within the ecliptic allows astronomers to predict and publish such occurrences years in advance. Note that occultations of BRIGHT stars, any of the "superior" planets (those more distant than Earth from the sun), ASTEROIDS and even deep sky (i.e, the Pleiades star cluster) are all published in Sky and Telescope magazine far ahead of time to alert you to these upcoming spectacular events.

Note that a star - because of its great distance, even those closest to us - "blinks" out behind the limb, or edge of the moon. A PLANET, on the other had, does not "blink out" because it is NOT a point source; its disk merely gradually diminishes in brightness (see Figures 4 & 5) as it progressively disappears behind the lunar limb (edge). Likewise, "multiple stars" - even those that cannot be separated into two or more components in you telescope -also have a double fading, with one component disappearing (the light suddenly diminishes) and suddenly the other, causing both stars (or more) to finally "blink out." Actually, several new double stars - and "double asteroids" - have been DISCOVERED using the occultation "double-blink" method!

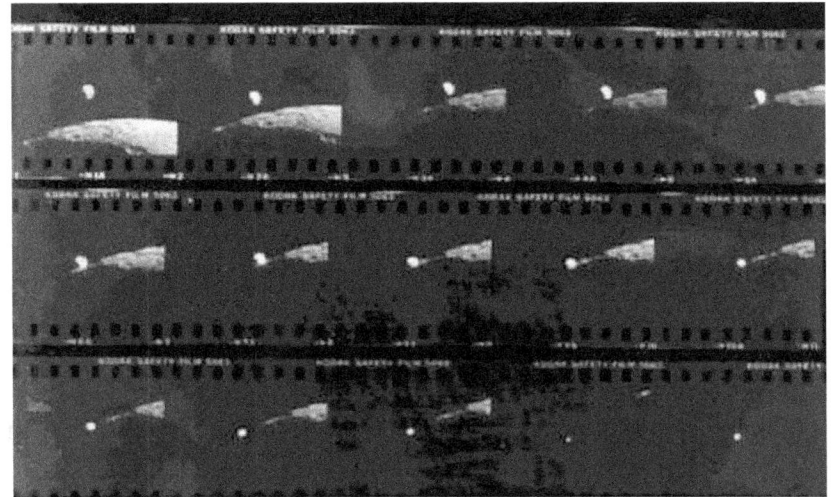

FIGURE 4 (above)
A Lunar Graze Occultation of the Planet VENUS This print shows the prime focus negatives showing the series as the moon crept eastward toward Venus, its southern cusp just "skimming" across the "quarter-lit" planet.

Star "blinking" is a result of two factors:
1) Because even the closest star is farther than 12 trillion miles distant, their apparent size, or angular size, is equal to that of a point of light, showing no diameter which can be slowly covered up by the edge of the moon; the largest stars may be over a million miles in diameter, yet their distances still show them to us on Earth as a tiny point of light.
2) The moon has no air surrounding its surface like the Earth's atmospheric "blanket." Hence, there is no gradual diminishing of light as would be expected as the star creeps deeper and deeper into an atmosphere. Such a dimming of starlight by the atmosphere is called "atmospheric extinction" and is particularly noticeable on Earth as bright stars grow dimmer and dimmer the closer they get to the western horizon during setting.

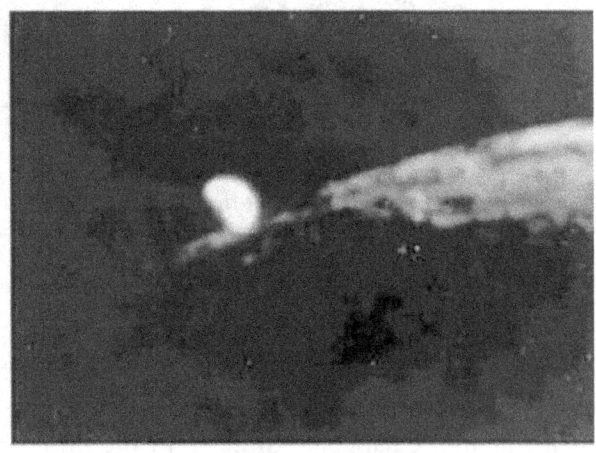

FIGURE 5 (right)
Note the rugged detail of the south lunar cusp as Venus skims past. The quarter-phase shape of the planet is clearly visible in this photograph.

Photographs by
Clay Sherrod, Dec. 26 '78
on Tri-x film, 1/30 second.
12" Clarke Refractor at
f/16 prime focus.
Central Missouri Methodist
College, Fayette, MO

Occultations are much easier to spot and record during waxing phases (Figure 1), those lunar phases between new moon through the next two weeks to full moon. The sunlight which illuminates the moon during such times is striking the western side of the moon, leaving the eastern or approaching side darkened. Even faint stars stand out in stark contrast to the dark limb of the moon.

Many times, particular close to either waxing or waning crescent moon, a bright reddish reflection of sunlight off the shiny atmosphere of Earth can be seen distinctly on the "dark side" of the moon. In addition to being a striking sight, it allows you to define the moon's edge and see the brightest craters even in full darkness. This gentle glow on the lunar surface is known as "earthshine," and is exactly the same phenomenon that Earthlings realize when walking about in the midst of brightened sky and ground of full moonlight.

Stars that normally would be visible without the bright full moon, are blocked by the scattering in our air of bright reflected sunlight from the moon. Because the moon, has no air, the faintest of stars can be clearly seen during "full Earth" (the Earth's phases as seen from the moon are exactly opposite those exhibited by the moon on the same given night! As previously mentioned, occultations of stars and planets by the moon during waning phases - or those that occur within the two weeks following a full moon - take place with the object disappearing on the bright limb of the moon. This means that the bright lunar edge greatly impairs your ability to view the star and thus its sudden disappearance. Indeed, during the waning phases of the moon, it is more desirable to watch the reappearance of a bright star as it suddenly "blinks" into view out from BEHIND the moon on the darkened eastern limb.

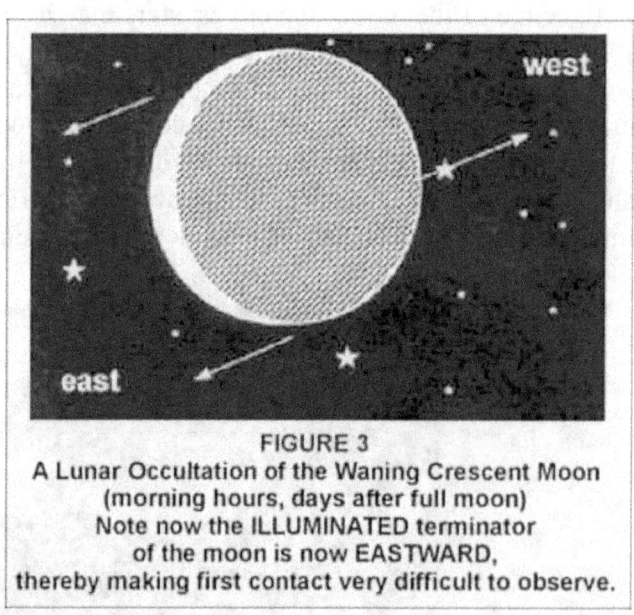

FIGURE 3
A Lunar Occultation of the Waning Crescent Moon
(morning hours, days after full moon)
Note now the ILLUMINATED terminator
of the moon is now EASTWARD,
thereby making first contact very difficult to observe.

Of course, your might wish to view BOTH occurrences - the disappearance and the subsequent reappearance of the same star. Because the moon moves its own diameter in one hour, a star that disappears directly in the "middle" of the lunar disk (as shown in Figure 1), will re-emerge from behind the opposite edge (limb) in about 50 minutes to one hour. Those that are occulted nearer the lunar poles will reappear in a much shorter interval that those near the lunar equator. The length of time, thus is dictated by what point relative to the moon's equator the star disappears, the farther from the moon's midsection (toward the lunar poles), the shorter time until it reappears on the opposite limb.

REMEMBER!! *ALL* stars and any object that are located or travel through the "ecliptic" are subject to occultation at one time or another!

PART II - APPEARANCE OF THE MOON FROM PLANET EARTH

The Phases of the Moon - Although modern astronomy has been a long time coming, held back to some degree by astrological beliefs, the simple concept behind the changing faces - phases - of the moon have been understood and documented as far back as the time of Greek philosopher Anaxagoras in 430 B.C.; it was Aristotle who first chronicled the explanation in his writings so time later.

Both realized that the changing shape (illumination) that we see of the moon's surface from earth varied because of the angle from which sunlight reflected from the moon as we see it from Earth (Figure 6).

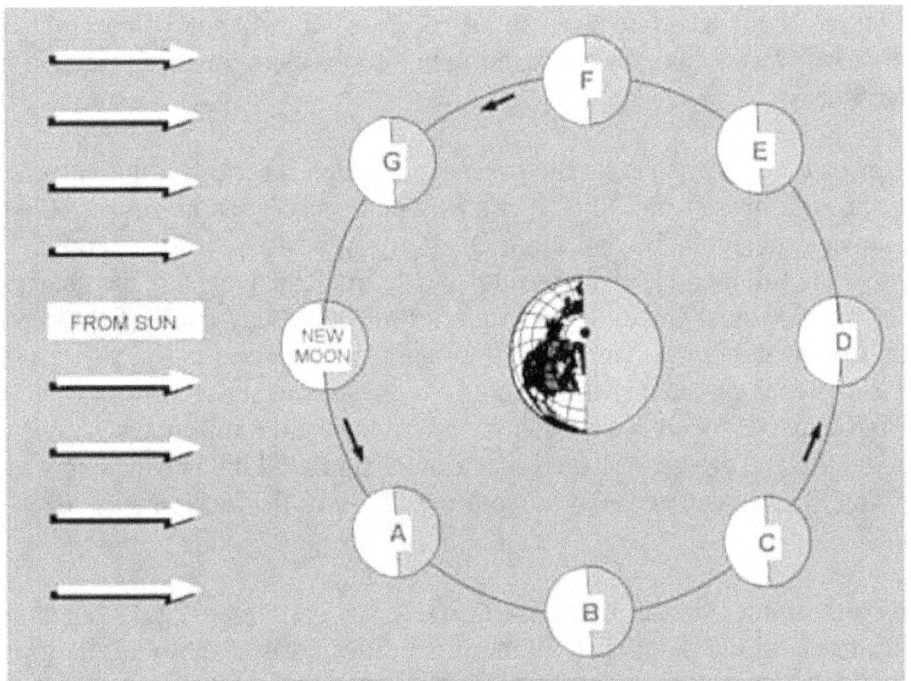

FIGURE 6 - PHASES OF THE MOON
Showing the Geometry of the phases of the moon as seen from Earth. Note that the letters in RED indicate a reference to FIGURE 7, our monthly Lunar Calendar.

Note in this figure that the phases are like timekeepers throughout nearly a month's period. In other words, from a new moon (between the Earth and sun, therefore its illuminated face is not facing us) to the first quarter (we are seeing only a quarter of its illuminated face) is almost exactly one week. Another week passes and we see a full moon, one which we see fully illuminated by sunlight. Thus, one can expect - on average - that the moon progresses from one crescent moon (say 3 days "old") to the next successive crescent moon of three day old in a period of right about 28 days. In addition to telling the progression of weeks using the moon, one may also tell approximate local time by remembering a few basic rules for each major phase (quarter) of the moon.

Local Time by the Moon - It is easiest to start with the full moon as a means of telling time through the nightly motions of our natural satellite. Remember, to be "fully illuminated," the moon must be opposite the sun from the Earth, as shown in Figure 6 above.

Consider if the moon was on the same side of the sky as seen from Earth, or a "new moon." Obviously in that arrangement, the moon and sun would "rise" and "set" very close together on that day. In a perfect world, at the equator on Earth, that would be rising about 6 a.m., crossing nearly overhead at noon, and setting as the sun sets, about 6 p.m.

Back to the full moon ("D" in Figures 6-7) located just the opposite of the new moon relative to the sun and Earth. If opposite, then the moon on THAT night would be rising and setting opposite the sun, right?

Therefore, the full moon rises at sunset, crosses overhead (the meridian) at midnight and proceeds to set about the time the sun rises. Thus, knowing the times of sunset and sunrise on any given evening and next morning, an observer can easily reckon his or her local mean time.

Not quite as easily to visualize as the opposite principle of the full moon, time keeping with both the first quarter and third quarter moons can be just as simple. First Quarter moon is shown as position "B" and Third Quarter is in position "F" in Figure 6-7.
Knowing that the full moon keeps opposite "clock time" than the sun, the quarter moon is off, simply by six hours, since it is one-quarter of the 28-day cycle (for the purpose of this discussion) of the moon. Thus, the first quarter moon rises in the sky at noon, local time, and is approaching the meridian (highest in the sky) at sunset.
Since the third quarter moon is a quarter of the cycle past the full moon, it rises about the time of midnight, and crosses the meridian around sunrise. This is why - on a very clear, crisp morning, it is sometimes easy to spot the moon in the western sky, even after the sun is up!

The Moon's Distance, Size and Apparent Size -
Like most objects in our solar system, the moon does not move around its gravitational center (the Earth) in a perfect circle, but rather an ellipse, or slightly flattened circle. Since the moon's distance is only is only 30 times the Earth's diameter (about 8,000 miles), it is easy to recognize a slight change in the size of the moon as this ellipse sometimes brings it closer (perigee) and moves it farther away (apogee) to the Earth.
We have noted that the apparent size of the moon, some 240,000 miles distant is about ½° of sky. The actual size of the moon is quite small - 2,160 miles, or about on-fourth the size of mother Earth.

The moon may approach the Earth as close as 221,463 miles and - in the same month - be as distant as 252,710 miles. It is this dramatic change in proximity to Earth and its location relative to the sun which causes our ocean tides.

Why do I see the same face of the Moon, no matter what day or what time I look?......Does this mean the moon doesn't turn, or rotate? -

On the contrary, the moon rotates on the lunar axis in what may be the most peculiar, though not coincidental from a "celestial mechanics" standpoint.
But the moon DOES keep the same face, with a little "rocking" variation, called libration.

Interestingly, if the moon did not rotate, we would see different features (i.e., the "man in the moon") at different times - and we would be able to see the features on the far side of the moon which have never been seen from Earth!

Remarkably, the moon rotates on its axis (its "day") in almost perfectly the same period of time that it takes for it to revolve around the Earth (a lunar-Earth "year'). Thus, as it moves eastward around the Earth, it is also turning synchronously on its axis at the same rate, keeping the same face exposed to the Earth.

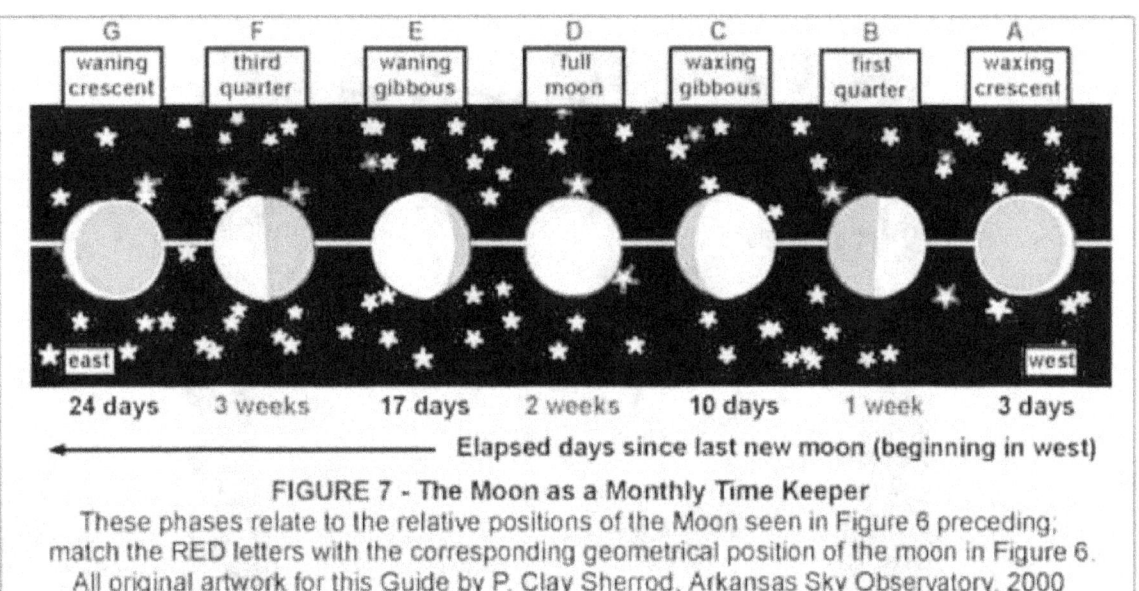

FIGURE 7 - The Moon as a Monthly Time Keeper
These phases relate to the relative positions of the Moon seen in Figure 6 preceding; match the RED letters with the corresponding geometrical position of the moon in Figure 6. All original artwork for this Guide by P. Clay Sherrod, Arkansas Sky Observatory, 2000

The Romance of the Moon –

The moon has been the source of novels, poetry and steamy love affairs throughout history. The definitive changes in appearance, the rapid movement through the sky, the quiet suspension of its bright orb against the curtain of distant stars....are all "stuff" that makes our romance of the heavens a profound one. Indeed, even the light of the moon - from kindling love to igniting battles of war - is a reminder to us that we depend on the starry sky and all that is in it. And....more profoundly...that we are but such a small part of it all.

The bright portion of the moon is reflected sunlight, but even the full moon shines back to earth only 1/400,000 of the intense light of the sun. Indeed, the reality of a three-dimensional space is perhaps best realized during the very thin "early" and "late" crescent moons, when the bright crescent meets the dark side at the terminator where day falls instantly to night with no dusk. With only the faint glowing reddish reflection of Earth is a sphere seemingly floating silently in space on the curtain of eternity.
Astronomy and romance perhaps, do have the common thread, and nowhere is that more evident than with our bright moon. Wars have been fought by its illumination, its silent splendor have kindled the fires of love, its eclipses have cast foreboding shadows of death and disease across ancient cultures.

If the true challenge of any romance is the "mystery" that surrounds it, then surely this nightly temptation of all the mysteries of the cosmos packaged neatly only a couple of hundred thousands of miles distant, affords us the ultimate in mystery....and a nightly romance as our minds travel through our telescopes to peer at mysteries locked away in a cold and silent world.

The "Super Moon" as photographed originally in color by Tom Walker

Chapter 27

OBSERVING THE PLANET VENUS
The System of Planets Throughout History

From the puzzled wonderment by ancient stargazers until ony a few centuries ago, the skies above us were thought to be fixed against the distant curtain of darkness except for a select few objects that seemed to either move a tiny bit....or in some cases: quite a bit.

Among the rapidly-moving were the "transient phenomena," like brilliant fireballs, or rich meteor showers, the occasional comet, the ghostly apparitions of a glowing curtain of aurorae, or perhaps a nova or "new star." And then - in the category of moving a "tiny bit" - were the *SEVEN WANDERERS* of the sky....the early Greek philosophers labeled them "*planetes*," the precursor of our modern title of "planets."

The meteors and comets were commonly associated with "atmospheric disturbances," much like rainbows, sun dogs and lightening. These occasional interlopers were not common, but were laden with superstition and myth even until recent times. Thus, such otherwise unexplainable happenings were blamed on "meteorological" causes.

On the other hand, the seven "wanderers" were well known since earliest mankind first looked skyward, moving slowly amongst the seemingly fixed points of light. They were somewhat predictable to early man, but certainly their motions were not understood. The "seven planets" consisted of those bodies outside of the earth and its atmosphere: the SUN, MOON, MERCURY, VENUS, MARS, JUPITER and SATURN.

VENUS IN EARLY HISTORY –

The planet Venus is referred to by many even today as both the "evening star" and the "morning star", depending on its ELONGATION (see following) relative to both the sun and the Earth. Even before the makeup and order of the solar system was known, the ancient Chaldean astronomers realized that this bright white planet appeared to "hover" close to the sun at all times and never attain an altitude such that it could be seen throughout the night. Of course, today we understand that Venus is a world not unlike the Earth, a planet orbiting the sun some 67 million miles distant. Only little Mercury is closer.

The striking white brilliance of Venus is unrivaled. The early Chinese astronomers called the body "*Tai-pe*," or "beautiful white one," and realized that whether seen low in the eastern morning sky at dawn or in the evening western sky at dusk, the appearance was of ONE planet, not two.

Not so for the Egyptian astrological priests who gave Venus in the morning sky the name "*Tioumoutiri*", and "*Ouaiti*" when it appeared in the evening; whether they assumed it was TWO separate planets, or merely signified its morning and evening names ritualistically different to "appease" the gods is unknown.

Later in time, Greek and Roman temples, statues and mythology centered around the bright heavenly orb; Polynesian tribes - as did the North American Pawnee - frequently sacrificed maidens to the morning appearance of this bright planet in special circumstances. Hardly fitting, it seems today, for the beautiful orb we commonly know as "the goddess of love."

As to the "planets," we of course no longer associate the Moon and sun among them, and have classified them into two major groups in regard to their relative positives to the Earth. The inferior planets - Mercury and Venus - are located in orbits between the sun and Earth. Their motions in our sky are noticeably different than the other group, the superior planets - Mars, Jupiter and Saturn (and now including the telescopic planets Uranus, Neptune, and Pluto - and perhaps others). The "superior" planets have orbits outside the Earths, or farther from the sun than the Earths orbit.

Because of their locations, the motions of the inferior planets - and hence what we SEE of them in our telescopes - are much different and perhaps a bit more difficult and confusing to keep up with than those outside of our orbit.

Like the moon, all planets have orbits that are elliptical, and all planets follow the pathway through the ecliptic, the 12 Zodiacal constellations which we observe from Earth. But since we watch them from EARTH that also goes around the sun, the planets move at different rates and in different patterns depending on how we see them at any given time.

SUPERIOR PLANETS appear to move slowly eastward month-after- month. Although their motions (particularly the closest to Earth: Mars) CAN be detected in telescopes in a matter of days or weeks as they creep slowly eastwardly relative to the stars, it takes some length of time for naked eye observers to realize the motion.

However, the motions of Venus and Mercury are much different and - at first - seem a little confusing for those not familiar with the makeup of the solar system and our vantage point - Earth - from which we must observe them.

MOTIONS OF VENUS - THE "INFERIOR" PLANETS –

INFERIOR PLANETS move from our vantage point in pathways that are not so neat and understandable as what we see from the more distant superior planets. This is only because of our vantage point and the fact that they are orbiting the sun closer than the Earth. Thus, we must take into account the position of the Earth as well as that of Venus and Mercury when examining their apparent locations in the sky.

Galileo Galilei, in 1610, first exposed the world to the peculiar motions of Venus and Mercury when - in his controversial publication Sidereus Nuncius - he proposed that Venus must travel around the sun, not the Earth as had been held up until that time. After observations through his telescope, with which he discovered the four major satellites of

Jupiter, the rotation of the sun from the movements of sunspots, and turned what appeared to be a smooth surface on the moon into an imperfect world of craters, mountains and valleys, he revealed that Venus exhibits phases, just like the moon. From this, he deduced that the only plausible explanation for him - on Earth - to see phases on Venus was that the planet was in orbit around the sun, and closer to it than Earth. (See Figure 1)

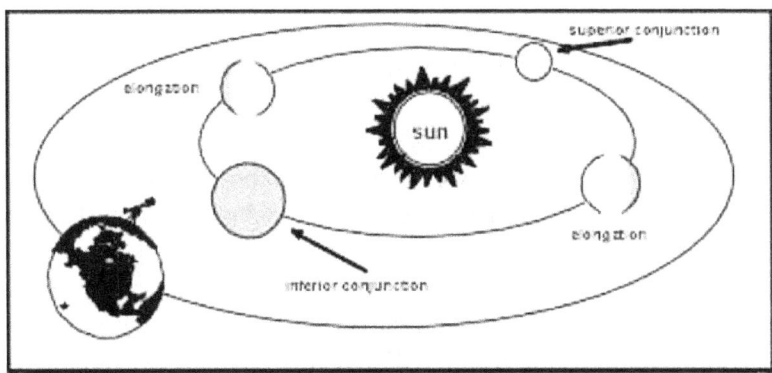

FIGURE 1

Figure 1 demonstrates the orbit of Venus (or Mercury) around the sun and the relationship to Earth, showing the explanation for the phases (more on this later) of inferior planets. Note that - because an inferior planet is seen ALWAYS in the direction of the sun from Earth - the planet can only be seen either in early evening after sunset, or at or near dawn in the morning. At dark, the Earth is always turned AWAY from the sun, of course, and thus the planet. Figure 2 demonstrates how Venus (and Mercury) will attain a "highest" point (or ELONGATION) as we see it from Earth. In this drawing this would be the "greatest eastward elongation" or the highest east of the sun that Venus reaches as we see it. This would be in the early EVENING SKY above the western horizon. Note that eventually, its path will take the planet "back down" to the horizon and eventually seem to disappear.

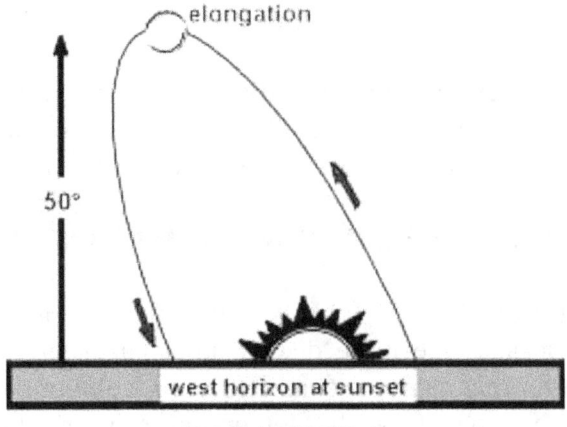

FIGURE 2

However, in only a very short interval of time - perhaps two weeks - Venus will RE-EMERGE as a morning object seen just prior to sunrise in the Eastern sky; each successive day will result in the planet attaining a higher and higher altitude above the east horizon until it reaches.....you guessed it: "greatest western elongation."

ELONGATIONS OF VENUS AND MERCURY –

Because Mercury and Venus are orbiting the sun closer than is Earth, they can never rise high in the sky (see Figure 2). Rather, they appear to slowly rise out of either the morning or evening sky higher and higher each subsequent morning, momentarily stop that rising path (elongation) then begin to sink ever so slowly back to the horizon.

The process is a slow one, and varies from "year" to "year" for the two planets in relation to the Earths orbit. Elongation in the western sky for both is known as greatest eastern elongation, while that in the early morning eastern sky is greatest western elongation.

The MAXIMUM elevation above the horizon for Mercury is only 28, whereas Venus can attain an elevation of almost 50. At elongation, both appear as "quarter illuminated," much as a quarter moon. The greatest brilliance occurs when Mercury and Venus are nearest inferior conjunction (Figure 1), when both present slim crescents and are very close to the Earth in their orbits.

Mercury (distance from sun average 35 million miles) takes only 88 days to completely circle the sun (its year). Being the closest planet to the sun, it circles inside the orbit of Venus (distance 67 million miles), and thus both exhibit this motion common to the inferior planets.

This motion results in Venus (and Mercury) exhibiting "phases" much like our moon as it encircles the Earth. However, the phases of Mercury are quite difficult to see, even in the largest telescopes. The main reason is that Mercury, being so close to the sun, is always seen in bright twilight sky, very low on the east or west horizon and never attaining the altitude as does Venus. The second reason that phases on Mercury are difficult to discern is that the planet is VERY small, only <u>3030 miles across</u> (barely larger than our moon!) compared to Venus' diameter of only 200 miles less than that of Earth.

As the inferior planets stretch their limits as high as possible in the eastern and western sky, we are reminded that we are but the third planet of the sun. The motions of the OTHER planets outside of OUR orbit are discussed here only for comparison.

THE APPARENT MOTIONS OF THE SUPERIOR PLANETS –

Those planets outside the orbit of the Earth exhibit apparent motions much different than Venus and Mercury. Because the more distant any planets orbit is from the sun, the longer time is required for that planet to complete one revolution, or "year." Earth does it in 365 days, and we have seen that speedy Mercury has a "year" of only 88 days.

By comparison, Saturn - at a remarkable distance about about 887,000,000 miles from the sun - completes its trip around the sun once every 29 year years! And Saturn is close compared to Uranus, Neptune and Pluto.

So, based on the table below, it is easy to see how the motion of, say, Mars is quite much more perceptible than that of even Jupiter and particularly of Saturn, the farthest naked eye planet of our solar system.

Since this GUIDE is intended to provide the information and tools necessary for the novice astronomer to get the most out of observing the brilliant planet Venus, the details of celestial mechanics are left out. We don't want to spoil the fun by overloading our cerebral databases.

So - for the most part - we'll leave our discussion of the "superior planets" with the fact that they creep ever-so-slowly EASTWARD each night as we observe them from Earth. Every year however, each planet outside the orbit of Earth will undergo "retrograde motion" in which the star will appear to STOP its motion against the background stars, then slowly REVERSE its course (begin moving westward) for a very short distance only to STOP again and end up resuming its eastward path. This is merely an optical "trick" that the orbit of Earth does to us as it "catches and passes" the planet, much like approaching, then passing and ultimately leaving "in our dust" a slower automobile on the highway. Figure 3 demonstrates a superior planet's pathway as it goes through its "retrograde motion." Venus and Mercury do NOT demonstrate retrograde from the Earth.

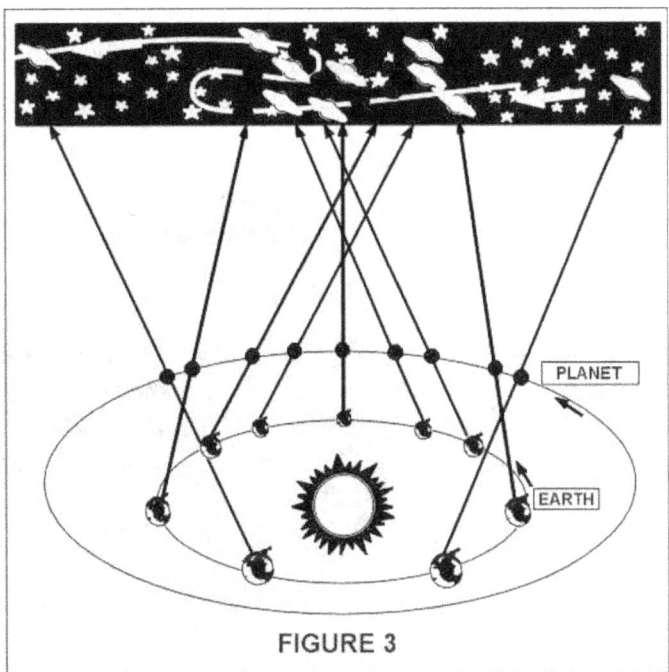

FIGURE 3

An interesting aside to this, however, is that if you were a Mercurian astronomer, Venus would be a "superior planet" and it, too, would exhibit retrograde motion as would the Earth!

In regard to the superior planets, once their locations have been learned by even the casual stargazer, it is difficult to forget where they will appear in the sky the next season, since their motions are minor when compared to the moon, or even Venus and Mercury. Even if their positions are forgotten or confused from one appearance to another, they can be quickly identified by eachs distinct brightness and color as discussed following.

Let us now examine the physical appearance of the brilliant Venus and many of the attributes - both telescopic and naked eye - that this "goddess of love reveals to our curious eyes and telescopes here on Earth.

BRIGHTNESS AND THE ORBIT OF VENUS –

Venus is a sight to be reckoned with....at magnitude **-4.2** it outshines all else in the sky save the sun and moon. It is so bright that it can be seen in full daylight. It is so bright that its brilliance can actually cast a distinct shadow into an otherwise darkened room.

Even when Venus is at "*inferior conjunction*" (Figure 1) and is showing only a sliver of a crescent to observers on Earth, it can attain its maximum brilliance and size. Venus appears huge by comparison to other planets in a common telescope.

A small 8 x 21 finder can easily spot Venus in the middle of the day. For this reason, GO TO computerized telescopes are particularly useful in observing the planet in daylight, when it can be seen overhead. Actually, observing Venus during daylight somewhat subdues its otherwise overwhelming brilliance. During such observations, it is common to see the irregular patterns of high Venusian clouds at the "terminator" (See Figure 4) and the crescent "cusps" extending far around the peripheral edge of the planet.

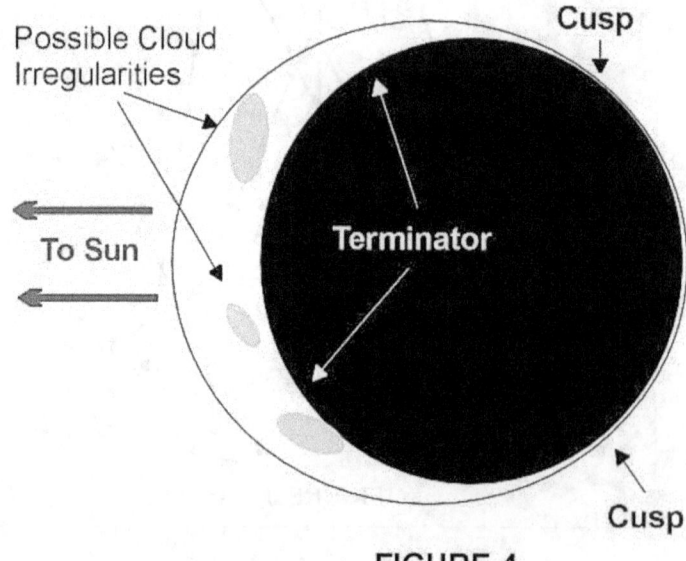

FIGURE 4
Nomenclature of the Appearance of Venus

To observe Venus in daylight (the best time by far) with a computerized telescope, follow these simple steps:

1) make sure your telescope is aligned with celestial north (sometimes it helps to leave the scope outdoors and aligned from the previous night if care is taken to cover it for protection);

2) place the telescope in "Home Position;"

3) whether in Alt-Az or Polar mounting mode, do an "easy" star alignment with the telescope; each time a star is selected for you to center, merely hit the "Enter" key and allow the computer to select your next alignment star; do NOT attempt to find the alignment star and center....it is useless;

4) after the alignment is done and the motors are engaged in the telescope, simply scroll to "Object" / "Solar System" / "Venus" and press enter; once located, press "GO TO" and the telescope will take off and get you fairly close to the bright planet;

5) remember I said you can easily see Venus in your 8 x 21 finder? Look carefully (it may take some time for your eye to adjust to the bright blue sky and find the planet, but once you find it, it is unmistakable) and locate Venus and center! You're there!

During daylight hours, you will realize that you can use more magnification to study Venus than you can during night to observe other planets. I commonly use around 410X to observe Venus and many variations of its clouds are easily noted. Note that during daylight hours the air surrounding your telescope is very unsteady, as heat currents will be rising directly from the telescope itself; hence the image is waiver frequently but many moments of excellent "seeing" are awaiting!

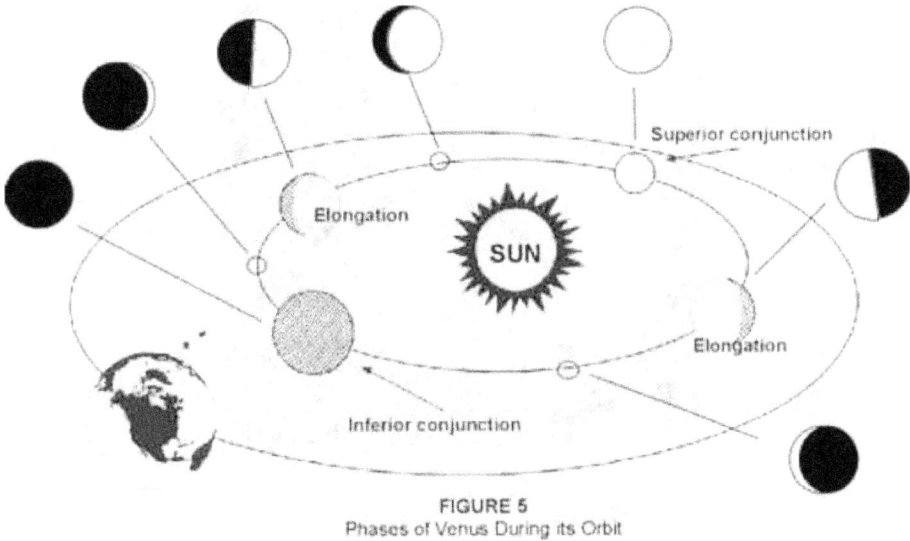

FIGURE 5
Phases of Venus During its Orbit

Venus is closest for observing (Figure 1) when it is near - but not at - "inferior conjunction" and presents a very thin sliver of a crescent. At such times the disk is very large, attaining a size almost 1/3 LARGER than mighty Jupiter. Frequently, the planet will attain an angular size of over 60 arc seconds when at its brightest.

At each "elongation", Venus is still very bright, but presents a "quarter-moon" shape, or seems half-illuminated as we see it from Earth. Its size will have diminished significantly from inferior conjunction. Figure 5 shows the apparent phases of Venus at its respective positions relative to Earth.

OBSERVING VENUS –

Venus presents a very difficult world to study. This is a strict rule not only for you and your telescope, but also for the many spacecraft that have ventured close for a look; it is a world cloaked in thick and uninviting clouds of corrosive acids and tornado-force winds.

Venus does not reveal her secrets easily, certainly not to terrestrial based telescopes of any size. But it is a world well worth visiting.....and a world worth a return trip every now and then. Figure 6 shows Venus in "visible light," or in the portion of the spectrum that our eyes observe. In this February 10, 1990 Galileo spacecraft photograph, the planet appears as it might at high magnification when just past ELONGATION toward SUPERIOR CONJUNCTION, presenting a recognizable "gibbous" phase.

In this photograph you can see some of the very subtle detail that MIGHT be visible under extremely steady conditions in your telescope, including the well-defined "terminator" and several wispy "streaks" of darkness within the otherwise very bright cloud tops.

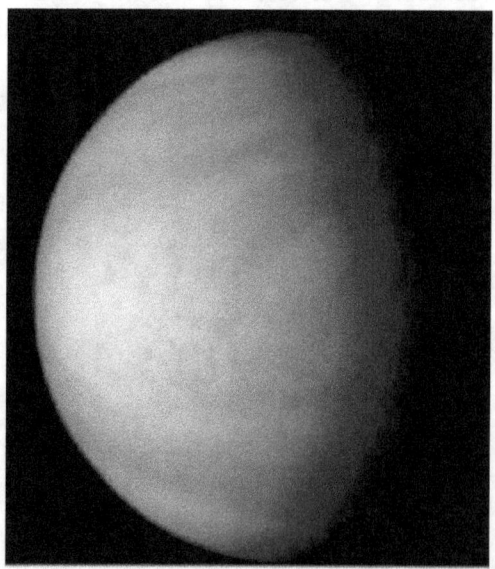

Venus in visible light
Photo Courtesy NASA

When at greatest elongation, Venus can set a full 3 HOURS after the sun. Beset visibility of the planet is during EVENING hours anytime Venus is visible in springtime in the

Northern Hemisphere, and during MORNING hours when the apparition occurs in Autumn.

If you do not observe Venus in the daylight hours, it is best to filter the brightness of the planet for two reasons: 1) a good filter of the right color enhances the penetration of your telescope into the thick "greenhouse" cloud layers of the planet allowing more detail to be seen; and 2) the filter eliminates glare and increases contrast by cutting down so much stray light from the bright planet.

There is not nearly as much detail to be studied on Venus as there might be on Mars, Jupiter, or Saturn. This is because we are not able to see the surface for the thick clouds, as we can with Mars, and unlike Jupiter and Saturn (on which all we can see are also cloud tops) the clouds are virtually featureless and uniformly bright.

Nonetheless, there IS detail to be seen on Venus. Try using a Wratten #47 violet filter to penetrate the clouds of the planet, and also reduce the brightness. You will see a purpleish discoloration for sure, but the reward in detail is worth it. Among features that you can look for are:

IRREGULARITY IN THE CLOUDS - Particularly along the terminator (Figure 4), there is much variation in the brightness of the Venusian clouds; there are many areas which appear to reflect less sunlight and thereby appear "mottled" to the keen-eyed observer. Frequently, some dark areas might even be glimpsed in the otherwise uniformly bright clouds as well, particularly when Venus is at elongation and demonstrates it "quarter phase."

ROTATION OF VENUS - Although not much can be seen to prove it, in 1890 G.V. Schiaparelli proposed that Venus must rotate on its axis (its "day") the same period in which it revolves around the sun (its "year"), much like the moon orbits the Earth. Indeed, he was correct in this assumption, with the planet rotating - and revolving - at about 224 days. Schiaparelli based his assumptions on direct visual observation of VERY faint cloud detail that moved as the planet turned. Can YOU make out any markings to watch day-after-day? If you observe enough....you can.

EXTENSIONS OF THE CUSPS - Note the elongated "cusps" in Figure 4. As Venus approaches toward inferior conjunction and enters its thin crescent phases from our vantage point, the bright sunlight actually PENETRATES the thick clouds of the planet and SCATTERS throughout many of the cloud layers. This scattering many times allows viewers to see the cusps extended like "embracing arms" around the otherwise dark (unilluminated) side of Venus. Many times it is actually possible (particularly when viewing in daylight) to witness illumination from light scattering of the section of Venus that is NOT directly lighted by sunlight, similar to "earthshine" we see on the moon.

CHANGING OF THE PHASES - Every observer should at least once in his or her "observing career" make a record of the date and phase of Venus through the course of one complete "Venus year" as observed from Earth. All it takes is merely a sketch showing the round ball of the planet, the dark vs. the illuminated portion as you see it, and any irregular/dark blotches that might appear on an otherwise "bright white" surface. Perhaps one of the nicest attributes of this beautiful world is our ability to watch these changes with such little magnification!

Like any lady, particularly one of "goddess stature," Venus does not give up her secrets easily or otherwise.

The story of Venus is one of a paradox in that we have, only 30 million miles away, a sister to planet Earth.....so similar and yet so disproportionately different. On Earth we savor the rich meadows of the wonder of biology and a world in seemingly static balance. And - over there, toward that very bright object in the sky - Venus exhibits the most hostile of all worlds for this glorious creation of life. Both worlds have landforms.....both have abundant water, volcanoes, land quakes and similar gases within their atmospheres, but each is so different than the other.

Venus is a world of the "greenhouse effect" gone wild, perhaps a lesson in nature of what can happen to this precious balance of our own world should we carelessly forget to nurture it.

It is the most mysterious still of all worlds of which we know, hiding the secrets behind a cloak of bright, shimmering disguise. Could it be that this shrouded world is a hint of the future of Earth?

If so....then WHAT must Venus have been like, long ago in ages passed?

* * *

"The sun, with all those planets revolving around it and dependent on it, can still ripen a bunch of grapes as if it had nothing else in the universe to do...."
— *Galileo Galilei.*

Chapter 28

Your Studies of the Red Planet Mars
Our Fascination with the God of War

Since the Viking (http://nssdc.gsfc.nasa.gov/planetary/viking.html) and Mariner (http://nssdc.gsfc.nasa.gov/planetary/mars/mariner.html) spacecraft visited the vicinity of the mysterious planet Mars, some of the wonder and mystique have disappeared from our attitudes toward the red planet. We have now closely examined the surface, and it appears that there are no creatures there. Indeed, not even the basic organics necessary for life exist in the Martian environment. Yet it appears that Mars at one time was quite active both geologically and meteorologically, considerably more so than at present.

Near the beginning of the twentieth century, astronomer **Percival Lowell** (http://www.nasm.si.edu/ceps/etp/mars/percival.html) had the theory, based largely on the ideas of Italian astronomer **Schiaparelli**, that Mars was a dying world, one in which its inhabitants were starving and dehydrating from the lack of water. Lowell theorized that - knowing that the Martian spring and winter would result in melting the Martian polar caps - the inhabitants had built a vast network of canals to take the life-giving moisture of those caps into the arid equatorial regions of Mars. We now know that Mars does not have canals and that there are no inhabitants who wait eagerly for the advent of spring. Yet Lowell was correct about one point of his theory - Mars is, indeed, a dying world. Sketch below is a rendering of Mars by Lowell.

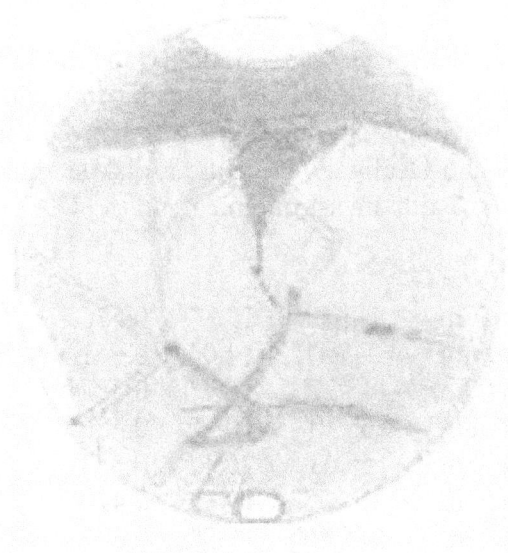

The processes we see today on Mars are the result of phenomena that occurred only yesterday in the life of our solar system; many of the processes continue to this day, only on a considerably smaller scale. Massive volcanoes not yet eroded by the winds of Mars still loom on the Martian surface. And still fresh canyons appear as the result of the activity in the recent past.

It is this lure of the similarities of features on this earth that keeps writers, poets, movie-makers and even astronomers captivated by the Red Planet.

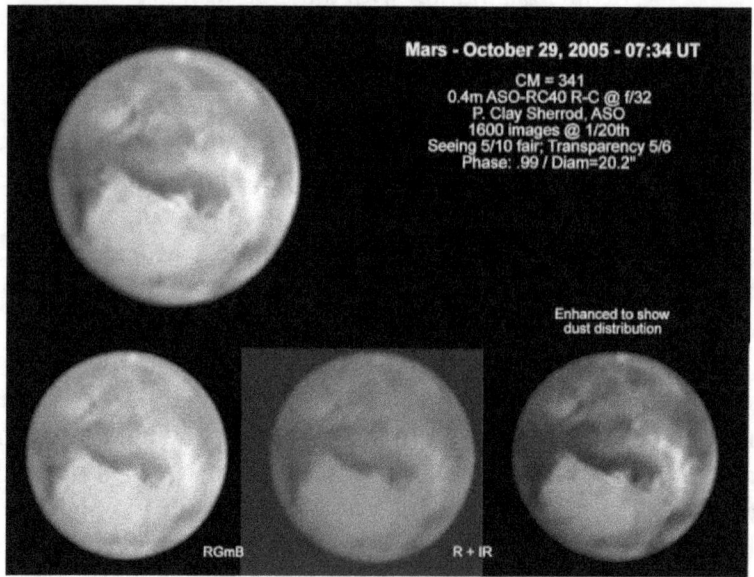

Dust storm activity captured by the author from ASO inn 2005

THE OUTSTANDING MARS OPPOSITIONS

Most favorable apparitions will place Mars similarly favorable to that of 1924....when folks on earth still longed to know the details about the "inhabitants" of that mysterious world. The following table (data courtesy N.A.S.A.) provides recent oppositions of Mars, its Right Ascension and Declination, apparent disk size in seconds of arc and the distance of the planet from earth in astronomical units (1 A.U. = about 92.7 million miles).

Date	RA	Dec	Diam	Dist (AU)
1995 Feb 12	09h 47m	+18° 11'	13.8"	0.676
1997 Mar 17	11h 54m	+04° 41'	14.0"	0.661
1999 Apr 24	14h 09m	-11° 37'	16.2"	0.583
2001 Jun 13	17h 28m	-26° 30'	20.5"	0.456
2003 Aug 28	22h 38m	-15° 48'	25.1"	0.373
2005 Nov 7	02h 51m	+15° 53'	19.8"	0.470
2007 Dec 28	06h 12m	+26° 46'	15.5"	0.600
2010 Jan 29	08h 54m	+22° 09'	14.0"	0.664
2012 Mar 3	11h 52m	+10° 17'	14.0"	0.674
2014 Apr 8	13h 14m	-05° 08'	15.1"	0.621
2016 May 22	15h 58m	-21° 39'	18.4"	0.509
2018 Jul 27	20h 33m	-25° 30'	24.1"	0.386
2020 Oct 13	01h 22m	+05 26	22.3"	0.419

Note that the apparitions of August 28, 2003 and July 27, 2018 are presenting Mars with nearly as large an extended disk as possible. The size of 25.1" for the outstanding 2003 apparition afforded a wonderful view of surface detail, cloud formations, melting of the polar caps and much more exciting developments.

ONE interesting note, however, separates the 2003 opposition to many other favorable ones before and after. Take a look at the <u>declination</u> of the planet in that year....minus 15 degrees, considerably higher in the sky for northern latitudes than any favorable placement since 1924, when the apparent size was nearly identical to that of 2002, yet it was slightly lower in southern skies for northern observers. Thus, if you are an observer in the Northern hemisphere, in such favorable oppositions, Mars shows a very large disk, fairly high in the sky and only about 35 million miles away from the base of your telescope.

AMATEUR STUDIES OF MARS

Because of the many uncertainties that still exist regarding the visible phenomena of the Martian surface and atmosphere, Mars continues to appeal to the amateur astronomer. In addition, we can rationalize in our human way that the spacecraft may "have missed something." For example, why do the maria, or dark areas, of the Martian plains appear to darken when the polar caps melt? What causes the mysterious "blue clearing" and the circulation of the Martian clouds? We still do not clearly understand why the darker features change shape and size erratically as we peer at them from earth. Spacecraft thus far have provided little additional knowledge of these phenomena, and in many cases have added to the Martian mysteries and our desires to study the planet. Mars comes into opposition with the earth every 2 years and 50 days on an average, yet each two successive oppositions are unique because the orbit of Mars is considerably more eccentric than that of earth.

Favorable oppositions of the Red Planet occur approximately every 2 years, either in January or February, or in late summer. Those that occur in January or February are known as aphelic oppositions (farthest distance for a close approach to earth and sun), and they show Mars as a very small disk of only about 13" arc. The oppositions that occur in summer, known as perihelic oppositions, bring Mars closer to the earth and sun so that the disk is twice the size as it is in the aphelic oppositions. From the Northern Hemisphere, Mars is seen quite high in the sky during aphelic oppositions and very low in the south during perihelic oppositions. Consequently, the conditions for observing the planet are improved during the winter months even though the disk is considerably smaller than during the summer oppositions.

Observations that the amateur astronomer (or should we say more aptly "non-professional astronomer?") can pursue are badly needed for our better understanding of Mars. Areas of study within a the scope of non-professional equipment should include the following:

1) Studies of the Martian polar regions and the two polar caps.

2) Long-term programs examining the Martian atmospheric phenomena.
3) "Patrol" observations for brightenings on the Martian maria or plains.
4) Studies of seasonal changes, including correlations with the melting of the polar caps.
5) Visual, photographic and CCD (webcam) studies of surface features.

Whatever program you undertake, it is better that you do not attempt to accomplish all objectives listed above. Concentrate fully on one or two closely related studies, such as the correlation of the appearance of the polar regions in conjunction with predicted seasonal changes. There is no substitute for dedicated systematic study of the planet. Try to overcome the urge to "put too many irons in the fire," and restrict your observations to a well-developed program that you know will lead to success in your studies. Basic physical knowledge of Mars is summarized in Table 8-1.

TABLE 8-1. **Martian physical data**.

Distance from the sun = 228 million (km) ave.
Orbital velocity = 24.1 km sec
Length of Martian year = 687 earth days
Equatorial diameter 6800 km
Maximum size of polar caps = 2200 km
Mass (x earth) = 0.11
Albedo = 0.15
Maximum magnitude = -2.8
Maximum apparent diameter = 26"
Atmospheric constituents mainly N, H_2O, CO_2, Ar

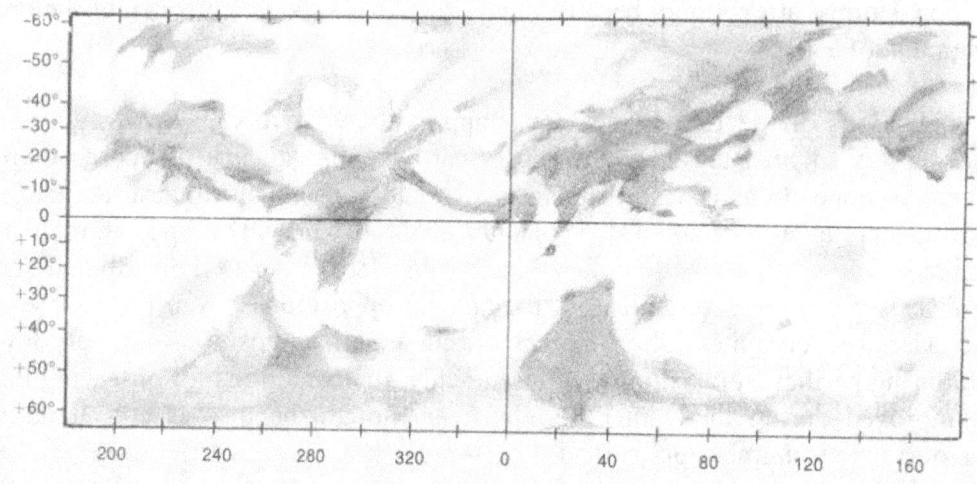

Figure 1 - A projection map compiled by P. Clay Sherrod, *Arkansas Sky Observatory* from 378 visual drawings of Mars in the year 1971

Table 2. The Principal Features of Mars

The principal features visible on Mars are described below, with their scientific name and Martian longitudes and latitudes given. Note that because of dust storms and actual morphological changes in the sizes and shapes of nearly all of these features, the given positions are only approximate. For the latest map from the Mars Section of the **Association of Lunar and Planetary Observers** (ALPO) click on: http://spider.seds.org/spider/Mars/Pics/B&WMarsmap.jpg (in alphabetical order with approximate CENTRAL Martian-centric longitude listed first) and latitude (both given in degrees)

Feature	Longitude	Latitude
Acidallum Mare	30	+45
Aeolis	215	-5
Aeria	310	+10
Aetheria	230	+40
Aethiopis	230	+1
Amazonis	140	0
Amenthes	250	+5
Aonius Sinus	105	-45
Arabia	330	+20
Araxes	115	-25
Arcadia	100	+45
Argyre	25	-45
Arnon	335	+48
Aurorae Sinus	50	-15
Ausonia	250	-40
Australe Mare	40	-65
Baltia	50	+60
Boreum Mare	90	+50
Boreosyrtis	290	+55
Candor	90	+10
Casius	260	+40
Cebrenla	210	+50
Cecropia	320	+60
Ceraunius	95	+20
Cerberus	205	+15
Chalce	0	-50
Chersonesus	260	-50
Chronium Mare	210	-58
Chryse	30	+10
Chrysokeras	110	-50
Cimmerium Mare	220	-20
Claritas	110	-35
Copals Palus	280	+55
Coprates	65	-15

Cyclopia	230	-5
Cydonia	0	+40
Deltoton Sinus	305	-4
Deucallonis Regio	345	-12
Deuteronilus	0	+35
Diacria	180	+50
Dioscuria	320	+50
Edom	345	0
Electris	190	-45
Elysium	215	+30
Erldania	220	-45
Erythraeum Mare	40	-25
Eunostos	220	+22
Euphrates	335	+20
Gehon	0	+15
Hadriacum Mare	270	-40
Hellespontica Depress.	340	-6
Hellespontus	325	-50
Hesperia	240	-20
Hiddekel	345	+15
Hyperboreus Lacus	60	+75
Iapigia	295	-20
Iscaria	30	-40
Isidis Regio	280	+20
Ismenius Lacus	330	+40
Jamuna	40	+10
Juventae Fons	63	-5
Laestrygon	200	0
Lemuria	200	+70
Libya	270	0
Lunae Lacus	65	+15
Margaritifer Sinus	30	-2
Memnonia	160	-2
Meroe	285	+35
Meridiani Sinus	0	-5
Moab	350	+20
Moeris Lacus	270	+8
Nectar	72	-28
Neith Regio	275	+35
Nepenthes	260	+20
Nereidum Fretum	55	-45
Niliacus Lacus	30	+30
Nilokeras	60	+25
Nilosyrtis	290	+42
Noachis	330	-4

Ogygis Regio	65	-45
Olympia	200	+80
Olympus Mons	133	+18
Ophir	68	-8
Ortygia	0	+60
Oxia Pilus	18	+8
Oxus	10	20
Panchaia	200	+60
Pandorae Fretum	340	-25
Phaethontis	155	-50
Phison	320	+20
Phlegra	190	+30
Phoenicis Lacus	110	-12
Phrixi Regio	70	-40
Promethei Sinus	280	-65
Propontis	85	+45
Protei Regio	5	-23
Protonilus	315	+42
Pyrrhae Regio	38	-15
Sabaeus Sinus	340	-8
Scandia	150	+60
Serpentis Mare	320	-30
Sinai	62	-25
Sirenum Mare	155	-30
Sithonius Lacus	245	+45
Solis Lacus	85	-35
Styx	200	+30
Syria	100	-20
Syrtis Major	298	+10
Tanais	70	+50
Tempe	70	+40
Thaumasia	75	-30
Thoth	256	+30
Thyle I	180	-70
Thyle II	230	-70
Thymlamata	10	+10
Tithonius Lacus	85	-5
Tractus Albus	80	+30
Trinacria	268	-25
Trivium Charontls	198	+20
Tyrrhenum Mare	255	-20
Uchronia	260	+70
Umbra	290	+50
Utopia	250	+50
Vulcani Pelagus	15	-35
Xanthe	50	+10

Yaonis Regia   320   -40
Zephyria   90   -12

TELESCOPES FOR OBSERVING THE MARTIAN SURFACE

The telescope required to observe the Martial phenomena depends on the type of program you plan to undertake. Obviously, if your objective is to patrol the Martian environment photographically, the telescope requirements would be different from those of a telescope used visually to monitor for Martian clouds, using selected color filters. For serious study of Mars, you should use a least a 4-inch refractor or a 6-inch reflector. The best telescope for an amateur to use to make best use of resolution, light gathering, and-perhaps above all-budget, is a 10-inch Newtonian reflector or popular Schmidt-Cassegrain (SCT) telescope. A 10-inch aperture is just about the size limit that has power capable of resolving the currents of air in our atmosphere. Anything larger may not be effective on nights of average air steadiness. Smaller telescopes sometimes can actually be more efficiently used on nights of poor seeing conditions than can larger telescopes. One prime requisite for your Mars telescope is that it be motor driven and have the capacity for accurate tracking.

Because Mars shows at best a disk less than half that of Jupiter, highest powers (perhaps 70x per inch of aperture) are necessary for observing its disk. The fine details present on the Martian surface requires your most attentive concentration. The necessity for making constant adjustments to a non tracking, or poorly tracking telescope will eventually frustrate even the most patient observer.

Magnification
The disk of Mars is quite small, even when the planet is closest to us, and thus very high magnification is required to view its subtle changes. The optimum magnifications, based on tests with a range of telescopes, all at the same location and operating under the same conditions, are recommended as follows:

Aperture	Magnification
6 inch (15cm)	300x
8 inch (20 cm)	400x
10 Inch (25 cm)	450x
12 inch (30 cm)	500x
14 inch (35 cm)	550x
16 inch (40 cm)	600x "

On nights with very poor seeing conditions, observations of scientific quality are impractical if not impossible. If the steadiness of the air changes rapidly and if there are some moments of average steadiness, magnifications of about half the amounts given above would permit some patrol work. The efficiency and the reliability of fine detail that are discerned at lower magnifications are not as great as with higher magnifications.

Eyepieces

Use eyepieces of the highest quality, preferably of Orthoscopic or Plossl design, no matter what type of telescope is used. Top quality eyepieces permit the finest color correction, improved eye relief, and the best resolution as opposed to that obtainable by using cheap, inexpensive makes and designs.

Filters for Observing Mars

Observation of Mars, more than any other planet, requires good filters. They often make the difference between being able to see some subtle marking and not being able to see it. A good set of optical glass Wratten (or equivalent) filters is a must for the serious observer, both to cut down unnecessary glare from the Martian disk and to accentuate fine detail. It is this detail that shows the start of many of the large-scale changes on Mars that are of scientific interest to the amateur.

Table 3 lists the various filters of interest to the amateur astronomer. Notice that not all filters can be used to any advantage when observing Mars whereas several others can be of considerable help. Most drawings of Mars should be made using either the orange (No. 21) or the red (No. 25A)

TABLE 3. <u>Types of glass filters and their uses.</u>
Wratten Filter Numbers Given

Number	Color

47   Violet
Not particularly good for observing Mars; can possibly use for heavy blue clearing.

80A   Blue 486.5nm
Use every night to search for blue clearing and to look for high (H2O) clouds in atmosphere.

58   Green 538nm
Use for observations at the "melt line" around polar caps and searching for yellow dust storms.

12   Yellow 583nm
Not much advantage when used for Mars; might accentuate the outline of polar caps and white clouds.

21   Orange 593nm
Very good filter for Mars because it reduces intensity of the reddish coloration and allows observers to note very fine delineation on the red plains; in addition it accentuates the maria, showing mottling within them.

25A   Red 617.2nm

Much the same advantages as for No. 21, but with much better color differentiation between red plains and dark maria. Red filters are quite dense and can be used to advantage only with lower powers and steady seeing.

(nm = nanometers)

Although the optical glass filters are superior, gelatin sheet filters may be cut and mounted in the standard cardboard frames available for mounting 35-mm color-slide film. Gelatin filters offer only the advantage of being cheaper than the glass. They will wear out and get scratched and stained when fine optical glass will not. Gelatin filters are usually simply held by hand (requiring the use of an otherwise free hand) between the eyepiece and your eye. Of course, the film should not be touched, because finger-prints are difficult to clean from the soft gelatin material. Glass filters are normally manufactured by commercial firms to thread directly onto the base of standard eyepieces. If the eyepiece is not threaded, the glass filter too may be simply held by hand between the eye and the front opening of the eyepiece.

It is helpful when observing Mars to run through quick overviews with all the filters to check for various features that might be visible through one filter but not through another, making careful notes of each. However, after you have made these checks (say for the blue clearing), insert either the orange or red filter for critical observations and drawings. It is best to use the orange filter for photography because the red filter is so dense that a factor of up to 8 times normal is necessary for the proper exposure.

Auxiliary Equipment for Observing Mars

Little else is needed for the systematic observation of Mars except a good telescope and steady skies. By using a few auxiliary items, your study of Mars can be greatly expanded. Some are:

* *filar micrometer* –
This device allows you to determine accurately the placement in latitude of Martian features as well as to determine precisely the changes in size of prominent areas (e.g., the diminishing polar cap as the Martian summer approaches).

The filar micrometer is becoming very difficult to find from commercial sources. The use of each instrument is dependent on the method by which it was made. Most measurements made with such an instrument are made in relation to the total disk size. For each date the American Ephemeris and Nautical Almanac gives a precise angular size of the Martian disk (in seconds of arc). Measure the Martian feature in proportion to the disk size for that date. Note also that there are excellent programs available on the Internet today that allow you to download CM calculator programs for any date, time and location of your observing station on earth. Two such programs are found at: http://sweiller.free.fr/links.html and also at: OnLine Tools from ASO available on this website. Both provide the precise angular size as seen on each date.

* *observing forms* –

See the Association of Lunar and Planetary Observers (ALPO) Mars Section at: http://www.alpo-astronomy.org/mars/marsfrm2.jpg for blank forms in downloadable format....a sample of a completed observing form is found at: http://www.alpo-astronomy.org/mars/marsfrm1.jpg

accurate time –
A watch accurate to 1 minute of WWV radio transmission from shortwave radio is necessary OR a direct time sync from: http://nist.time.gov/timezone.cgi?Central/d/-6/java (this example is for central time) can be used to update either your watch or your clock on the PC for daily precision checks. This aids in the accurate determination of Martian longitudes of various features, and can be used in indirect determination of the east-west extent of a feature.

The American Ephemeris and Nautical Almanac (or good PC program)
gives physical data for each date throughout the year, such as longitude of central meridian for 00h 00m Universal Time, phase angle of Mars, size, magnitude, and so on. Note that for each upcoming apparition, this same data is easily obtained through the Mars section of the ALPO website: http://www.alpo-astronomy.org/marsblog/. You can also use the outstanding Mars meridian program onto your desktop for instant retrieval of Mars CM longitudes for any given moment by going to the ON LINE TOOLS section of this website.

OBSERVATION OF THE POLAR CAPS

Like earth, Mars has two polar caps, although neither is as densely deposited on the Martian poles as on those of earth. A tilt of up to 24° of Mars as it approaches and recedes from the earth allows you to view one pole well, although the northern cap is easily visible only during the aphelic oppositions. Thus it is somewhat more difficult to study.

Prior to the Viking spacecraft visits to Mars, it was predicted that the polar caps of Mars were probably composed of frozen carbon dioxide (CO_2) -or "dry ice" -with traces of water as well. This hypothesis was first made by astronomers **R.B. Leighton** and **B.C. Murray** in 1966, and it was confirmed partially by the 1969 visit to Mars by the Mariner spacecraft. Beginning with the spectroscope's use in planetary astronomy, scientists have been at odds as to the nature of the caps, whether they were water ice or carbon dioxide ice. As recently as 1952 photoelectric infrared studies by **G.P. Kuiper** suggested that the caps were composed predominantly of water vapor deposited thinly on the Martian poles, with little or no carbon dioxide being present. By contrast, the *Mariner* flyby in 1969 revealed a cap composed predominantly of frozen CO_2, and having little or no water vapor. More recent trips by the Viking probes, which landed on the surface of Mars in 1976, proved both schools of thought to be correct.

As Mars recedes from the sun and the Martian autumn and winter approach, the little water vapor present in the atmosphere is slowly deposited at the poles, because the freezing point of H_2O is higher than CO_2, Then, as Mars rapidly cools below the freezing, or sublimation, point of CO2 (-110°F), the CO2 is also deposited over the thin sheet of water ice. Therefore, neither Kuiper nor the Mariner craft was wrong. If Kuiper's observations were made during the Martian early spring or late autumn, he would have seen an ice cap, composed chiefly of water, from which the CO2 had sublimated because of the higher temperature. Similarly if the temperature was below -110°F, the Mariner craft may have recorded a cap composed of CO2 that covered any sign of the water ice beneath.

Observations of the North Polar Cap
https://en.wikipedia.org/wiki/Martian_polar_ice_caps

The northern polar cap of Mars apparently never completely disappears during the aphelic summers. E.C. Slipher, of the Lowell Observatory, determined that the cap never shrinks below an average width of 6°, usually greater, and may reach a maximum expanse of 72°. During the advent of autumn on Mars, a haze stemming from the polar regions develops, and any further shrinking of the northern cap ceases. Even during late summer, when one might expect the maximum recession of the cap, you can sometimes see that it stops shrinking and actually increases in size. Interestingly, the growth of the cap is always associated with the reappearance of the region.

The unpredictable nature of this cap can perhaps be explained by the following three factors:

1) The topography of the northern polar region may allow for greater deposits of ice or CO2 in the higher altitudes. Exposure to Martian winds may also increase the effective chill.
2) Average temperatures of the polar regions, depending on planet-wide meteorological conditions, may vary during certain seasons, and the distribution of temperature gradients may change from 'Martian year to year.
3) The density of the deposits at the caps may also vary in the amount of mass of either the CO2 or H2O vapor.

Further study of the degree of melting of the northern polar cap are important and desirable, and the effect and cause or the polar haze associated with this cap are equally important.

Observations of the South Polar Cap
(http://www.nasa.gov/centers/ames/multimedia/images/2005/marscap.html)

During perihelic oppositions of Mars, the large south polar cap is visible to observers on earth. The south polar cap seems to go through more changes than its northern counterpart. During approaches to Martian summer at perihelic oppositions, the tilt of Mars' axis of rotation can change from 5° to 24°, giving excellent views of the cap. And the closer proximity to the earth during these close approaches helps astronomers make more detailed studies of the southern cap. Considerable meteorology is associated with the thawing and melting of the southern ice cap on Mars. It is this defrosting that enables

Mars to warm considerably as a result of decreased reflection of sunlight from the large cap. The many cloud formations that are discussed following are the result, either directly or indirectly, of the thawing of this large cap.

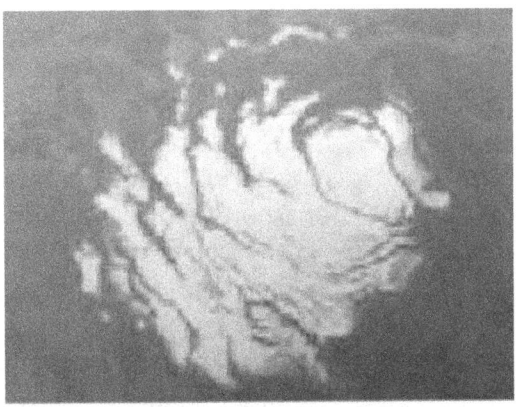

FIGURE 6
Mars' South Polar Cap
(Mars Orbiter)

As spring on Mars advances, the southern polar cap appears to split into two segments, the result of rapid thawing of a feature known as Novus Mons, or the "Mountains of Mitchell" (see http://www.cosmographica.com/gallery/portfolio/portfolio001/pages/004-TheMountainsOf%20Mitchell.htm). This large mountain range protrudes high above the cap remnant and is the first visible delineation of the cap seen at each opposition. Around the edge of the southern cap you might also often see finely detailed rifts that emanate from the melting cap and stretch delicately into the peripheral plains. The meteorological conditions in the Novus Mons area are of particular importance to astronomers.

Procedures for Observing Martian Polar Caps

Both the northern and southern polar caps are studied similarly, although each cap has its own characteristics. At least FOUR areas of detailed studies enable the amateur astronomer to provide important patrol information, as follows:

    1. *Polar Meteorology* - Examine the caps carefully for such meteorological phenomena as the formation of clouds, mists, or fogs when the polar caps begin to melt and again when they begin to refreeze. Make examinations for such meteorological phenomena throughout Martian spring and midsummer, and again throughout the autumn when the cap is reforming. The cloud formation will usually be confined to the hemisphere in which the cap under observation is located. So, during perihelic oppositions (every 13 years, roughly), the southern cap should instigate meteorological activity in the southern hemisphere. Likewise, at aphelic opposition, the northern cap will be the precursor of activity in the northern hemisphere. Predictable meteorological activity that might occur is the formation of the north polar haze during Martian summer

and the formation of white clouds in the Hellas basin during July and August when Mars is at perihelic opposition. Green, yellow, and orange filters aid in the study of such developments.

2. *Melting/Freezing Curves* - Draw carefully, or measure on photographs, the size and shape of the polar cap as it begins melting. Continue your drawing and/or measuring until Mars is no longer visible. Measurements of the polar caps can be made in three ways: (a) by using a filar micrometer, (b) by using a reticle eyepiece, and (c) direct from a good digital or ccd image while displayed on your computer monitor. First take a measurement of the size of Mars from north pole to south pole to determine the micrometric extent for any given day. Then convert it into seconds of arc (" arc) by finding the angular extent of Mars on that date from the American Ephemeris and Nautical Almanac (AENA) or via one of the many fine computer Central Meridian and Mars data programs mentioned previously. Then measure the polar cap width and convert it similarly. With the advent of good digital and CCD imaging available to the amateur astronomer at a reasonable cost, many contributions have been and continue to be made direct from high resolution images.

Measurements can also be attempted (although with a likely considerable reduction in accuracy) via a reticle eyepiece. By using a graduated reticle inserted into an eyepiece whose field is well known, you can extrapolate those gradations of the reticle in fractions of the field of view. Consequently, by determining the angular extent of Mars' disk on the night of determination (using the AENA or your PC programs), you can compute the division-per-degree ratio of the planet's surface. Then it is a simple matter to determine proportionately the total angular extent of the width of the cap in relation to the pole-to-pole extent (angular size in "arc) of the planet.

Such measurements - regardless of which method you are using - enable you to compile a "melting" or "freezing" curve that graphically displays the changes in size of a cap throughout its appearance. Because the amount of atmospheric activity is thought to be attributable to the degree of melting, the changes in size give you the percentage of the polar cap that is melting or freezing.

3. *Topography* - Examine the caps under high magnification during moments of very steady conditions to search for any significant features within the caps, such as Novus Mons in the southern cap. It is important to note the time and date when you first saw the feature, any changes in intensity and character, and the date when you last see it. You can obtain valuable information concerning the temperature, the terrain of the land, and the relation to atmospheric phenomena from the visibility of features of polar caps.

4. *The "Melt Line"* - Measurements and drawings of the polar cap melt line (the perimeter surround each cap) are of interest to the amateur astronomer. This dark band surrounding the periphery of the polar cap develops as the cap melts in early Martian spring, and it is visible until summer. These bands are probably caused by the melting of the water ice and the dispersion of liquid from that portion of the cap. Because the amount of H2O that may be locked up in a cap can vary depending on conditions

previously described, the melt line also can vary. Studies comparing the rapidity of polar cap recession and the width of the melt line are valuable. Record the time when you first see the melt line and when it disappears.

STUDIES OF MARTIAN ATMOSPHERIC PHENOMENA
(https://en.wikipedia.org/wiki/Climate_of_Mars)

Mars is not a static world; it is a planet still alive both geologically and meteorologically. The atmosphere of this small planet is representative of the climate, the seasons, the chemical constituents of the air and ground, and-of course-the melting of the polar caps. There is no doubt that the seasonal variations in Mars' atmosphere are directly related to the melting and the sublimation of the polar caps as Martian spring progresses. As the atmosphere condenses at the poles, the formation of clouds and transient veils diminishes, and you can monitor these events readily. Other than a telescope of moderate aperture, what is most needed for the studies of Martian atmospheric phenomena is a good working knowledge of the Martian terrain as it normally appears. Good maps of the Martian features aid you in quickly identifying transient clouds superimposed over major known areas.

Five types of atmospheric phenomena are suited for studies by amateur astronomers. Except for the efforts of these nonprofessional astronomers, very little investigation from earth-based observatories is being made into these phenomena of Martian meteorology, which are:

(1) whitish blue and white clouds;
(2) yellow clouds,
(3) dust storms,
(4) the blue clearing, and
(5) the W-shaped clouds.

Whitish-Blue and White Clouds (http://www.alpo-astronomy.org/mars/discrete.htm)
The whitish-blue and the white clouds are attributed to near-surface fogs and mists or perhaps to actual deposits of frost in sheltered depressions on the Martian surface. The very white clouds appear to increase in number and the area they cover at about the time the polar caps melt, and they are not necessarily restricted to one hemisphere. Thus, the white clouds are thought to be seasonal occurrences, beginning in Martian spring and ending in early autumn. In addition to being seasonal occurrences, it is possible these whitish clouds are daily occurrences as well, forming in early morning on the Martian terminator and disappearing in the heat of day. It is important for amateurs to monitor for such changes to help establish the theory of the possible daily formation of low ground fogs and frost deposits.

You should record the appearance of any cloud, whether the cloud was visible on the morning terminator or the evening terminator of Mars, or whether it was visible all day. The areas given in Table 4 are known to exhibit the seasonal forming of white clouds. Most of the reported white clouds are confined to the mid-temperate and equatorial latitudes, and they are more predominant in the southern hemisphere than in the northern. This may be misleading, because the true cause for such distribution may be the favorable circumstances under which we observe the southern hemisphere. Not only is the planet of greater angular size when its southern hemisphere is visible, but the angle of reflected sunlight from features such as the white clouds is considerably improved over that seen in oppositions of a northern tilt.

Your observation of the whitish clouds can be considerably improved if you use the Wratten No. 58 green filter for the maria or the Wratten No. 12 yellow filter for the plains. These filters enhance the brightness of the clouds on the maria and plains.

TABLE 4. AREAS OF SEASONAL WHITE CLOUDS
(in degrees latitude and longitude, in order of increasing longitude)

Feature	Latitude	Longitude
Aram	0	15
Sinai	-25	62
Ophir	-08	68
Thaumasia	-30	75
Tharsis	+02	100
Nix Lux	-08	112
Olympus Mons	+18	133
Memnonia	-20	160
Zephyria	-12	190
Elysium	+30	215
Isidis Regio	+20	280
Neith Regio	+35	275
Nymphaeum	+08	305
Hammonis Cornu	-10	315
Deucalionis Regio	-12	345

The Yellow Clouds of Mars
http://www.uapress.arizona.edu/onlinebks/MARS/CHAP10.HTM
Precursors of the great dust storms of Mars, the yellow clouds form readily during years when Mars is near perihelion and at the time of the Martian summer solstice. Because of the timing of such developments, it is speculated that the yellow clouds as well as the dust storms are raised by the rapid transfer of heat in the thin Martian atmosphere, which causes violent winds during this interval. The winds carry the clouds of dust and distribute it all across the planet. Yellow clouds form quite rapidly, and sometimes they spread equally fast. They can turn the planet into a vague diffuse globe, with no sign of even the brightest or darkest feature beneath the blowing dust. Although the clouds appear quickly, they can take weeks, even months, to disappear. It is their longevity that allows amateur and professional astronomers to map the transient circulation patterns of Mars.

The origin of the dusty yellow clouds is highly localized, coming from the Serpentis-Noachis-Hellas basin area, centered at latitude -28°, longitude 320°. With few exceptions, these clouds, as well as the dust storms, occur quite low in the Martian atmosphere, skirting the landscape of Mars with dust-laden clouds. You will get best results if you monitor the yellow clouds and the extents of the dust storms using a Wratten No. 12 (yellow), No. 21 (orange), or No. 25 (red) filter.

The Martian Blue Clearing
https://www.researchgate.net/publication/283855760_MARS_BLUE_CLEARING_AND_ALLAIS_EFFECT
If you view Mars in blue light, you will usually see very little detail. The planet appears to have a uniformly smooth surface with a bright polar cap. The farther you go into the blue region of the spectrum (toward the ultraviolet) in your viewing of Mars, the less you will be able to see. You might glimpse some detail if you use the Wratten No. 80A (medium blue) filter, but all detail vanishes if you use violet light (No. 47 filter). Because of the nature of the materials that make up the Martian landscape and portions of its atmosphere, more light is absorbed in the blue end of the spectrum, whereas most of the light of longer wavelengths - in yellow, orange, and red - is reflected.

A little-understood phenomenon occasionally occurs during which the Martian surface is very poorly delineated if you use filters of the longer wavelengths that you would normally use in viewing the planet. If you use a blue (No. 80A) or violet (No. 47) filter, however, the surface features again become visible. Such conditions, known as the blue clearing, may last for several days and are of great interest to astronomers. This clearing in blue light was originally suspected to be caused by a blocking layer of ultraviolet clouds, high in the Martian atmosphere. Recent studies, however, indicate that this might not be the case, or that another phenomenon-the albedo of various surface features, as well as the polarization of light from those features-might be a partial cause of the blue clearing.

To search for the blue clearing, merely look first with the blue filter on your telescope and then look without the aid of a blue filter. It is best. to search for the clearing using the medium blue (No. 80A) filter. If you suspect the presence of the blue clearing move to the more dense No. 47 filter for further scrutiny. You should become suspicious of the presence of the blue clearing if, when viewed through white or unfiltered light, the planet looks like a featureless, orange disk. Make observations of a blue clearing on every possible date, noting the location (whether the clearing is restricted to a small area or is planet-wide), the date and time you first saw the clearing, and the results of your filter observations. This unusual phenomenon is not usually restricted to small areas, as are some of the other atmospheric phenomena on Mars; rather, the clearing seems to affect the entire planet on most occasions.

The Curious W-shaped Clouds http://www.ast.cam.ac.uk/HST/press/mars95.html
The most curious of all Martian phenomena-the "W-shaped" clouds-form in the vicinity of massive volcanic peaks. Reported first by Earl Slipher of the Lowell Observatory, and confirmed later as a recurring phenomenon by Charles Capen in 1966, the clouds (usually quite large) apparently are associated with reflections in longer wavelengths of light. They are therefore best seen in the medium blue to violet range of the spectrum (Wratten

No. 80A filter). It is possible that these unusual clouds occur less frequently in the southern hemisphere than in the northern. They are also seen more often in the Martian summer and early fall. Because the W-shaped clouds move fairly rapidly, it is important that you record them on the date you first see them. Record also their motions across the planet relative to known features; if possible, longitudes on each date should be determined, and the latitudes of these clouds estimated. It is important that you record and report the point of origin of the cloud.

The W-shaped clouds form near the following features (in order of increasing longitude):

Feature	Origin Latitude	Longitude
Ascraeus Lacus	+110	1040
Pavonis Lacus	+010	1120
Arsia Silva	-090	1200
Olympus Mons	+180	1330

It is interesting that all the features listed above are large, volcanic peaks and are located near one another. Perhaps the origin of these clouds is not nearly as simple as might be explained by atmospheric conditions.

The High, Blue Clouds
http://www.msss.com/mars_images/moc/6_12_98_color_wa_release/p345/
High in the Martian atmosphere are thin clouds not visible to the naked eye, or even to the eye aided by a telescope. Photographs taken in far-violet or ultraviolet light reveal the existence of these small clouds; ultraviolet-sensitive film is often used on steady nights by amateurs who possess large instruments [i.e., 32 cm (12-1/2 inches) and larger] capable of taking long photographic exposures at high magnification. I emphasize that the would-be Mars observer should become familiar with the Martian surface as it appears without clouds, so that atmospheric phenomena can be recognized for what they are.

Recording such information can be of great help to the professional planetary astronomer. It also will enable you, the amateur, to compile all your observations and derive a cloud map of Mars, showing the locations of the most frequent areas of cloud formation and their circulation after being formed. All clouds and any other atmosphere-related features should be recorded as follows:

1. The date and time (UT) on a feature is first seen.
2. The location of the feature relative to known features on the planet.
3. The approximate size of the feature, in proportion to .the total size of the apparent disk.
4. The longitude of the feature and an estimate of its latitude.
5. A disk drawing of the planet, showing the feature (optional).
6. Date on which the feature is last seen.
7. The amount and direction of the feature's drift on the planet.

PATROL OBSERVATION OF MARS

You can make patrol observations of the Martian surface either telescopically or photographically, but you should make them as consistently as possible. Basically, by setting up a patrol, you scan the surface on every available date so that you are sure to see any rapid changes. Report such changes immediately so that others may study them. Such changes include cloud formation, blue clearing, dust storms, changes in the size and shape of the maria, the appearance of bright spots on the plains or maria, new features in the polar regions, and so forth. Unless some new phenomenon is noted, you need not make drawings unless you have the time and the desire to do so.

Seasonal changes....the Maria: Because Mars has an axial tilt of 24.9 degrees, it has a seasonal cycle, much like that of earth....except not to the predictable levels of our planet. On Mars the seasons are much more extreme and last twice as long. Some of the effects of seasonal changes have been discussed via the polar caps and these basic changes cause most of the other seasonal changes on Mars. For example, the density of the atmosphere increases proportionately with the melting of the polar caps. The density increases as the caps melt and sublimate, and it decreases as they refreeze in Martian autumn and winter. With the melting of the ice caps, the amount of sunlight reflected from the planet decreases, which causes more heat absorption by the planet's surface. This in turn, results in increased atmospheric convection.

From your standpoint as an amateur observer, the most notable seasonal changes, other than those of the caps themselves, occur in the maria, which are seen as dark regions. The changes were formerly thought to be caused by growing vegetation that made these regions darken as the Martian summer progressed lighten as autumn approached. We now know that these are related to the albedo, or the amount of sun- light absorbed compared to the amount reflected. The more light that is reflected, the brighter the feature appears to us.

The wave of darkening: http://www.msss.com/http/ps/life/life.html as it has been termed, begins in midspring on Mars and continues until most of the polar ice cap is gone. It is more predominant in perihelic oppositions than in aphelic ones, possibly as a result of the differing densities of the north and south caps. As the caps melt or sublimate, the darkening progresses from pole to equator. An increase in atmospheric H2O slowly disseminating from the pole and drifting toward the equator could possibly explain such sequential darkening. Not only do the features darken, their size also often increases greatly over their size in the early spring, and they even change their shape and their position on the planet. Again, your familiarity with the normal appearance of the maria is essential. Only when you know how they are supposed to appear, can you detect the subtle seasonal changes that always occur. Table 5 is only a small sampling of areas that exhibit changes such as those described.

You can determine the relative degree of darkening of the maria by using a scale of 1 to 10, noting the relative contrast between a feature and its surroundings, with 1 being the lowest contrast and 10 the highest contrast. All the maria can be better differentiated by using color filters; the orange (Wratten No. 21) or the red (No. 25) are the best for such studies.

TABLE 5. Notable areas of seasonal changes
http://humbabe.arc.nasa.gov/mgcm/faq/climate_study.html

Feature	Latitude	Longitude
Margaritifer Sinus	-02	30
Hydrae Sinus	-02	30
Mare Australe	-65	40
Acidalius Fons/Tempe	+58	60
Nilokeras/Lunae Lacus	+25	60
Solis Lacus	-35	85
Candor/Tharsis	+10	90
Aonius Sinus	-45	105
Amenthes	+05	250
Thoana Palus	+35	256
Thoth	+3	256
Nepenthes	+20	260
Moeris Lacus	+08	270
Antigones Fons/Astaboras	+22	290
Syrtis Major	+10	298
Aeria	+10	310

(in order of increasing longitude)

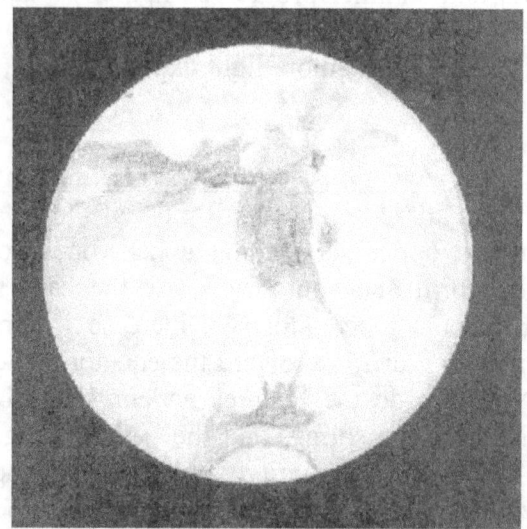

Sketch of Mars made by the author, Feb. 25, 1980. The prominent feature, Syrtis Major, is seen near the central meridian (see following); north polar cap is at the bottom, as it should be in all correctly oriented astronomical views.

TELESCOPIC AND PHOTOGRAPHIC RECORDS OF MARS

To compile a history of the appearance of Mars is a rewarding experience. You can compile the history by drawing on a standard disk form the appearance of Mars as you see it in the telescope, or you can photograph or record via CCD the planet. Whichever way is chosen, adhere to it. Do not switch back and forth to one or the other method. Photographic records of the planets, particularly Mars, can actually be less desirable than high-resolution drawings. With the aid of your 10- to 12-inch (25- to 30-cm) telescope, you can detect nearly as much as can professional astronomers making photographs through the largest telescopes on earth. However, the great advantage of a photographic program is that the photo verifies itself, so to speak. Modern webcam-type astro cameras are ideally suited for incredibly high resolution results of stacked images.

A photograph of a W-shaped cloud would be of great value, for example, and would virtually substantiate itself even if observers elsewhere had not seen this transient feature. For photographing Mars via film photography, choose a high- contrast, fine-grain, and red-sensitive film. The Kodak 2415 film, developed in Kodak HC-110 developer, dilution D, is best suited for photography of the red planet in black and white format. For CCD imaging, very high resolution, small field CCD chips are desired. Remarkably clear and quick results can be obtained via modern digital cameras which can be operated in manual mode.

Visual Studies of Mars

Visual studies of Mars are better suited for the endeavors of the amateur astronomer than are photographic studies, unless on is equipped with high resolution CCD equipment. Telescopic studies reveal low contrasts and subtle detail that cannot be recorded by the camera. You can record all the Martian phenomena thus far discussed quite accurately by making highly detailed drawings on which you draw and label precisely what you see in the eyepiece. During moments of steady seeing, you will be able to see extraordinary detail crisscrossing the Martian surface. Your first efforts at drawing the tiny disk of Mars may seem somewhat comical, with strange abstract figures and geometrical patterns drawn on a white circle. With a little practice and the experience of only a few drawings behind you, however, the very low contrasts and subtle details begin to form, appearing much as they do in the map shown in Figure 1. Both Mars and earth have atmospheres, and it is only when both atmospheres are simultaneously steady that the telescope can penetrate to the Martian surface to view the fine network of detail.

The trained Mars observer knows to wait for such moments. Drawings of the Martian disk provide a permanent record of your impression of the planet. No drawing is without a certain amount of bias, showing features that result from some preconceived notion of what you think "should" be there. Draw only what you really see - not what you think you "should" see. If you are not familiar with the Martian features, particularly the very small ones, you might not record some fine linear marking that you saw only for a moment, simply because you do not trust what your eyes have actually seen.

When seeing conditions permit, attempt to make observations at least every other night, preferably at about the same time each night. Ideally, make two drawings each night, when possible, to allow for greater coverage of Martian longitude. These drawings can make a composite record of all changes seen over a long period as well as show the expansion of very large features. If you make a second drawing 4 hours after the first, almost 600 change in longitude on Mars will be shown as a result of the rotation of the planet. Use a standard form, part of which is a circle 2 to 3 inches in diameter, as a standard for making visual observations of Mars. The form should have places where additional data can also be recorded, including the following:

1. The observer's name, the date, and time of the observation.
2. Telescope used, magnification, and any filters used.
3. The steadiness of the air on a scale of 1 to 5 (5 best, 1 worst).
4. Transparency of the sky (1 to 6, representing the magnitude of the faintest star seen to the naked eye).
5. Any transient or unusual features on the planet.
6. Martian longitude on the central meridian at the beginning of the drawing.

Make copies of each drawing as accurately as possible from the original, retaining the original and filing it in some systematic way. Send the copies to: MARS SECTION of The Association of Lunar and Planetary Observers or e-mailed direct to: rjvmd@hughes.net

The ALPO, http://www.alpo-astronomy.org/mars/marsfrm2.jpg provides link to the standard observing form used by the Mars section of the Association of Lunar and Planetary Observers (ALPO). Also, we have an excellent form in PDF format made by Carlos Hernandez and graciously provided to ASO for use by anyone by going to the ASO On-Line Tools section www.arksky.org .

For your drawings, it is best to follow a strict sequence in which features are recorded to provide accuracy and consistency in your records. The following sequence is recommended:

1. Draw the polar caps first. This allows for better accuracy in the placement of other features, and sets up a basic north-south orientation for you.
2. Draw all prominent detail in the center of the disk, using the maria as guides for small, indistinct detail.
3. Draw details on the preceding limb (eastern, or left on the form shown reference above). Notice that south is at the top on this form, as it should be in all drawings.
4. Sketch in all fine details, looking carefully for clouds, "canals," bright spots, and so on.
5. Add details seen in filters that were not seen without them, but be sure to note (by a small number) which were seen this way, and through what filter.

Adhering to this routine helps you to place features in their proper positions, rather than too far north, south, east, or west. After all the most prominent details have been drawn in their approximate positions, then draw the finer details on the drawing in relation to the obvious features. A normal Mars drawing should require a full hour. You can take the time to wait for moments of optimum steadiness in order to discern the finest detail.

DETERMINING LONGITUDE ON MARS

To make your observations more scientifically valuable, it is necessary that you determine the central meridian of the Martian globe at the time of any particular observation. For each drawing made, or photograph taken, determine the longitude of the central meridian. The central meridian (CM) is merely an imaginary line passing from the north pole to the south pole of the planet, perpendicular to the equator. Record the central meridian at the time when the drawing was begun, and again when the drawing was finished.

An excellent downloadable computer program will assist you in determining exact longitude for any time, date or earth location; go to the Meridian program in the ASO Tools Section for instant longitude readings for the red planet.

NOTE: If it took you one hour to render the disk drawing, then 14.6 degrees of longitude will have rotated past the CM. For a photograph, the CM recorded should be that at the instant the drawing or photograph was made.

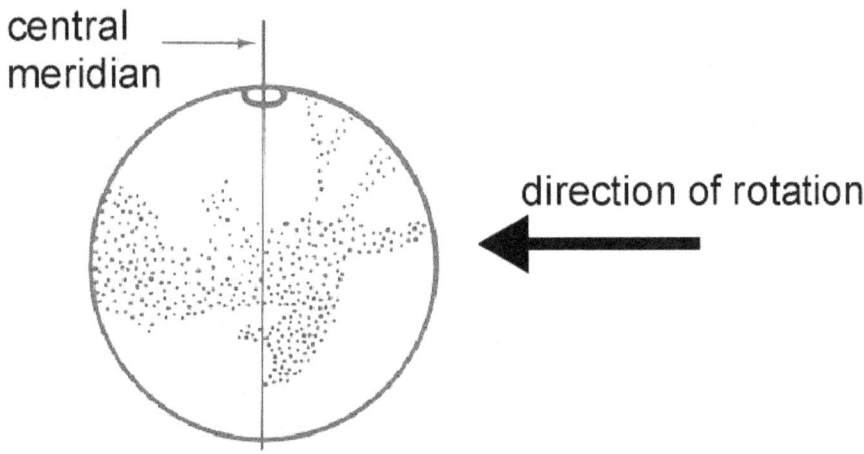

FIGURE 3. Drawing showing the central meridian

Because the rotational rates of Mars and the earth are almost the same, the same face of Mars will be visible with only slight deviation for several successive nights. Over a period of 36 days, you can view an entire rotation of Mars through 3600 of longitude if you make your observations each night at approximately the same time.

Another important advantage in determining Martian longitude is that it allows you to determine the correct longitudinal placement of features, as well as their angular size horizontally. The longitudes of Mars' surface features are determined by noting the time, to the nearest minute, at which that feature is exactly centered on the Martian disk (the central meridian would then cut symmetrically through the feature). To determine angular longitudinal expanse, one simply times the instant the preceding edge of the feature first crosses the CM, and the instant the following edge crosses some moments later.

After determining the longitudes for each timing, the difference of the two is the angular extent in longitude of that feature. The process of determining longitudinal placement of features on Mars is much easier than for those on Jupiter or Saturn, because of the much slower rate of rotation of Mars. Closer precision may be achieved if a vertical wire in a finely threaded cross-hair eyepiece is positioned to exactly coincide with a north-south line.

An even more accurate determination of both latitude and longitude of any Martian feature will be afforded through direct measurements from quality high resolution CCD or photographic images.

Reducing the Timings To Determine the Longitude

Every planet has 360° of longitude starting with 0° and progressing all the way through 359.9°. On Mars the 0° point -the starting point for all subsequent measures- was arbitrarily set at the center of a prominent feature, Sinus Meridiani, which is also located near the Martian equator (-05°). All subsequent markings are measured in increasing longitude westward from that point, as they are on earth, as we measure westward and eastward from the 0 degree longitude of Greenwich, England. The use of either an on-line Mars ephemeris database to determine longitude (and thus exactly what features you might be viewing, or interesting new clouds or developments over existing features) or the AENA will result in much greater scientific value to your observations.

It has been said that Mars may well hold the key to the development and evolution....and perhaps even the HISTORY.... of our solar system and our planet earth. Perhaps Percival Lowell was not so wrong after all about his hypotheses regarding Mars....maybe at one time there was enough water to sustain life of some type, and we know that Mars is a drying and dying world. Perhaps Mr. Lowell was merely speculating a bit too late in the prehistoric history of a primitive world of mystery.

Often I wonder what EARTH might appear to be if all life suddenly ceased upon its surface....no more skyscrapers, no bridges nor highways....no city dumps or great dams upon the waterways. If our planet were left alone - void of the force that we call "biology" - for perhaps three million years, what signs would be visible upon this planet to the visitors from some world only 36 million miles away? The greater forces of geology, meteorology, and the force of sunlight itself may well render our world to appear "drying and dying" to those yet to come.

* * *

Chapter 29

JUPITER

An Overview of Observing Programs for the Jovian World

Part One - Overview

Jupiter is a dominant object of our skies. Not only can it be seen throughout much of every year as one of, if not THE, brightest objects in the sky, but it also the largest and the one that demonstrates to viewers with even small telescopes "the most bang for the buck" when it comes to visible changes and events.

Jupiter is the largest of the gas giants, and indeed all planets of our solar system. To observers on Earth, its yellowish disk is the largest of the distant planets, except for rare approaches of Venus as it attains a slightly larger disk.

We cannot see any surface detail on Jupiter, just like with any gas giant and there likely is NOT a surface to be seen. What is seen in the visible spectrum through our telescopes are the many bright cloud tops....the dark chasms where clouds plummet downward....violent storms that can attain sizes more than THREE TIMES the entire diameter of Earth.

It is truly a violent world, ever-changing. You can witness first-hand much of this turbulent and active world in even small telescopes. In addition the four large satellites of Jupiter - the Galilean "moons" - can be monitored every night as bright starlike disks that appear to dance around, across and behind the bright mother planet.

There is a lot to be seen on Jupiter, but it takes a trained eye to see most of it. Too many times novice observers have been misled into thinking that subtle detail like the Great Red Spot is "easy pickings" on the planet. Nothing can be further from the truth, as the Jovian contrast is very low and detail is subtle to the unfamiliar.

Fortunately time is on our side. Anything you miss at one moment on the great planet will be back around in less than 10 hours! That's right, even with its huge mass, the 88,000 mile plus planet rotates on its axis quite fast, at 9h 55 minutes in mid-to-polar latitudes and only **9h 50 minutes** around the equator! This "differential rotation" causes the "air" of Jupiter to be "stirred" violently as the slower-moving cloud tops resist the faster moving currents.

The result is a feeding frenzy for amateur astronomers who can witness great storms seen as white ovals (like cyclones) or dark spots (as the Great Red Spot) and as wispy trailing clouds and downwinds.

Being a gaseous giant planet, there are NO surface markings to be visible on Jupiter, only faint "belts" which are downward convection currents. Like canyons in the clouds, sunlight cannot reflect from these areas and hence, they appear darker to us in our telescopes than do the surrounding high clouds, or what we call "zones" on the gas planets. The striking photographs from Rick Krejci which accompany this observing guide demonstrate clearly the remarkable number of belts and zones on Jupiter that can - at one time or another - be seen even in a 3" or larger telescope.

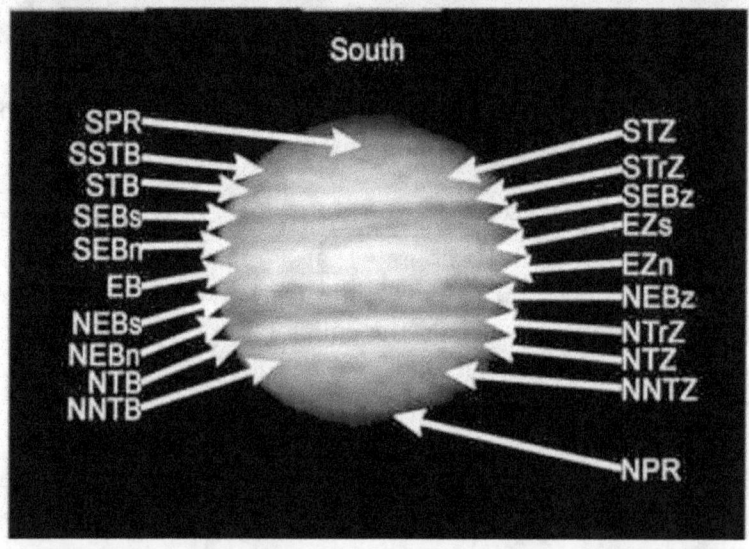

Occasionally, a bright white cloud will be seen on Jupiter, these usually forming just north of the North Equatorial Belt (NEB) in the North Tropical Zone (NTrZ) or immediately south of the South Equatorial Belt (SEB) in the South Tropical Zone (STrZ) - both shown clearly in the attached photos and clearly visible in your telescope. These white clouds are cyclonic storms which force cloudtops high into the Jovian atmosphere and, hence increase their bright reflectivity.

Even as beginners in astronomy, we are used to hearing references to *"the Great Red Spot"* (GRS) on Jupiter.....festoons, white spots, swirls, zones, belts and shadows of its four *Galilean satellites* as they transit between the giant planet and the sun, casting dark disks onto the clouds tops far below.

Therefore, I have compiled a description of Jupiter and its major features for viewing and a brief overview of what IS visible in common telescopes and what you might expect to see. Note that this discussion is painfully realistic and does not exaggerate what is actually visible in telescopes of various sizes.

Attached are numerous beautiful high resolution DIGITAL composite photographs supplied graciously by *Rick Krejci*, taken with only an ETX 90 telescope(!). These clearly show a variety of markings on the giant planet; the reproductions (by the time they reach YOU) are very close to the way that Jupiter can look in a very good 5" to 10" telescope under incredibly steady seeing conditions; smaller telescopes can reveal just about the same detail, but perhaps with not as good contrast and color. Note, that because of the contrast and color enhancement available through Mr. Krejci's imaging system and computer techniques, you WILL NOT be able to see the detail as "vividly" VISUALLY IN ANY MODERATE-SIZED TELESCOPE (size up to about 16"), but the detail is there in very subtle contrast and low color. Careful observation, however, CAN eventually result in much of this detail being seen even in small telescopes.

Keep in mind that the following description details features which can be seen under OPTIMAL viewing conditions: 1) when the air is very steady; 2) when Jupiter is high in the sky; 3) when your telescope has cooled down (1-2 hours minimum) and 4) under planetary viewing magnification, or about 50x per inch aperture minimum.

BELTS AND ZONES OF THE JOVIAN CLOUDS

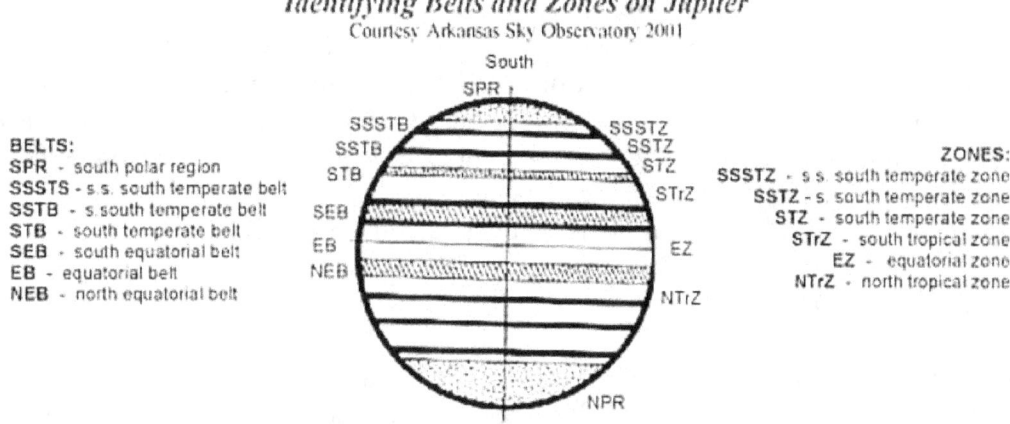

Identifying Belts and Zones on Jupiter
Courtesy Arkansas Sky Observatory 2001

BELTS:
SPR - south polar region
SSSTS - s.s. south temperate belt
SSTB - s. south temperate belt
STB - south temperate belt
SEB - south equatorial belt
EB - equatorial belt
NEB - north equatorial belt

ZONES:
SSSTZ - s.s. south temperate zone
SSTZ - s. south temperate zone
STZ - south temperate zone
STrZ - south tropical zone
EZ - equatorial zone
NTrZ - north tropical zone

NOTE: Northern Nomenclature is Identical to Southern Nomenclature

We have already discussed that the *ZONES* of Jupiter are seen as bright areas and represent cloud tops of gases forced high into the Jovian atmosphere by upward convection currents.

By contrast, the dark *BELTS* are cooler areas between the zones which represent plummeting clouds forming deep and narrow canyons between zones. Many times, as can be seen in the accompanying photographs, the belts will appear "double" or separated into two components with a narrow but definitely discernable zone between the two components. This is CLEARLY seen in the South Equatorial Belt (SEB) and the North Equatorial Belt (NEB) in the photographs; when such a zone appears it is termed either a north or south "equatorial belt zone."

The nomenclature for these belts and zones is standard for all gas giant planets: Jupiter, Saturn, Uranus and Neptune, but only Jupiter shows the number and the intensity demonstrated in the accompanying chart. Here is a drawing (from "*A Complete Manual of Amateur Astronomy*", P. Clay Sherrod, Prentice-Hall 1982, used by permission) that demonstrates the astronomical nomenclature for all gaseous planets: Jupiter, Saturn, Uranus and Neptune. It is the same for all planets. Note that, ASTRONOMICALLY, south is always shown at the TOP of photos and diagrams. Now photos and the drawing can be compared; although the electronic files of the photos are certainly not as clear as the originals, try to make out which "belts" and "zones" you can make out on Jupiter

You should become very familiar with these markings, and learn to differentiate between NORTH and SOUTH when viewing in your telescope; remember, a telescope with a right angle prism gives images with NORTH UP, and the image reversed right-to-left. However, the photographs shown here always have SOUTH UP, as this is the correct and accepted orientation in the astronomical community.

A couple of quick notes regarding belts and zones:

1) the equatorial belt (EB) is neither north nor south;
2) the polar regions are NOT "belts", they are "regions;"
3) there are NO "tropical belts," **only** "tropical zones;"
4) the belts are not always seen the same SIZE nor INTENSITY, sometimes much more vivid than others, and many times wider or even double from the night or week before.
5) be sure to make observing notes concerning the COLOR of the belts and zones, as they change over time; you will see yellows, white, blue, orange and tan in the zones and dark tan, brown, black, red and orange in the belts.
6) REMEMBER: "belts" are always dark; "zones" are always bright.

THE GREAT RED SPOT

To amateur astronomers perhaps the best known feature of Jupiter is the "*Great Red Spot*" (GRS), a very prominent and darkened oval in the southern hemisphere of the planet. Now known to be an immense storm of cyclonic nature, detailed photographs from the Hubble Space Telescope (HST) show a spinning nature to the GRS, with considerable central detail within the spot itself.

The spot, as well as some conspicuous darkening within it, can be clearly seen in the pair drawings by the author; the GRS is in the center of the sketch. Note that the spot is "embedded" (actually it "plows through" with its own motion) in the SEB, south component, and is encroaching on the STrZ; a very dark center can be clearly seen in the drawing.

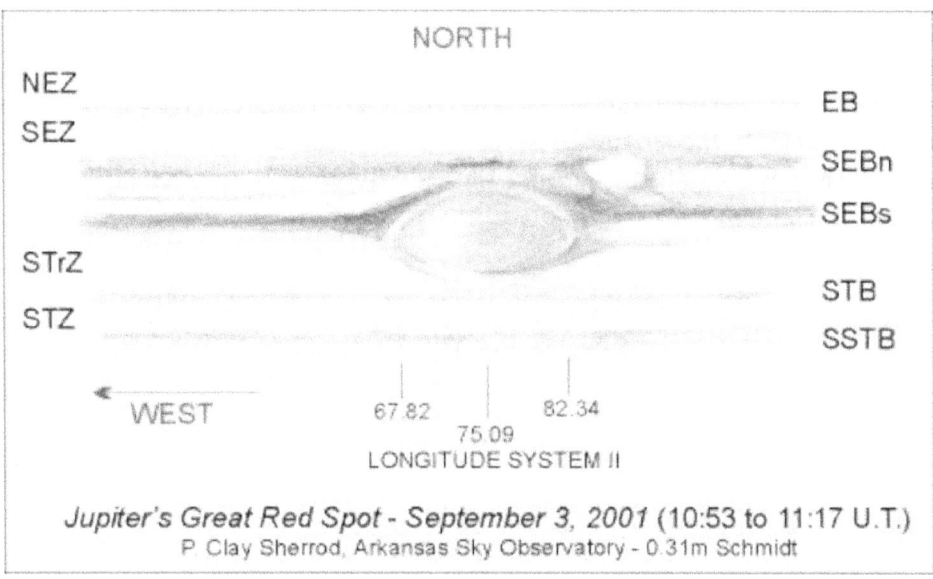

The influence of the GRS on other Jovian features is great as demonstrated in this drawing. Note how the SEBs has "curved" to conform to the shape of the GRS, and the trailing (left) edge of the SEBs is considerably darkened and displaced to the south behind the Red Spot.

Above the GRS note the duplicity of the SEB and the very distinct and bright South Equatorial Belt Zone (SEBz). This appears to "begin" immediately following the GRS (lower photo) but is absent preceding the GRS. Such tiny phenomena are the real captivation of the planet to amateur astronomers. Modern multi-frame webcam captures to high resolution images reveal cyclonic swirls within the span of the GRS.

The GRS is clearly visible in small and large telescopes. During recent years, the GRS has experienced a significant "fading" to our line of sight and is a bit more difficult to spot and certainly less vividly red than in some years. For those who persevere, though, the spot will eventually regain its majesty. As of 2016, the spot is re-emerging and is displayed in considerable structure by amateur planetary imagers.

Although it can FAINTLY be detected (particularly in the lower photograph), the GRS displays a very thin ring around it at the present time; I have seen this for several years now, the ring outlining the pinkish spot and appearing very thin and only somewhat darker than the spot itself. My ETX 125 shows it during very steady conditions, but the ring - and the dark central spot within the GRS - are not visible in smaller telescopes right now.

Check the listings in "*Sky & Telescope*" magazine for transit times (look in the "calendar" section) of the GRS. Most uniformed viewers seem to think that the spot should be visible any time they set out to observer, but this is not the case, the spot rotating on the planet once around every 10 hours or so. ALSO note that the ASO Tools section has a computational program for predicting each passage of any feature on Jupiter.

STORMS AND FESTOONS

Much of the fun in observing Jupiter is watching the nightly changes that take place in the visible clouds. Even the smallest telescope can reveal changes from one night to the next....even during one night's observing, the fast spinning of Jupiter allows you to see many "faces" as the planet turns yet more of its "surface" toward us.

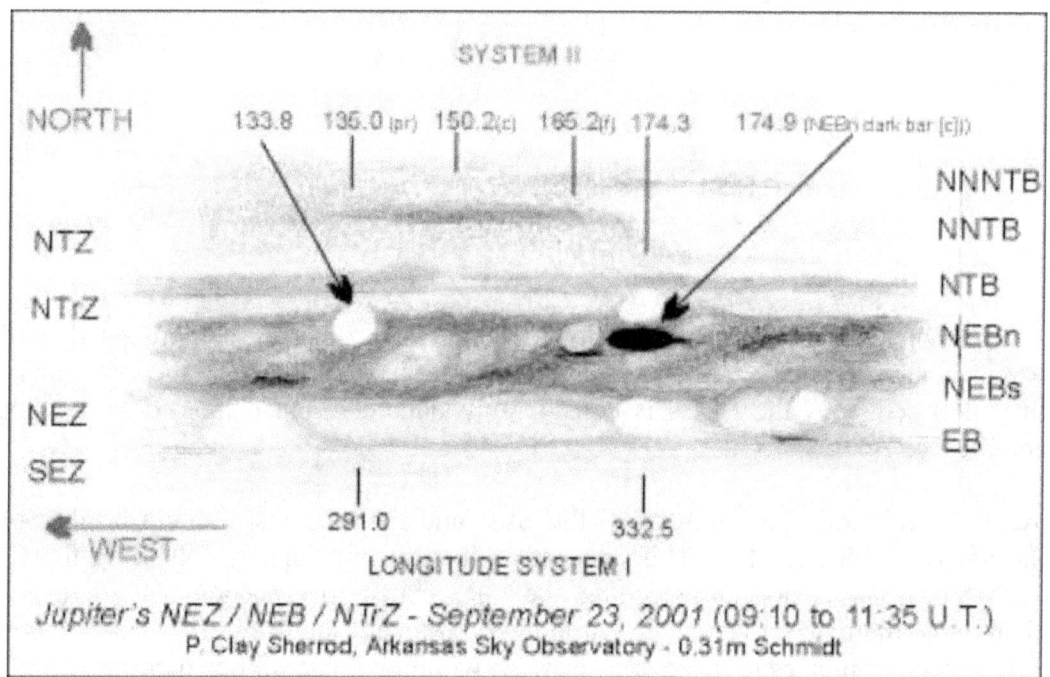

Among the many interesting features that can be monitored in small and medium telescopes are the "white ovals" and "dark ovals", storms similar in nature (cyclonic) to the GRS, but with vortex motion that forces the cloud tops HIGHER (as opposed to lower with the Red Spot) in the atmosphere than the surrounding ZONES. Hence, the higher clouds reflect more sunlight and appear brighter.

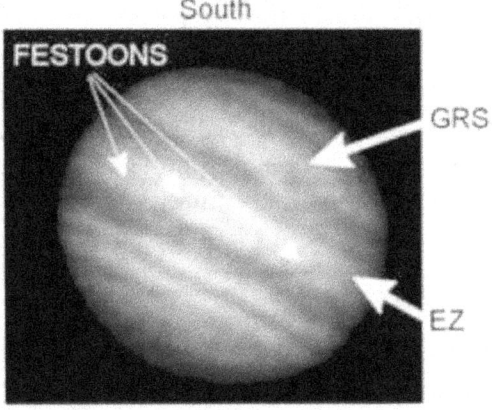

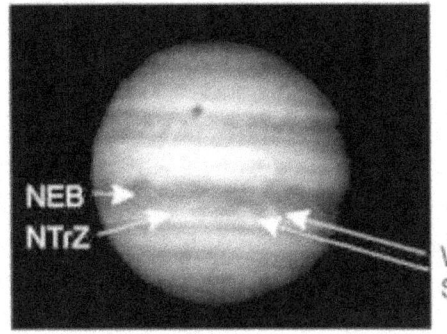

Festoons & White Ovals
PHOTOS BY
RICK KREJCI
ETX 90 - Dec. 4, 00 (top)
MEADE LX 90 - Feb. 5, 01 (bottom)

In **Rick Krejci's** two photos above (used to demonstrate results with small aperture telescopes!), you can clearly see some rapidly-changing features that are visible to you. Look at the lower photograph first, demonstrating two very bright white ovals in the North Tropical Zone (NTrZ) and encroaching into the north component of the North Equatorial Belt (NEBn). These are bright cloud tops.

Not seen right now, as far as my observations go, are any "dark ovals" which are also cyclonic storms of about the same size as the white ovals, but are representative of "downdraft cyclones" in which clouds are pulled inward rather than thrust high into the atmosphere.

NOTE: The dark spot at the top (south) in the lower photograph is the SHADOW of the Jovian satellite "Io", discussed in the following section.

The storm spots are not long-lived, but some last for many years. Indeed, the Great Red Spot has been seen for well over 300 years, even though in fades in and out of intensity from year to year. The white ovals appear to last longer than the dark ovals, and in some cases both only last a few weeks and quickly dissipate. Their motion is independent, but they do move very similar in rotation to the belts around them.

Observers with small telescopes can easily detect the bright and dark ovals provided that the observer is viewing a the time the planet has turned them to our direction here on Earth. Since the ovals can suddenly appear overnight (as can much transient and dramatic outbursts), observers are advised to watch nightly and attempt to spot these outbreaks early in their infancy. Who knows....you might discover a new disturbance on the giant planet!

Now look closely at the top image, examining the region between the two Equatorial Belts. The white arrows point to three clearly-visible "FESTOONS" that gently curve in the Equatorial Zone (EZ). If you look carefully at the top photo, you will see that the festoons appear to originate at the south edge of the North Equatorial Belt (NEB) and curve southward (toward the top) and to the right before reaching the equator of Jupiter.

Likely, within only a few days after Mr. Krejci took this fine image, these festoons were either gone, or greatly altered from the form in which you see them in his photograph. The festoons are "streaks" of swirling gases intermixing rapidly in the turbulent equatorial zone of the planet. It is a very active region and change can be seen on a night-to-night basis.

OBSERVING THE GALILEAN SATELLITES

In amateur telescopes (indeed, in ANY telescope less than 36" diameter), only FOUR of Jupiter's satellites can be detected visually. These are the same four that were discovered by **Galileo** in 1610, and hence entitled the "Galilean Satellites. Charts are published in "*Sky & Telescope*" magazine each month detailing the daily motion of these satellites as they orbit Jupiter.

This photo demonstrates what you can expect to see during one of the many weekly "transits" (when the satellite passes IN FRONT of Jupiter) of a Jovian moon. Here, we see the SHADOW (yes...it is a solar eclipse on Jupiter!) of the satellite IO, the closest of the Galilean satellites to the planet. Although not visible in this photo, Mr. Krejci has been able to capture BOTH the satellite AND the shadow simultaneously in animated time-lapse motion in the ETX 90 – a three-inch aperture telescope.

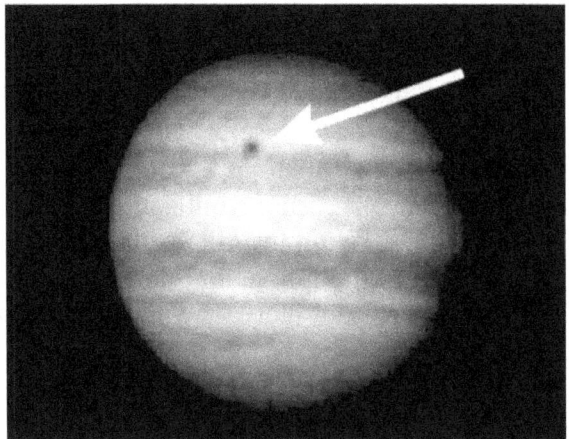

South

This is a phenomenon that YOU can see with your telescope! Clearly visible are both the disk of the small satellite AND the shadow that it MAY cast as it transits the planet (not all transits lead to eclipses). Another satellite fancy would be for you to observe one of the four satellites as it DISAPPEARS BEHIND the giant gas globe of Jupiter. The telescope (and/or camera) will clearly show the small satellite as it slips behind the planet, and in many cases, when one might re-emerge on the opposite side!

OBSERVING TIPS FOR JUPITER

As mentioned, the detail is there to be seen on Jupiter, but the contrast on the planet is VERY low, making the detail difficult to discern. To see all that we have discussed requires two things: 1) your familiarity of WHERE the features are located and what to look for; and, 2) understanding what techniques and tools you can use to bring out the subtle detail.

MAGNIFICATION - As with all planets, keep in mind that "magnification" is necessary, but not too much. (see the observing guide on "Seeing and Transparency" on this web site) Magnification is limited by the steadiness of the night air, and if the stars twinkle and images of planets are not sharp....you WILL NOT see the detail no matter what scope you are using! Whatever telescope you use to observe Jupiter, use this rule for magnification: when the steadiness is "come and go" and rather average, limit your magnification to 50x per inch of aperture; this means about 175x in a 90mm scope and about 250x in a 125mm scope. A 200mm (8") scope MIGHT stretch out the 400x, but that power will be pushing the limit even on a good night. On a night exceptional seeing, up the power-per-inch rule to 75x or 80x per inch to see even finer detail. Moonlight will not interfere with planetary viewing. When choosing the perfect magnification for any night, you might do what I follow: get an eyepiece (do not use zooms for planetary detail....they do not show fine detail like eyepieces alone) and magnification that shows a wealth of detail in your opinion; then go a little "higher" and see if that detail holds; if it doesn't the first eyepiece is your limit. If high magnification continues to show the same fine detail (always judge by the most difficult and tiny feature you can make out) you see at lower power, keep raising the magnification until you begin to see breakdown of detail. You have reached above your limit at that point!

FILTERS FOR JUPITER - I have mentioned time and again that detail is very subtle on Jupiter. There are TWO filters that you can screw onto the base of your eyepieces to really bring out considerable more detail than you can see without them: 1) the Wratten #58 green filter is incredible for increasing contrast between the belts and zones, revealing dark and bright spots, intensifying the GRS, and showing fainter belts than you might see without it; 2) a good "light pollution" or "nebula" filter is also recommended for Jupiter work; it does not provide much increase for other planets, but is excellent for Jupiter providing an overall increase in contrast planet-wide.

AVERTED VISION - this is an astronomical trick, not used too much anymore since most people never "look" through telescopes. When you look DIRECTLY at any bright object, particularly a small one like a planet, you are viewing with the worst part of your eye. In this position you view only with the RODS of your retina which - the older you get - deteriorate with age and use. Try this: with Jupiter centered, attempt to move your eye so that you never actually stare RIGHT AT the planet but rather skim past it, or look just to one side. This activates the "CONES" of your retina which are more color and light sensitive. It is amazing what you can see with this "averted vision" than without it! Try it also on double stars and deep sky objects.

The mighty planet Jupiter puts on a new show for those ready to find its features every night. It may not be as spectacular at first sight as the ringed majesty of Saturn, but its secrets are far more obvious and merely awaiting the trained eye and inquisitive mind to unravel.

Part Two –
Physical Characteristics of the Giant Planet

BACKGROUND

Detailed observations of Jupiter's Great Red Spot (GRS) in 1975, nearly five decades ago, at the Arkansas Sky Observatory revealed a crimson-red and large oval, the shape and color normally associated with this wonderful cyclonic feature first noted by **Cassini** nearly 400 years ago in 1660. It is believed that this is the famous spot through which Cassini actually first determined the reasonably accurate rotational rate of the gas giant. However, it was not long after observations in 1975 when this famous celestial bookmark began to show signs of fading both in intensity and color; within only a few years the spot had all but disappeared from view, noted ONLY though is absence in an indentation in the South Equatorial Belt (SEB) that is known as the "*Red Spot Hollow*" (RSH). [P.C. Sherrod, *Project Publication 003*, Midsouth Astronomical Research Society, 1975).

Thus, for the next thirty years during which time the GRS was all but invisible, the exact rotational rate and drifting in both latitude and longitude was measurable through where it was "supposed to be" rather than where we actually "saw it to be." Such remarkable fading of the GRS is not at all unusual and, in time, the spot will eventually return to its characteristic intensity and color.

The GRS, like most features on both north and south latitudes from the North Equatorial Belt and South Equatorial belts, respectively, moves in a slightly slower rotational rate - 9 hours 55 minutes 40 seconds - than the Jovian equator itself - 9 hours 50 minutes 30 seconds. Since the GRS is actually positioned latitudinally in System II, but intrudes northward into the SEB which is in System I (see latest drawing, below) you can image how it "stirs the waters" so to speak as it attempts to retard the motion of clouds, storms and anomalies located in the faster equatorial stream.

The faster-moving System I is confined to a narrow band centered on the Equatorial Belt that spans a total of about 10,000 to 15,000 miles north-to-south, all moving about 200 miles per hour FASTER than the clouds and features seen north and south of it. In the second figure below more of the planet is exhibited from the following morning's observations which reveal the "opposite side" of Jupiter, nearly 180 degrees from the GRS shown in the more narrow scale view in the first drawing. NOtice the remarkable intensity of the North Equatorial belt (both components).

LATEST NEWS

Since reappearing in morning twilight in fall of 2001, the GRS has shown a slight intensity change that has continued until today. A sample drawing of this region is shown below. It was made on the morning of September 3, from 10:53 Universal Time to 11:20 UT, with the steadiness of the air rated at "8" on a scale of 1-10 with "10" best. It was drawn using the 0.31m (12") Meade Schmidt-Cass at the Observatory at 625x using no filters. NOTE that drawings of Jovian features MUST be made very quickly (albeit very accurately as well!) since the rotational rate of the visible features results in very rapid movement across the center of the planet (Central Meridian - "CM"). This drawing was made within minutes of the CM crossing, which allows very accurate determination of Jovian longitude.

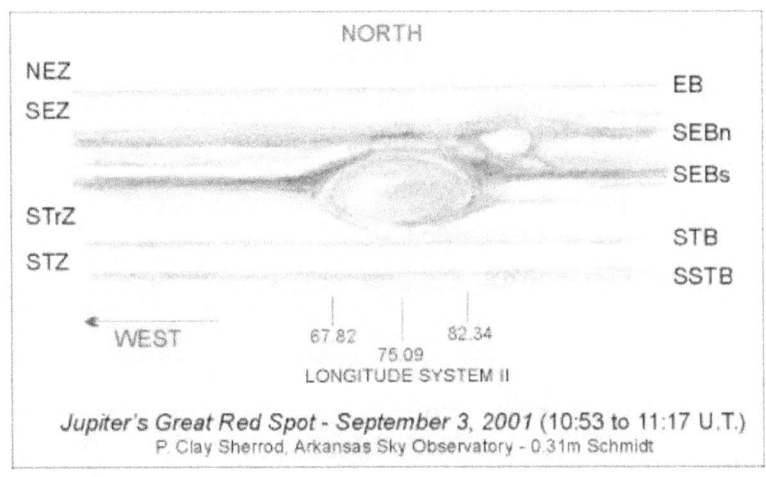

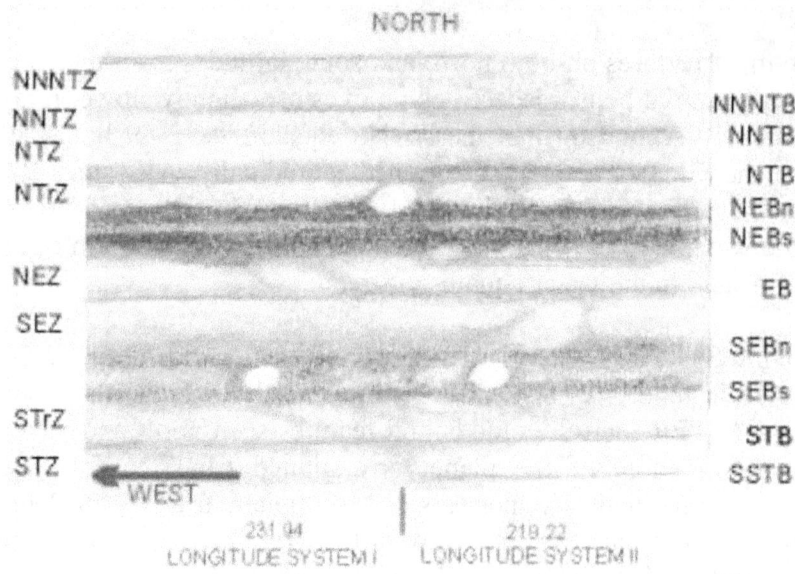

Drawing by author, September 1, 2001 showing belts and zones

INDEX TO NOMENCLATURE:

Zones (brighter areas)
NNNTZ - North-North-North Temperate Zone
NNTZ - North-North Temperate Zone
NTZ - North Temperate Zone
NTrZ - North Tropical Zone
NEZ - North Equatorial Zone
SEZ - South Equatorial Zone
STrZ - South Tropical Zone
STZ - South Temperate Zone

Belts (dark "stripes") –
NNNTB - North North North Temperate Belt
NNTB - North-North Temperate Belt
NTB - North Temperate Belt
EB - Equatorial Belt
SEBn - South Equatorial Belt, north
SEBs - South Equatorial Belt, south
STB - South Temperate Belt
SSTB - South-South Temperate Belt

1. <u>Red Spot Hollow</u>: you can clearly see the GRS hollow indenting the SEBs, and thus creating the Great Spot Hollow. In smaller telescopes, the GRS will likely ONLY be noted by observing this hollow at least early in the 2001 apparition, since the spot itself is

very vague and lacks much contrast or color at this point. Note from the drawing how the SEBs preceding (westward) the GRS appears to be disturbed by the intruding motion of the GRS itself; indeed, a small "festoon" can be seen just west of the GRS intruding into the STrZ as though "splintered" or pushed southward from the SEBs.

2. <u>South Equatorial Belt</u>: Note in the above drawing both "Systems" are rotating from right to left (east to west); however, it is important to note that System I (usually the SEB - particularly the north component - , the Equatorial Zones and the North Equatorial Belt (not shown) is moving FASTER from right to left than the GRS and the rest of System II; this is interesting in light of the disturbance following (east) of the GRS within the space between the SEBs and SEBn. Such activity is common where this differential rotation interaction occurs and many times will lead to major disturbances following the GRS and extending far south into the STrZ. The color of the SEB (both north and south components) is currently a bluish-brown and nearly equal to both intensity and color to that seen in late 2000.

3. <u>The Great Red Spot</u>: The GRS itself presents a somewhat "football shape", not at all uncommon. The size is an incredible 14.5 degrees in width. The actual size of any Jovian feature can be determined by the simple formula: $x = (E) \times 1187$, where "X" equals the actual size in kilometers of the feature (in this case the GRS), "(E)" is the longitudinal expanse (14.5 degrees) and "1187" is the number of kilometers per degree of longitude seen on Jupiter. [from: *A Complete Manual of Amateur Astronomy*, P. Clay Sherrod, Prentice Hall, 1981). Hence, the present size of the GRS is some **17,211 kilometers** east-to-west....easily large enough for the earth to nestle within with plenty of room to spare! Although this is very large by any standard (10,671 miles), the GRS has been recorded as large as 30,000 miles in width - enough to encompass THREE Earths!

At present, do NOT look for a "RED" spot.....look for a large pink oval that intrudes northward into the very conspicuous SEBs. As seen in the drawing above, the GRS is not at all uniformly dark nor colored. The west, or preceding, end of it is much lighter and without much color, while the eastern (right in the drawing) side is most definitely a ruddy pink color. An unusual aspect of this spot right now is the wonderfully defined pink "outline" that completely inscribes the perimeter of the GRS! This is NOT seen in telescopes smaller than 8" (0.2m) aperture, but is unmistakable in larger instruments.

4. <u>North Equatorial Belt</u>: Although not shown in the strip drawing above, the North Equatorial Belt (NEB) is the darkest feature by far on the planet, exhibiting a very intense orange-brown color; it - like the SEB - is showing TWO components that seem to be nearly perfectly parallel for the entire 360 degree expanse of Jupiter. The vivid darkness and color, as well as this duplicity, is something that cannot be missed in common telescopes of all sizes. The color, particularly in higher longitudes as shown in the second drawing above, is clearly a very reddish brown; note the white oval at 219 degrees System II that is shown in this drawing and extending into the North Tropical Zone

This drawing, done on September 5, 2001, demonstrates a very dark and large elongated oval, or "Bar" in the tiny zone between the North and South Components of the North

Equatorial Belt. Because of this location, more observations will be required to determine if this feature is moving in System I or System II (see text). This appears the same intensity as a dark shadow cast from a satellite would, but it is clearly elongated and smaller in N-S width, yet is visible easily in a 5" telescope under high magnification and good seeing.

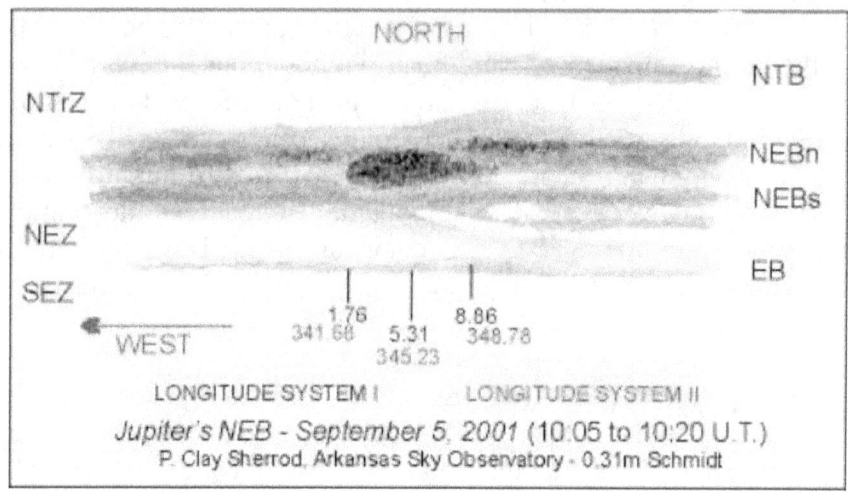

OBSERVING JUPITER

Observers who wish to determine the exact time of the transit of the Great Red Spot and other features, need only to download onto their computers the fabulous "Meridian" program by *Claude Dulpessis*, of Quebec. This will provide ephemeris data of EVERY planet, and for Jupiter will determine for your exact location and times the passage of the GRS across the central meridian. This freeware is available at:
http://pages.infinit.net/merid/index.html . Note that there is also a CUSTOM central meridian passage tool available here at ASO under the ONLINE TOOLS tab. In addition nearly all modern computer sky programs have tabs for instant calculations of central meridian, GRS passage times and other physical data on the giant planet.

Jupiter will dominate the evening skies of winter in the northern hemisphere and will be ideally located for both north and south of the equator for nearly nine months in 2001 and 2002. Observing Jupiter will allow you to move from a "casual" observer to a contributing "avocational astronomer" through your drawings, timings of features and general patrol observations that can lead to discoveries of storm outbreaks and changes on the giant planet. For more vital information regarding Jupiter and your contributions that are valued for this planet, go
to: http://www.alpoastronomy.org/jupiter/sendobs_0708.html for information concerning the Association of Lunar and Planetary Observers (ALPO), forms for recording and drawing and much valuable information and links to YOUR observing success and satisfaction with the greatest planet of our Sun's remarkable neighborhood.

Chapter 30

Observing the Ringed Planet Saturn

Because of its prominence of the planet Saturn in our skies nearly every year the number of inquiries regarding what can (or should) be seen on the ringed planet are many.

Saturn is a favorite: majestically presenting its rings (or "ears" as **Galileo** first referred to them) toward us at a sharp 26-degree angle (the southern hemisphere of the planet is tilted our way right now). The colors of the planet are splendid, with vivid yellows, oranges and whites contrasting nicely in both the rings and the planet itself.

Being a gaseous giant planet like Jupiter, there are NO surface markings to be visible, only faint "belts" which are downward convection currents. Like canyons in the clouds, sunlight cannot reflect from these areas and hence, they appear darker to us in our telescopes than do the surrounding high clouds, or what we call "zones" on the gas planets.

Occasionally, but not as often as on Jupiter, a bright white cloud will be seen on Saturn, these usually forming just north of the North Equatorial Belt (NEB) or immediately south of the South Equatorial Belt (SEB - shown in the attached photo and clearly visible in your telescope). These white clouds are cyclonic storms which force cloudtops high into the Saturnian atmosphere and, hence increase their bright reflectivity. Commonly, telescope users hear us refer to features on Saturn such as "*Cassini's Division*," the "*Crepe Ring*", "*Encke's Division*," and so on. Sometimes in our haste to explain what an observer SHOULD be able to see in a particular telescope, we forget that the jargon may not be familiar to all telescope users.

Therefore, I have compiled a description of Saturn and its "components" for viewing and a brief overview of what IS visible in common telescopes and what you might expect to see.

Those of you who have read many of my philosophical wanderings understand that I truly believe that ANY observer's FIRST view of Saturn in his or her own telescope in "real time" is an event never forgotten no matter what age we attain. So, at least in my opinion, it is important to know WHAT we can see and something ABOUT that which we do see.

Above is a NASA/Hubble photograph which I selected primarily because this reproduction (by the time it reaches YOU) is very close to the way that Saturn should look in a very good 5" telescope; smaller telescopes can reveal just about the same detail, but perhaps with not as good contrast and color.

Keep in mind that the following description details features which can be seen under OPTIMAL viewing conditions: 1) when the air is very steady; 2) when Saturn is high in the sky; 3) when your telescope has cooled down (1-2 hours minimum) and 4) under planetary viewing magnification, or about 50x per inch aperture minimum.

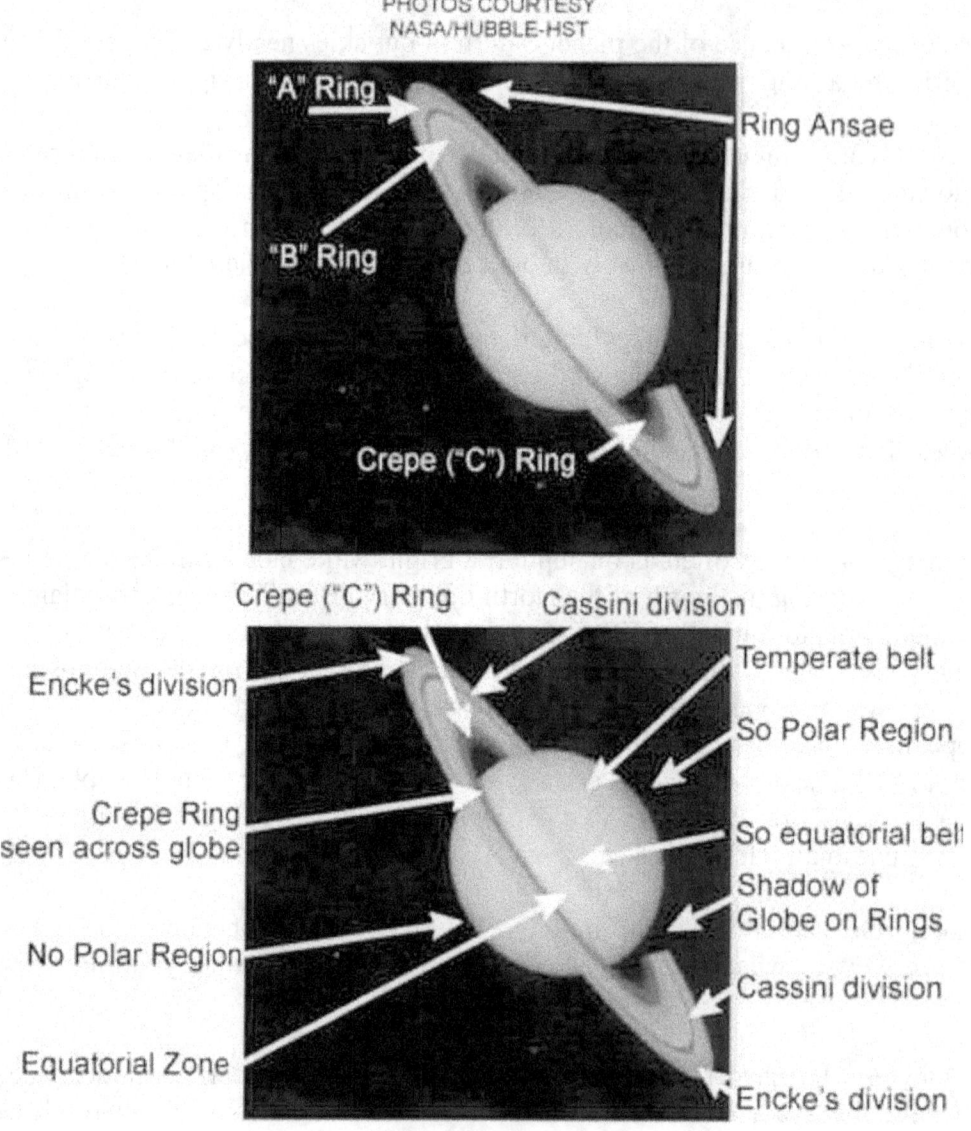

SATURN'S RINGS

Saturn, as we all know by now from Hubble and the many spacecraft which have penetrated the far solar neighborhood, has many more rings (and features) than we can see here on Earth; that applies to even the largest telescopes available.

With common telescopes under good conditions, THREE (3) primary rings are clearly visible on Saturn (see photo description): 1) the **OUTERMOST RING** (ring "A"); the **MIDDLE** and largest ring (ring "B"); and, 3) the **CREPE** or innermost (ring "C").

The Crepe Ring is VERY transparent and sometimes is difficult unless very high magnification is used. Look for the Crepe ring easiest where the ring appears to curve the sharpest (the "Ansae" at each side of the rings) or where the "C-ring" passes directly IN FRONT of the globe of Saturn; indeed, where the ring passes in front of the globe, it is actually possible even with a 3-inch scope to see the glove THROUGH the Crepe ring! Try it and impress your friends and neighbors!

The "B-ring" is the brightest and usually appears a whitish yellow color; on nights of extremely steady viewing a 5" or larger telescope can depict faint detail within this ring.

Outermost from the globe of the planet is the "A-ring", not as bright as the "B-ring" but most definitely easy and brighter than the elusive Crepe ring. It is always a dull yellow-orange color compared to "Ring B", and it is within this ring that the VERY difficult *ENCKE'S DIVISION* is seen on perfect nights.

A fourth, very faint, ring can sometimes be seen at very high power on perfect nights, even with smaller telescopes. However, apertures of greater than 12-14" are recommended and even then unskilled observers normally miss the "D-ring" which is found INSIDE the Crepe Ring and just OUTSIDE the ball of the planet itself.

RING DIVISIONS

There are two primary ring divisions that we are interested in common telescopes: *CASSINI'S DIVISION*; and *ENCKE'S DIVISION*.

Cassini's Division - this is an easy test for your 3" telescope and should be held steady at magnifications of 100x or greater. A MINIMUM of 40-50x in nearly all telescopes should reveal the dark gap which appears like an inky line separating Rings A and B. On nights when the seeing is particularly troublesome, the image of the faint curved line will sometimes come and go, but do not be discouraged...it is NOT your telescope!

Encke's Division - probably not possible to discern in a 3-inch telescope, but a test for a 5", I have clearly seen Encke's division with my ETX 125 on numerous very steady nights; on nights of average seeing, it evades observation. Likewise, observations with a 40" Boller & Chivens Richey-Cretien Cassegrain failed to disclose this very faint ring gap on one night in a Nebraska corn field, but the next night it was sharply distinct, with three other minor ring gaps visible! Those using 8" telescopes and larger should make this a priority test; look for Encke's in the "A-ring", about one-third of the width of that ring from the OUTSIDE edge. For the purists out there, Encke's division is a gap that is only .35" arc, thereby a real test for amateur telescopes; smaller telescopes, although they cannot resolve to that small a measure, can "see" Encke's through the CONTRAST of the "A-ring" on each side of it!

To prove the aforementioned point, note that CASSINI'S division is clearly visible in telescopes of 3" and above (even a good 2.4" refractor). Yet the ACTUAL width of this object is ONLY 0.6 arc seconds! In arc", the outer edge of "Ring B" is 16.9", while the inner edge of "Ring A" is 17.5" from the outer edge of the planet's globe. So, even though your telescope may not be able to actually "resolve" Cassini's, it "sees" it through the bright contrast between the "A" and "B" rings.

DETAILS ON SATURN'S GLOBE

Here is a drawing (from "*A Complete Manual of Amateur Astronomy*", P. Clay Sherrod, Prentice-Hall 1982, used by permission) that demonstrates the astronomical nomenclature for all gaseous planets: Jupiter, Saturn, Uranus and Neptune. It is the same for all planets. Note that, ASTRONOMICALLY, south is always shown at the TOP of photos and diagrams. Now, the Hubble photos I have provided and the drawing can be compared; although the electronic file of the photo is certainly not as clear as the original, try to make out which "belts" and "zones" you can make out in the southern hemisphere of Saturn.

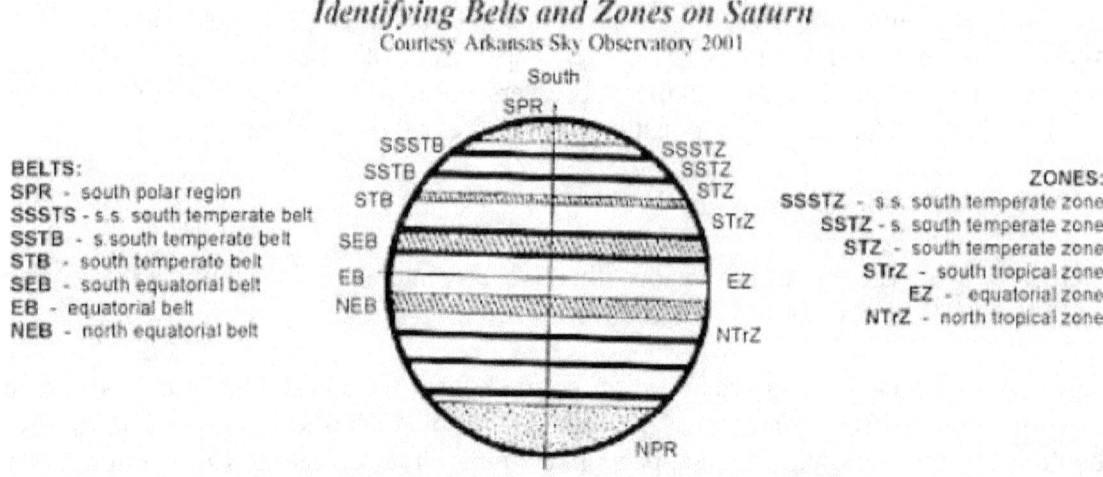

I can clearly (from a print of this image) make out (from the equator progressive SOUTH): the equatorial zone, the SEB, the South Tropical Zone (STrZ), the South Temperate Belt (STB), the South Temperate Zone (STZ), and the South South Temperate Belt (SSTB), as well as the South Polar Region (SPR).

This is exactly what you can expect to see on a very steady night - even with a 90mm telescope! If you cannot, here are a couple of hints:

1) <u>AVERTED VISION</u> - this is an astronomical trick, not used too much anymore since most people never "look" through telescopes. When you look DIRECTLY at any bright object, particularly a small one like a planet, you are viewing with the worst part of your eye. In this position you view only with the RODS of your retina which - the older you get - deteriorate with age and use. Try this: with Saturn centered, attempt to move your eye so that you never actually stare RIGHT AT the planet but rather skim past it, or look just to one side. This activates the "CONES" of your retina which are more color and

light sensitive. It is amazing what you can see with this "averted vision" than without it! Try it also on double stars and deep sky objects.

2) FILTERS - if you think that the planets look "washed out" and your skies are perfect yet little detail is visible, use a filter. The actual filter used depends on which planet you are observing, but we will concentrate here on Saturn. A Wratten #80A light blue filter is perfect for viewing Saturn, providing optimum contrast between the darker belts and zones. Try a Wratten #12 (yellow) to view the rings and provide the very best viewing of Cassini's and Encke's divisions.

FOR YOUR VIEWING PLEASURE....

A couple of quick recommendations for your viewing pleasure while looking at Saturn (refer to the photograph):

1) Look for the delicate "Crepe Ring" (it is seemingly transparent) as it crosses IN FRONT of the globe; you will clearly see it as a darkened streak across the body of Saturn (clearly visible in the photo);

2) The globe of Saturn casts a magnificent SHADOW onto the rings behind it (see photo), providing one of the most spectacular and believable "3-D" images you can have in a telescope!

Enjoy Saturn, and learn to understand what you see when you gaze 800 million miles away. The thrill of your first look at this ringed marvel will never be equaled.....but I can guarantee you that seeing it EVERY OTHER TIME from now until your last glimpse of days....

....this thrill will last your entire lifetime. Enjoy, it is a gift.

*   *   *

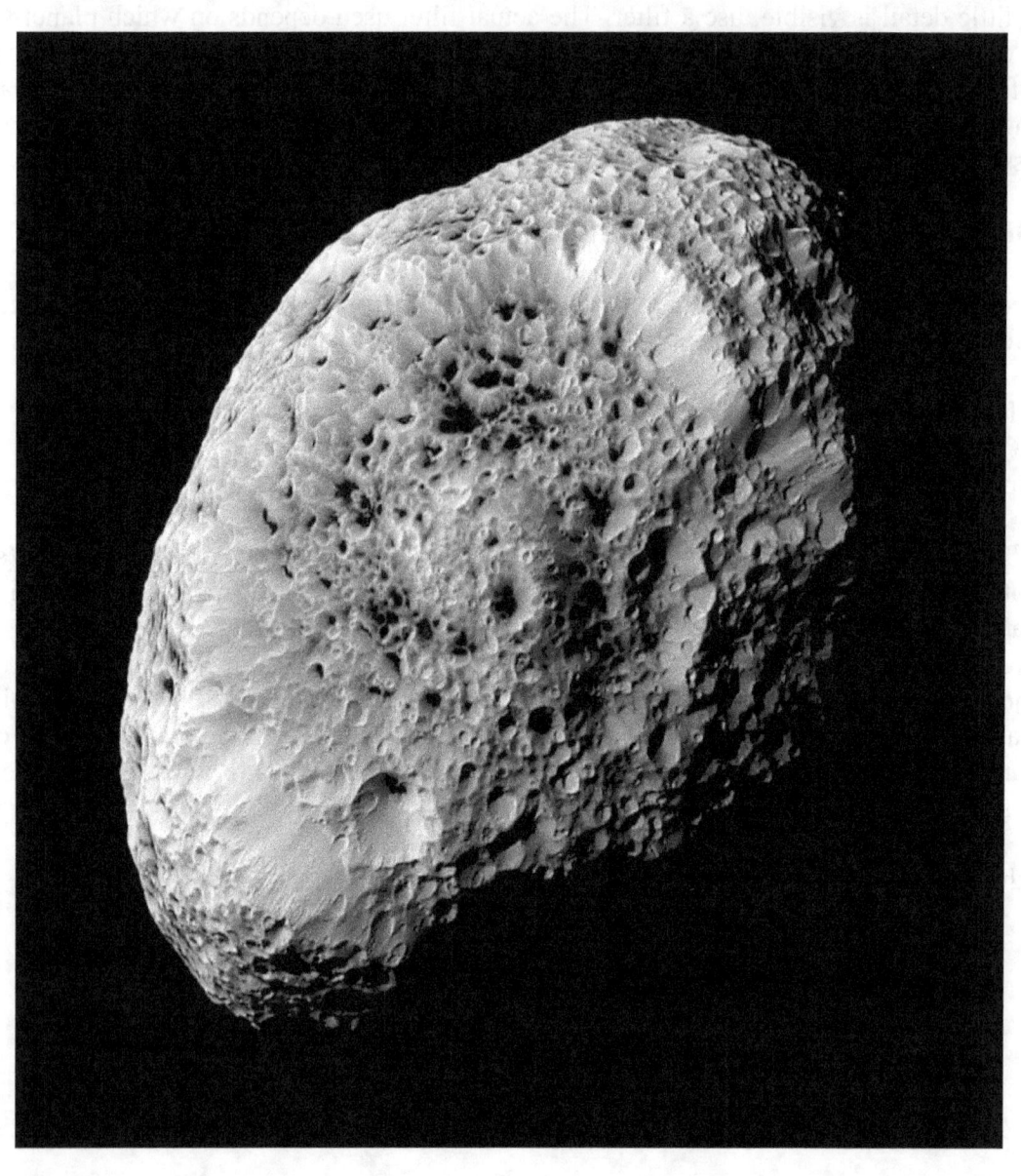

Near-Earth Asteroid APOPHIS approaching the Earth
For a close encounter in 2029
Courtesy NASA

Chapter 31

Comets and Asteroids
THE FASCINATING WORLD OF INTERPLANETARY OBJECTS

The primary objects within the sun's family - the large and bright planets - afford telescope users a plethora of observing opportunities, not to mention a gallery of unusual "out-of-this-world sights. Guides for telescopic observing of Jupiter (see ASO Observing Guides) and Saturn (see ASO Observing Guides) and other celestial wonders are found on this web site. Learning the tips and techniques, as well some pertinent information about the object itself, lends itself to a rewarding and challenging lifetime in amateur astronomy.

But in addition to the major planets and stars, there is an overwhelming wealth of other - even MORE unusual - material just within our "neighborhood": the solar system. Much of this unfortunately goes unnoticed and, for the most part - unobserved by amateur. This "stuff" comprises the *ASTEROID BELT* containing thousands of "minor planets," the APOLLO close-approach objects that many times pass even between the Earth and moon, and other "interplanetary objects", including the fascinating *COMETS*.

Fortunately for those of us with GO TO telescopes and PC sky programs, the capability exists to locate even the faintest telescopic asteroids and comets that we might otherwise miss if it were not for the computer's capability to "take us to them." Indeed, your modern computer-driven telescope is capable of accessing HUNDREDS of such objects and taking you to them, perhaps for the first time in your life!

Once the telescope has found a COMET for you, typically it is immediately recognizable...it may not be "bright" or have a "tail" and appear as ominous as the very bright naked-eye comets, but it probably will have a distinctly "fuzzy" look, like a very faint nebula.

On the other hand, even though computer acquisition, you can place your telescope dead center on an ASTEROID, unless you have a good set of star charts (the commercially-available CD or diskette Sky programs are adequate, to magnitude 7.5) you will never REALIZE that you are, indeed, looking at an asteroid. They merely look like faint stars; even the brightest asteroid, CERES, is a mere blip in a good pair of binoculars and looks only like a moderately bright star in a telescope!

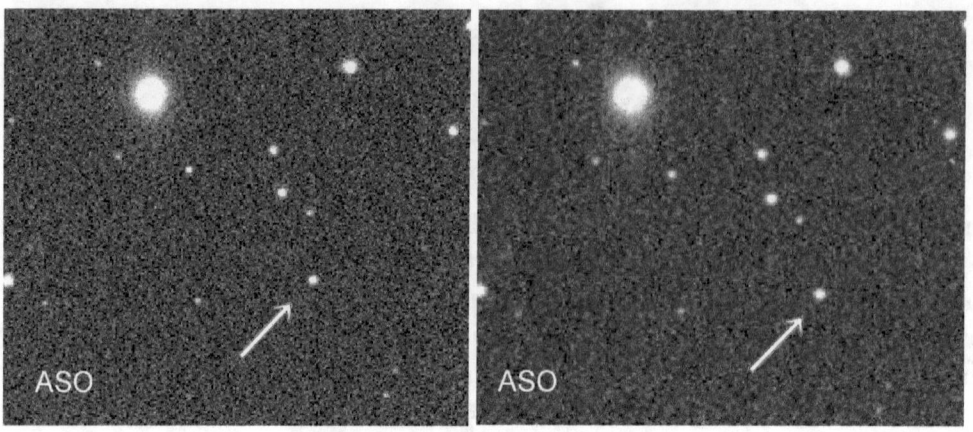

Motion NE (toward upper left)
of minor planet 4179 Toutatis in only 6 minutes

Figure 1 demonstrates the starlike appearance of #4179 Toutatis; note also it rapid motion in less han 10 minutes as it speeds across the sky at opposition.

The jury is still out, and likely will be for some time to come, as to what this "material" that makes up the comets or asteroids really is. Among the interplanetary matter are small and not-so-small objects. Some of it, indeed most of it, are objects of little regular shape, the majority appearing as chunks of either stony or icy materials. All, just like the planets, orbit the sun - but for the most part each of their orbits are very distinct from anything else out there. For example, the comets seem to be large blocks of frozen gases mixed with particles of cosmic dust, while the asteroids - or "minor planets" - are hard and often metallic bodies pocked with craters. With the February, 2001 "NEAR-Shoemaker" satellite surface landing onto the asteroid EROS, it was confirmed that this little minor planet is "potato-shaped" and has a large crater one-third from one end of it.....a shape AND a characteristic (the crater) that was mapped out nearly 25 years ago by AMATEUR ASTRONOMERS using small telescopes to record ever-so-slight variations in reflected sunlight as the elongated minor planet turned on its axis.

A History of The Asteroids, or Minor Planets

Just beyond the orbit of Mars, about midway to that of Jupiter is a fascinating "belt" of debris, predicted over 250 years ago in a mathematical progression known as "*Bode's Law.*" In this prediction, it was held that there SHOULD be a planet somewhere between Mars and Jupiter, or about 300 million miles away, three times more distant from the sun than is Earth.

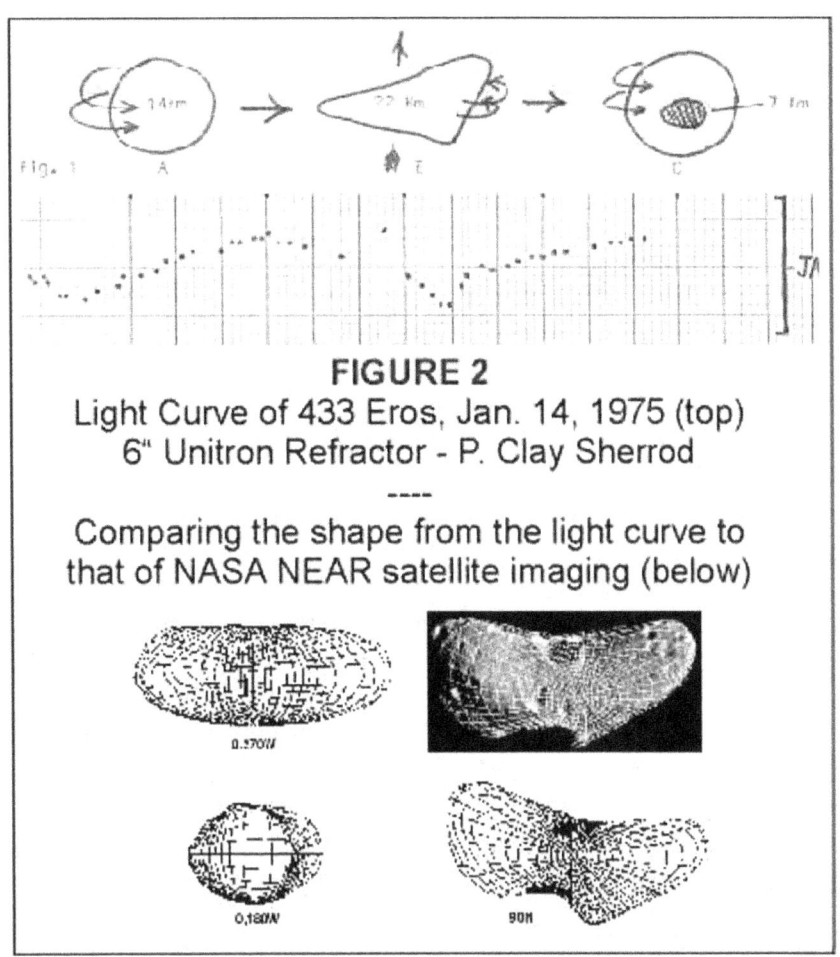

FIGURE 2
Light Curve of 433 Eros, Jan. 14, 1975 (top)
6" Unitron Refractor - P. Clay Sherrod

Comparing the shape from the light curve to that of NASA NEAR satellite imaging (below)

Note that the "light curve" as plotted about every 10 to 20 minutes refers to aspects of the asteroid as drawn from over 2,000 observations with a 6" telescope. Note how remarkably close the deduced shape (even to the large crater) was 25 years before the satellite actually took the NASA images! Smaller telescopes CAN make a contribution to science!

In "*Bode's Law*" there was also room for ANOTHER "missing" planet at that time, one that the mathematical progression deemed should be just outside of the orbit of Saturn, the most distant planet know in the 17th century.

Indeed, the discovery of Uranus in 1871 by **William Herschel** showed that a planet did exist just where Bode's progression said this trans-Saturnian planet should be - 1.8 billion miles from the sun!

Soon after - on the very first night of 1801 - the Sicilian astronomer **Giuseppe Piazzi** located a "star" that appeared telescopically to change its position against other stars. Later, it was confirmed that this object "Minor Planet 1" - the asteroid *CERES* (appropriately named by others, meaning "*goddess of Sicily*") - was orbiting exactly where the other and closer "missing" planet should have been as predicted by Bode's law - some 300 million miles from the sun, between the orbits of Mars and Jupiter.

Within the next six years, the largest of all the minor planets had been discovered: *Ceres* (1801), *Pallas* (1802), *Juno* (1804) and *Vesta* (1807).

Incredibly, because the asteroids are so small, and thus very faint, it was not until 1845 that the fifth minor planet was discovered. In 1970 over 300 were known; today, there are over 60,000 documented!

Unfortunately for the naked eye observer, the asteroids cannot be seen. Careful planning and plotting on star charts will reveal the brightest four in a pair of 10 x 50 binoculars, but only very faintly. Even though the aforementioned four are the largest asteroids, they are still tiny when considering that we are viewing them from a distance of over 200 million miles!

Looking at the sizes, consider that mighty Jupiter, with its diameter of over 87,000 miles is seemingly small in telescopes. Thus, the minor planets - even with the largest telescopes on Earth - are but pinpoints of light and cannot be magnified to exhibit a "disk" of their irregular bodies. The sizes of the 9 largest are given in Figure 3, below as well as their BRIGHTEST ATTAINABLE magnitude, usually during "opposition," when the minor planet is opposite the sun as seen from Earth. Although some can reach naked-eye brightness, picking them out amidst the thousands of tiny stars on a moonless night is very difficult....hence, the computer can show you the way!

FIGURE 3
The Ten Largest Minor Planets

MINOR PLANET	SIZE (miles)	MAGNITUDE
Ceres	352	4.0 (maximum)
Pallas	217	5.1
Juno	61	6.3
Vesta	177	4.2
Astraea	72	6.6
Iris	68	677
Flora	-	9.0
Metis	-	9.8
Hygiea	79	6.4

Orbiting spacecraft as well as the Hubble Telescope have been able to obtain close-pass images that reveal highly non-uniform shapes and substantial cratering from impacts, perhaps of meteors or even other asteroids in the asteroid belt as well as seemingly random *"Near Earth Objects"* (NEOs) that can come dangerously close to Earth on some passes, past and future. At present there are now over 50,000 Near Earth Objects known.

Being outside the Earth's orbit, like superior planets, the asteroids orbit with similar characteristics; nonetheless, the asteroid belt is very wide and the motions of the bodies within it are regulated by their masses and relative distances from the sun. Also perturbations - one object pulling on another through its forces of gravity - force many of the asteroids to vary in the orbits from time to time as the small planets are affected by the gravity of other passing bodies, particularly Jupiter. Hence, there is a HUGE effort now by amateur astronomers to monitor NEOs regularly to refine their orbits and chances of posing danger to our Earth.

Like all superior planets, the minor planets exhibit retrograde motion. As might be expected, the average passing across the sky (its "year") of an asteroid takes longer than does Mars', but not so long as Jupiter.

Scientific Studies of Minor Planets: NEAR EARTH OBJECTS:

A major contribution to the science of astronomy is now being done by many non-professional astronomers in the studies and reports of both ASTROMETRY (extremely accurate orbital postion measuring) as well as PHOTOMETRY (precise computer measures of brightness changes, the results of spinning or multiplicity of an asteroid) of Near Earth Objects. For a complete GUIDE to understanding how YOU can contribute to this science, please refer to the excellent "How To" guide at the Minor Planet Center: http://www.minorplanetcenter.net/iau/info/Astrometry.html
similarly, COMETS (following) are a very serious effort today by non-professional astronomers using modest equipment and the reports of observations are discussed at that same link.

I urge you to seriously consider working toward making YOUR observing location a Minor Planet Center Observing Station, complete with your own earned "ObsCode", where you can confirm, monitor, and study the motions and changes in this very serious and requested contribution to the science of astronomy. Your observations can make a huge difference.

All that is needed is an accurate GO TO computerized telescope, a telescope program such as *Astrometrica*, TheSky, *Guide* (Charon) or similar program, and a strong work ethic to monitor dozens or hundreds of dangerous asteroids and comets annually. NOTE that the very useful "**MPEC Sort**" program (tab at Home Page of this website) allows users to custom-sort the enormous list of NEOs and comets every day to update for Right Ascension, Declination, magnitude, NEO type, and location, all filtered for YOUR desired listing of target objects. For example, you can download and display all NEOs/comets in the span of right ascension 07 hours to 12 hours, with a magnitude threshold of "19". and higher than declination say, "-10 deg." for any specific day. All current targets for those parameters will be automatically downloaded and sorted for you. The **MPEC Sort** link is found on our ASO Home Page and is free for all observers. The link above, tells you exactly how to qualify for your ObsCode (ASO has three: H41, H43 and H45) and begin making observations.

At *Arkansas Sky Observatories*, thousands of such observations of NEOs and comets are made yearly and have been since 1971, now numbering over 100,000 submitted observations, all archived and recorded at ASO.

The Mysterious Comets –

There was a time, even less than 200 years ago in many cultures, that the sudden appearance and passing of a bright comet foretold impending doom or disaster. William Shakespeare wrote of comets and their appearances heralding the deaths of rulers, and other historians attributed bright comets in the sky to such dramatic events as: the fall of Constantinople, the miscarriage of pregnant cattle, and even hens laying strange eggs!

Part of the fear of comets is the UNKNOWN - their unpredictable nature. Before they were fully understood - at least as to their motions in the sky - they suddenly just seemed to "appear" like ghosts hovering in dark skies. They did not behave like the "planetes," or the "seven wandering stars" (Mercury, Venus, Moon, Mars, Jupiter, Saturn and the sun) moving predictably through the field of stars, and their motions many times could be seen within a night's time as they rapidly approached and receded. Surely something so mysterious and unpredictable has to be the work of the Devil!

Comet Hale-Bopp – April 30, 1997 by the author.
ASA 400 Ektachrome film @ 4 minutes, 80mm Canon Lens

Today it's easy to brush off the "comet scare" of times gone buy.....until you see a bright one - a really big, bright one. Even to those who have seen comets come and go, and even now that we are better able to understand when and why they appear, bright comets strike awe in the hearts of even the most seasoned stargazers.

On a dark night, with no moon in the sky, and the blanket of stars so deep that the motionless heavens seem three-dimensional, the suspension of a bright, ghostly and silent partner among those stars is a sight to behold.

So, some comets come rapidly and brush by the Earth so close that they seem to stretch across our dark skies, only to be gone in a few days. Normally those comets are sungrazers, interlopers that either have not come by the sun before, or some that approach in prehistoric antiquity. As such sungrazers are pulled into the sun, the pass remarkably close to the sun, and hence sometimes the Earth. Indeed, comets like Ikeya-Seki in 1965 skimmed through the nuclear fiery atmosphere of the sun, arcing around it in fierce speed, and exploding to such brightness that it could be seen at high noon in broad daylight!

But not all comets are bright, nor so large.... not so spectacular. On ANY given night, there are literally hundreds of comets strewn about the sky, most in sight of only the world's largest telescopes. BUT - there are also hundreds of comets visible in YOUR telescope with the aid of the computer comet directory! They are easy to GO TO and fun to observe!

Such comets that are cataloged in the PC programs are the periodic comets, those which have been by the sun previously and are now locked into a closed orbit, highly elliptical (the flattened circle, like Mars), coming back around every set number of years. Halley's comet, the famous comet predicted by **Edmund Halley** is a comet that, in earliest times, brought fear and repentance among all men. Being a periodic comet, Halley's returns to the vicinity of the sun every 76 years or so, close enough that the sun's energy can warm the frozen gases and vaporize enough for naked eye view of its coma (the bright "head" of a comet) and tail (a stream of bright ionized gases and dust that are pushed away from the comet's head). In the preceding photograph, you can clearly see the beautiful development within the tail of comet 73P on May 11, 2006, taken by the author.

Yet, like the automobile running out of gas, it appears that Halley's comet is literally "running out of steam." During its last pass, the sun's energy seemed to break through the icy barrier and break apart large chunks of the estimated 10-mile wide "dirty iceball." Astronomers with large telescopes noted in April, 1986 that several nuclei (the nucleus of a comet is the bright star-like point of light that is the comet itself, usually seen central amidst the comet's head (see Figure 6).

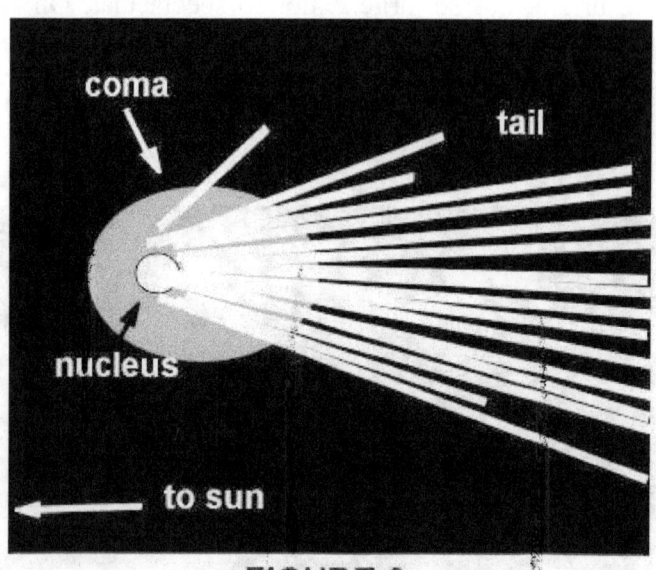

FIGURE 6
Graphic demonstrating the many features
of a bright comet. Note: tail is opposite sun;
Note: tail streamers at 1:00 & 2:00 positions
are dust trailing tails (anti-tail).

PERIODIC COMETS - First we will discuss the motions of the periodic comets. As mentioned all comets ultimately - once passing the gravitational pull of the sun and being "perturbed" into an orbit (Figure 7) - are "periodic." Some have regular and predictable orbits which bring them by on a relatively short basis.

One of the most famous periodic comets is *Comet Encke* with an orbital period around the sun of only 3.3 years, compared to Halley's Comet at 76 years. Once a very spectacular and bright object, Encke's comet, at every pass by the sun (perihelion), was gravitationally tugged into a less and less elliptical orbit and its outer limit (aphelion) gradually grew closer and closer to the sun until it was locked into nearly a circular orbit of only 3.3 years.

There are thousands of periodic comets throughout the sky and each year a few of them pass close enough to the sun and Earth that they become easily visible in small telescopes, binoculars or even the naked eye. Observing hyperbolic comets offers the most spectacular and unexpected rewards within the comet family.

HYPERBOLIC COMETS - The most spectacular comets, with few exceptions, are the newly discovered "hyperbolic comets," objects that enter our solar system and appear to make their first treks around the sun. Recent observations of such comets reveal that - once passing the sun - they do not "escape" and disappear forever away from its gravitational forces. Hence, the name is somewhat outdated but sticks nonetheless.

Essentially, once passing perihelion, such comets become extremely lengthy periodic comets, and indeed will return at some future data. It is important to note here the obvious difference in the motions of the short period comets and these relatively new interlopers. There is actually reason to suspect that there are truly NO "hyperbolic orbits" to any comets.

If you are fortunate to see one of these "hyperbolic comets" it is likely the first time it has been seen by mankind and likely will not be seen again in your lifetime. They are the ones referred to by ancient historians as causing great calamities and the miscarriages of pregnant cattle. As well, they show up very unexpected for the unprepared. Prior to the invention and use of the modern telescope, comets went totally unnoticed until they were so bright as to call immediate attention by the lay public. Many times the suddenness of their appearance would cause panic and terror, as many of these comets attain tails stretching from the horizon to nearly overhead, and many take on mysterious shapes, appendages and unusual characteristics.

As we have seen the periodic comets are not so bright, so it are the new comets to which we refer when speaking of "spectacular" comets. And there is ample reason that they should be most spectacular.

A comet approaching the sun for the first time enters from space, perhaps at a distance of 12 trillion miles, being nudged by some gravitational force by another object and "pushed" into a very slow (at first) trip into the center of the solar system.

This "*fresh comet*" is packed full of frozen gases and dusty particles left from the actual creation of our solar system, well over 5 billion years ago. Until reaching about the distance of the orbit of the planet Mars, or about 60 to 100 million miles from Earth, the new comet remains for the most part unchanged, slowly evaporating its outer layers as the effect of sunlight and solar wind gently warm this hitherto frozen and unexposed object.

Today, many new comets are discovered as much as a year or two prior to becoming a public spectacle. At even this early discovery, the comet may only be a few hundred million miles distant, as opposed to its starting point nearly one-third the distance to our nearest star, Alpha Centauri.

The trip from its "birthplace" to our region where we might first glimpse the comet has taken millions of years. Comets are traveling at a snail's pace when they leave the Orrt Cloud; by the time they reach perihelion, many are speeding by in excess of 25 miles per second.

At such a fast pace it is easy to understand why the comets seem to suddenly appear, change shape and brightness radically and as quickly....disappear.

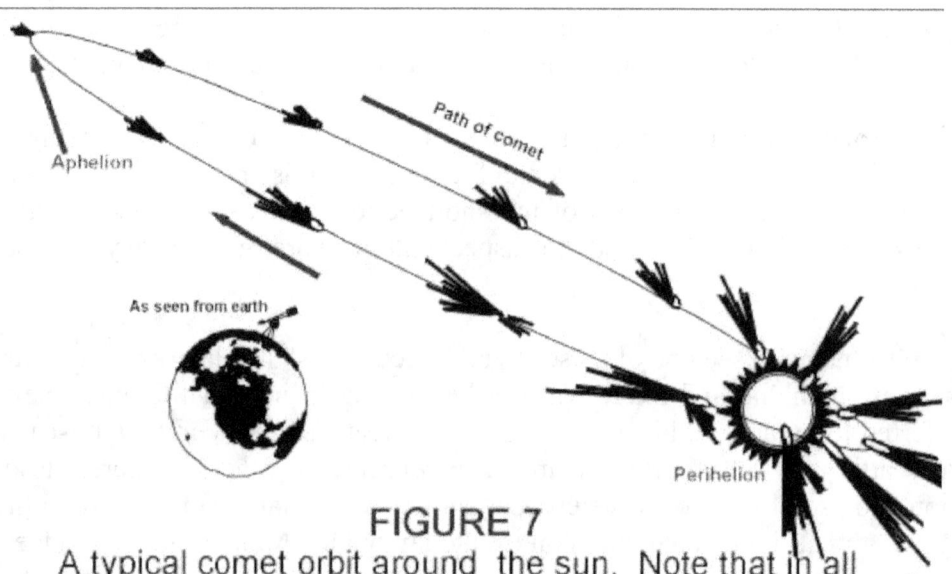

FIGURE 7
A typical comet orbit around the sun. Note that in all positions the main "tail" of the comet is pushed AWAY from the sun, rather than "trailing" the comet!

For the casual observer viewing a very bright comet is much like observing Venus or Mercury, although there are some exceptions. When the comet is close enough to become naked eye, it obviously is very close to the sun, the sun's energy interacting with the volatile materials of the comet, causing ionization of gases (like a glowing fluorescent light) and rapid melting of hence frozen dust and solid particles which stream away from the center of the comet away from the sun. Notice in Figure 7, how the "tail" seems to follow the comet as it approaches the sun; however, after passing its closest approach to the sun ("PERIHELION") the tail now clearly "leads" the comet, being propelled away by the force of solar wind.

During such close approach the long and famous "tail" develops and is pushed by the solar wind away from the sun like the dust particles. Only the heaviest particles (see actually "follow" the comet in its orbital path as shown with the photograph of Halley's Comet in April of 1986 after it had passed the sun (Figure 8). In this photograph you can clearly see the dust tail (from a "clock position of 12:00 to about 2:00") making a fan-shape of the main tail (from comet to upper right).

FIGURE 8
Halley's Comet after Perihelion April 16, 1986
Tri-X @ ASA2300 Magnitude = 1.7
7" f/2.5 Aero-Ektar

OBSERVING THE COMETS WITH YOUR TELESCOPE AND NAKED EYE –

Because the comet is so close to the sun, it will usually be seen best either before sunrise in the eastern sky, or after sunset in the west. Since the very brightest comets are those which pass the closest to the sun (thereby more heat and energy), it stands to reason that the very close ones must be viewed very close to dusk in the evening or dawn in the morning.

As the comet moves farther away from the sun after perihelion, it will rise higher and higher in the sky when compared to the sun, but its brightness from Earth also diminishes for two reasons:

1) the comet is getting less and less energy from the sun, and therefore is not as luminescent as when receiving maximum energy;

2) as the comet recedes, its brightness decreases proportionally to its distance. Thus, when a bright comet appears, the best words of advice are: "see it while you can and as often as you can! Indeed, the comets afford a different "show" every night, with

many gas jets erupting as they pass closely to the sun; the tail grows and diminishes as the comet approaches and then recedes from the sun.

Watching a comet over its brief appearance is a rewarding experience to share. For starters:

1) note (plot if you like) the changing POSITION of the "tail" relative to the comet's head; comparing your plots with a star chart will reveal that the tail does indeed point "away" from the sun! Remember that the tail is being PUSHED by the solar wind, not actually FOLLOWING the comet; only the heavier dust grains follow in the comet's path, and those are seen by REFLECTED SUNLIGHT (see Halley's Comet Figure 8).

2) compare the magnitude of the head of the comet with bright stars in the sky, either with your telescope if the comet is reasonably faint, or with binoculars and the naked eye; if using binoculars or a telescope, try "defocusing" the image in the optics to make the same size as the comet is seen IN FOCUS and this will give you a more accurate magnitude comparison. The HEAD, or COMA of the comet can be as large as the distance from the Earth to the moon, or greater than 250,000 miles across. The coma can be seen because of the IONIZATION and FLUORESCENCE of the rare gases in the nucleus, a result of exposure to the sun's energy.

3) look nightly for the bright NUCLEUS in your telescope; some times, as in the last pass of the famous HALLEY'S COMET, the tiny nucleus can split apart from solar radiation and become a multiple fragmented body! The nucleus of a comet is a very small object, when you consider the coma may be over a quarter-million miles across! Indeed, even the largest nuclei of comets range no greater than 15-20 miles across (Figure 6).

4) under very dark skies examine carefully for the "anti-tail", the tail which can be seen actually FOLLOWING THE COMET in its path; this anti-tail is actually heavier particles of dust and grit that the comet is leaving behind in its wake, unlike the bright gas and luminous cloud which is repelled by the solar wind. Usually, the best time to pick up the "anti-tail" is immediately after the comet passes PERIHELION, or closest pass to the sun, as seen in Figure 7).

PIGGYBACK PHOTOGRAPHY OF COMETS –

The very best photographs of comets are usually taken by "piggyback photography" whereby a camera "rides" atop a tracking telescope; the camera with a standard lens can cover wide expanses of sky; with a bright comet (see Hale-Bopp, Figure 9) exposures of only 5 to 60 seconds with modern CCD and digital DSLR cameras are required! Faint comets to magnitude 20 can be captured via CCD with amateur telescopes in 120 sec. or less.

Beautiful photographs such as the one above can be obtained easily with a piggybacked camera on any telescope, or with very short exposures via CCD (to keep star images from trailing as the Earth turns). Use modern DSLR cameras 400 and expose for only about 30 seconds with camera lens using a "cable release" to keep the camera's shutter open. With piggyback photography, lenses of all focal lengths (see Figure 8, taken with a 7" piggybacked lens) can be used for very long exposures while the telescope actually tracks for the motion of the Earth and comet and the camera/lens combination does all the work!

There is much literature available on the physical nature of comets and the remarkable detail that can be seen within them. Using any optical aid will enhance the views of the inner secrets of the comet.....outgassing of geyser-like spews of rare gases ("jets" that emanate from the nucleus and curve gracefully into the coma.....faint "tail components", extensions from the nucleus that combine into one large bright tail...and sometimes a nucleus that divides, or erupts in brightness unexpectedly.

All await your yearning for discovery in your telescope. Yet, the true splendor of these as-yet still mysterious ghost-like bodies can best be seen with the unaided eye on a dark night as they hover silently over the eastern horizon at dawn or the western horizon at sunset. They follow the sun like sentries that soon will fall into the shadows of the great star of our solar system.

* * *

Component "B" of Schwassmann-Wachmann 73P - May 4 02:49 UT - ASO-H45 / 0.41m RC @ f/3.3, Kodak KAF 0402me CCD
Field = 17.8 x 12', 1x60s / South is Up / P. Clay Sherrod

Chapter 32

Comet Magnitude Estimates via CCD Imaging
Assuring consistency and accuracy with saturated vs. unsaturated images

Although this comet appears very bright and total brightness (m1) was estimated at mag. 7.1 on this date, CCD measurements of the exact nuclear magnitude (m2) revealed that the actual core brightness was 11.2 via CCD estimates

Although there are visual observers in the ranks that do very quality work on comets the magnitude estimates for publication (for MPC or other agencies who serve as repositories): http://www.minorplanetcenter.net/iau/lists/LastCometObs.html
are subject to very high variability and inaccuracies when comparing estimates for identical comets made under similar conditions.

It is very difficult, if not impossible to get comparable and reliably research quality estimates from a visual observer next to one obtain via CCD.

Several parameters come into play:

1) the experience of the visual observer;
2) the size of the telescope and type of instrument used visually
3) the length exposure (saturation) with the CCD observer
4) the quantum efficiency of the CCD camera/sensor being used for the estimate.

But there should be consistency and there CAN be in today's modern equipment between observers using CCD for photometric measurements.

I follow a variable aperture technique for all comets, including many that appear to be stellar; many years nearly 70% of my comet analysis is on comets which exhibit no nebulosity (faint comets) and thus reduction of magnitude many times can be similar to that of measuring stellar, provided that a reference index star is always used for proper adjustment and calibration (this is done here at H45).

However on comets with an extended coma the aperture used to do the actual densitometry reduction MUST be used consistently in proportion to the overall coma size of the comet.

Put simply: the larger the comet, the larger the CCD "aperture frame" must be to accept the overall diameter of the comet's coma, or head. Using the old method of assigning a measuring aperture of "11" or 13" is simply not practical.

There is a huge difference in result, for example, in the measurement of a tiny comet at threshold (say, mag. 19.1 which clearly shows NO coma nor nebulosity) and one that extends a huge coma with considerable scattering. For the faint comet, and aperture of "13" is far too large, allowing sky brightness to fill the field being photometrically measured, whereas the "13" size aperture setting for the photo above would lead to eliminating more than 50% of its overall size and outer brightness.

A very small aperture (I believe that "7" is appropriate for stellar comets, while a variable aperture for large comets depending on overall coma size) is needed for faint stellar comets and an appropriately larger aperture for large comets.

By "aperture" I am not referring to the telescope aperture - rather the aperture of the computer program IRIS being used to measure brightness; all measuring CCD programs have this and the size of aperture is variable and can be changed from 3" to larger than 27"

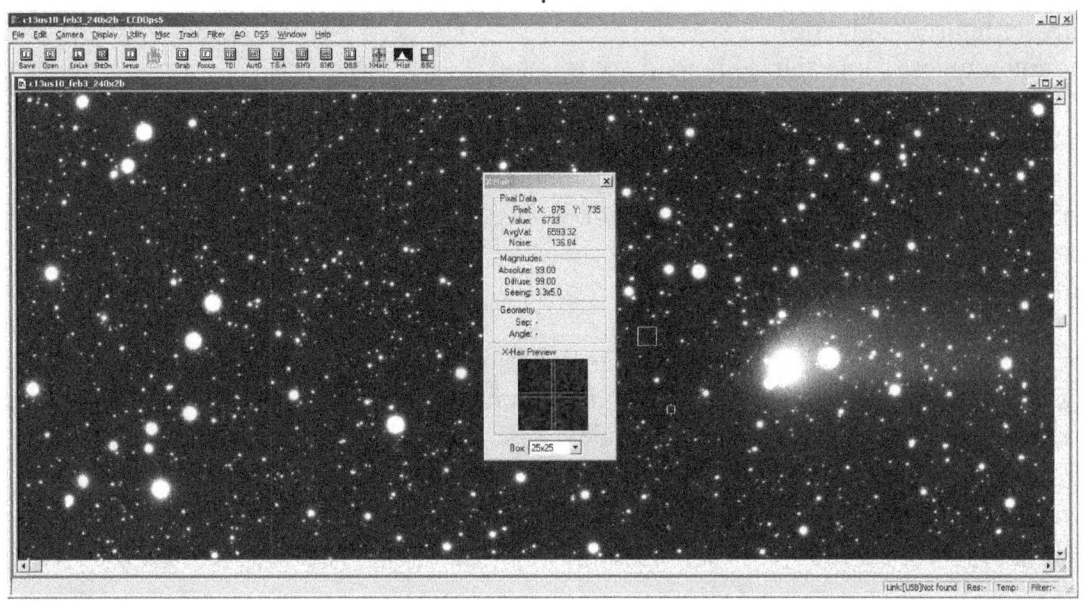

Above: From a screen shot of the same comet on the same date, here is the CCDops Crosshairs program for precise measuring of the comet's magnitude. *Note the small white box just to the right of the Crosshair window; that is set to aperture "25 x 25" (see selection just below red crosshairs); that is the IRIS setting for the box and you can clearly see that this aperture is still too small for the overall Coma size of the bright comet to its right. My measurements are made using the same box, with size to FIT the brighter stars that I might pick to the left and their magnitudes adjusted to coincide with the* UCAC-4 U-Mass *star catalog; once that calibration is done, the box is moved to the COMET and size adjusted (IRIS) to fit the comet's coma. The magnitude is provided via the program once that size is selected. CCDops allows the box to move with the user's cursor.*

* * *

The assignment of aperture is relatively subjective at this point, based on the overall witnessed extent of the coma from a resulting CCD image of good saturation. I strive to include the "major extent" of a coma rather than the "total coma" since on many, many comets (28p, Pons-Gambart, etc.) you have actually two gradients of coma: outer and inner. If the inner coma is distinctly segregated from a large and very veiled outer coma, that inner coma ONLY is used and the aperture on the densitometer (digital) is adjusted for that size, NOT the overall larger extended coma.

Keep in mind **one primary rule** about selecting IRIS diameter on your measuring program:
- For TOTAL (m1) the IRIS setting must be large enough to cover the entire saturated image of the COMA of the comet, but not so large as to let the SKY BACKGROUND be

an influence on the final reading. Hence, for a tiny stellar 19th magnitude comet, the IRIS setting should be near
minimum unless some sign of nebulosity is noted.

- For **NUCLEAR** (m2) magnitudes, you want to isolate the relative brightness of just the actual hard component, nucleus, of the comet; hence stopping down the IRIS setting to a standard "13" is recommended, unless the comet has no nebulosity whatsoever....in that case
the aperture should be minimal.

The reason for this is that if we use the entire extended outer coma, we are beginning to change our relative measurements of sky background vs coma intensity and the results become muddy. Perhaps years of practice of doing this (first by hand and now by computer) have resulted in me being able to quickly assess exactly where the aperture of the densitometer needs to be set for each particular comet.

Once that aperture is set, I use that size to measure a base value of my background sky, the move the aperture to a known stable star on the UCAC-4 catalog and calibrate the given magnitude accordingly; once set, the aperture is then moved and centered over the extended inner coma (or just coma if it is condensed and very well delineated) and my value is obtained accordingly.

This method has proved incredibly consistent and accurate; I simply do not know what all the fuss and error is about among comet observers. The method for CCD should be quite simple and consistent if everyone would measure this way. In fact, one of the most common CCD camera operating systems, CCDops by SBIG, provides somewhat this very concept on their "Crosshairs" measuring device. It is a well kept secret for some reason, but their program is amazingly quick and quite accurate when comparing with my custom programs that I use.

The trick is to use the system consistently for all comets and for all observations. I have found in my 50 years in astronomy that there are many "wolf-criers" out there who like to see their names in print and in the Internet circles. Many of these are professional astronomers and many of them are amateurs held in high regard. Some of these folks will semi-intentionally put out values that get our attention on a comet, simply so that their observations will cause a knee-jerk reaction. Such is the case among the horribly scattered data from which assessments such as comet **c2013 US 10** (above) were being made.

Chapter 33

Observing the Companion Star to the Brightest Star: *SIRIUS*

In our telescopes large and small the bright star **Sirius** is a breathtaking sight, filling the eyepiece with its brilliant bluish light, overshadowing all else which surrounds it. Indeed, the star field encircling the brightest star of the heavens is one of the prettiest that can be found in low power telescopic viewing. This star field comprises just a small part of the beautifully delicate winter Milky Way, with brilliant Sirius, at magnitude -1.58, serving as a beacon toward the southern winter sky.

Sirius is so bright in a medium power telescopic field of view - indeed, NINE TIMES brighter than a "regular" first magnitude star - that it is briefly a rather painful sight to behold. It is one of the few stars (or even objects, for that matter!) that can be easily seen in daylight, visible in any telescope at low power. Any computerized telescope can GO TO this object by merely "dummy-aligning" from near north (pressing enter at each alignment star instead of centering, of course) and then doing a "GO TO" under "Star" to Sirius.

Sirius is our fifth closest star, only *8.7 light years* from our sun and slightly more luminous; in fact as recently as only 2000 years ago, there are historic Chinese and Greek references to Babylonian cuneiform text telling of Sirius as "*the ruddy [reddish] star*". Even as recently as 140 AD, the Greek philosopher Ptolemy grouped Sirius among all of the "*fiery red*" stars of the northern hemisphere.

Today we know it as brilliant blue-white, and we also have known since 1862 by the discovery of famous telescope-maker **Alvan Clark**, that it has a very, very close companion, about 5 arc seconds distant (during recent times) from the brilliant primary star. At magnitude 8.6, and a full 5" arc distant.....WHY have more skilled observers NOT seen this interesting "double star?"

A TOUGH DOUBLE TO SPLIT

I have known of - and seen many times - the duplicity of Sirius since 1966. Recently, it's one of those astronomical statistics/facts that I had just stored away and forgotten until a keen ETX 125 user posted an inquiry on the *Weasner Mighty ETX Site* perplexed because he could not see the companion.

And perplexed for all the right reasons...he was RIGHT: after all! He had a 5" aperture telescope with the finest of optics (a 3" telescope should be able to see it, according to "resolution" and "light gathering" and all those empirical values that we hold so dear) and he seemingly was doing "everything right."

So WHY could he not see it? A star of magnitude 8.6 so far from its "mother star" should be seen in a 2" target spotting scope (you would think).

In 1993 to 1995 the star was only about 2.5" arc from Sirius and almost invisible to all telescopes (even though, on a good night at medium power the 3-inch could easily separate any OTHER star of that separation!). At that time I would have advised observers to "forget trying to split Sirius" and worry about some other celestial challenge. But now, things are changing; for the next three decades, beginning after about 1999, the faint companion will be moving (as seen from Earth) farther and farther away from the bright primary in its 49.9-year orbit around Sirius.

So the time is right....we have telescopes that SHOULD be able to resolve this star.....let's go look at "**Sirius B**", the orbiting companion star to its more famous big brother, "**Sirius A**".

SPECIAL OBSERVING TECHNIQUES

You do not need a special, or even a larger, telescope to view Sirius B for the coming years....what you need is "technique." The problem in observing this double is the overwhelming brightness (and glare in the optics you use) of Sirius A. It literally FILLS the eyepiece field of view with brightness, some of it scattered through the Earth's air through which you are looking, and a lot of it scattered within the optical elements of your telescope and its eyepiece.

Most of you are familiar with my "star test" techniques in which I emphasize that the "Airy Rings" - faint concentric rings of light which seem to encircle a bright star - indicate excellence in both optical quality and alignment.

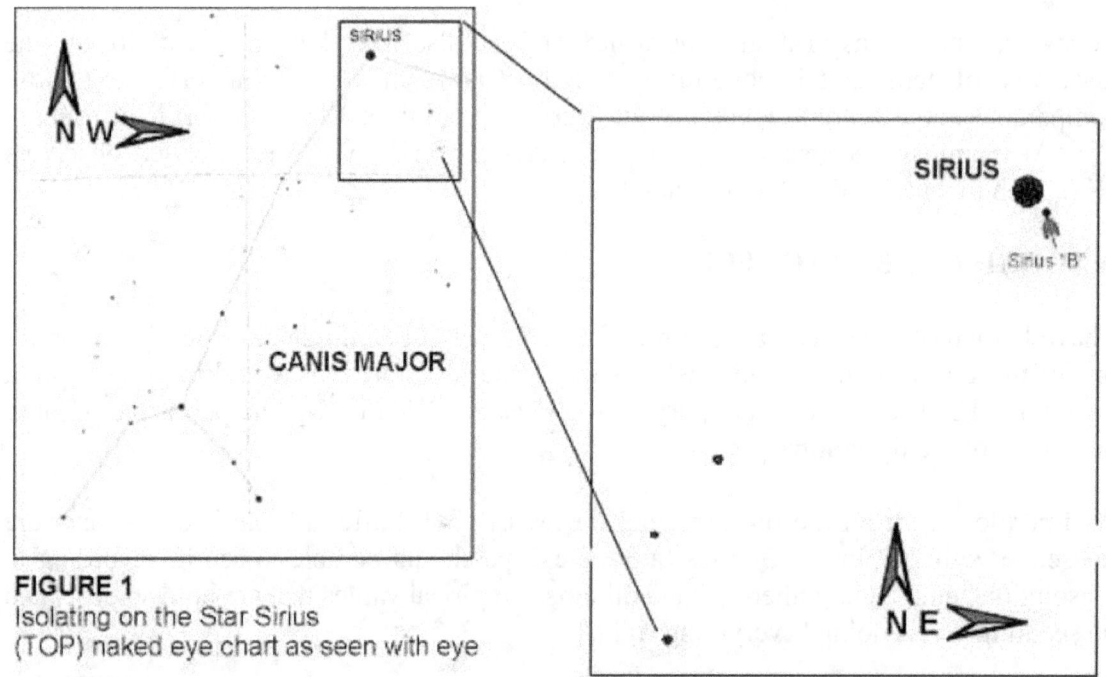

FIGURE 1
Isolating on the Star Sirius
(TOP) naked eye chart as seen with eye
(RIGHT) Reversed chart as seen with ETX or LX with medium or high power

However, when attempting to separate bright, and close, double stars, the Airy Rings can also "get in the way" and that is exactly what happens when attempting to view the elusive Sirius B. Note in Figure 1 (TOP) the naked-eye chart indicating Sirius and a little triplet of stars to its west (lower right); now move to the chart on the right which is "a closer view" and REVERSED, like images appear in your catadioptic telescope with the right angle prism for the eyepiece. You will see the same three (3) reference stars in both Figure 1 diagrams, but those in the reversed field will be a mirror image of what the naked eye sees. This has been included to show RELATIVE to Sirius A where the faint companion is located.

Don't get any ideas that it is this "easy" to see. Not by a long shot.

Now, look at Figure 2, showing a "telescopic" appearance that you might see of Sirius IF there were no diffraction rings, scattered light nor brightness from bright Sirius. Notice the current position of Sirius B (denoted by the YEARS of each position in its orbit) relative to Sirius. This field would be a VERY high (i.e., 80x to 100x per inch aperture) magnification and only usable on the steadiest of all nights.

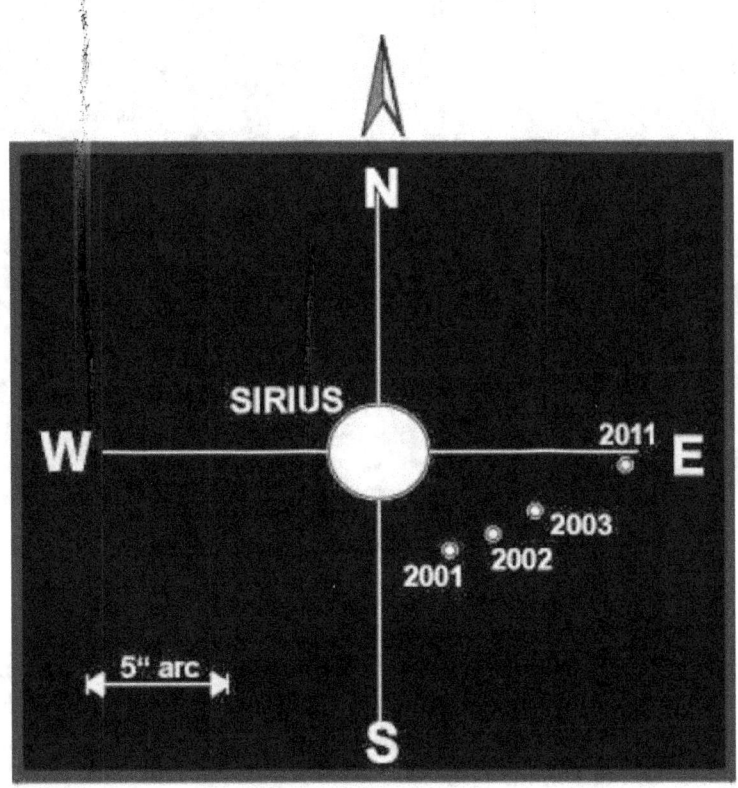

FIGURE 2
Sirius "B" in Relation to Sirius "A"

Indeed, it is important to note here that ONLY THE PERFECTLY STEADY NIGHTS when you can "defocus" Sirius and observe its broad bright Airy "disk" to have NO perceptible motion whatsoever, can you expect to see the fainter companion star. It takes, quite simply, the most steady nights you have ever witnessed; hazy nights with very high clouds in late winter are many times very steady.

Figure 3, is like Figure 2, except all of the dates, titles and N,S,E,W cardinal axes have been removed as it might look under very high magnification. Have you learned WHERE the star is now? Chances are, you STILL cannot find it....there are likely diffraction rings covering it up, or there is so much scattered light from Sirius A that you might be STARING RIGHT AT IT, and never see it! It happens all the time.

It's there....you just can't see it.

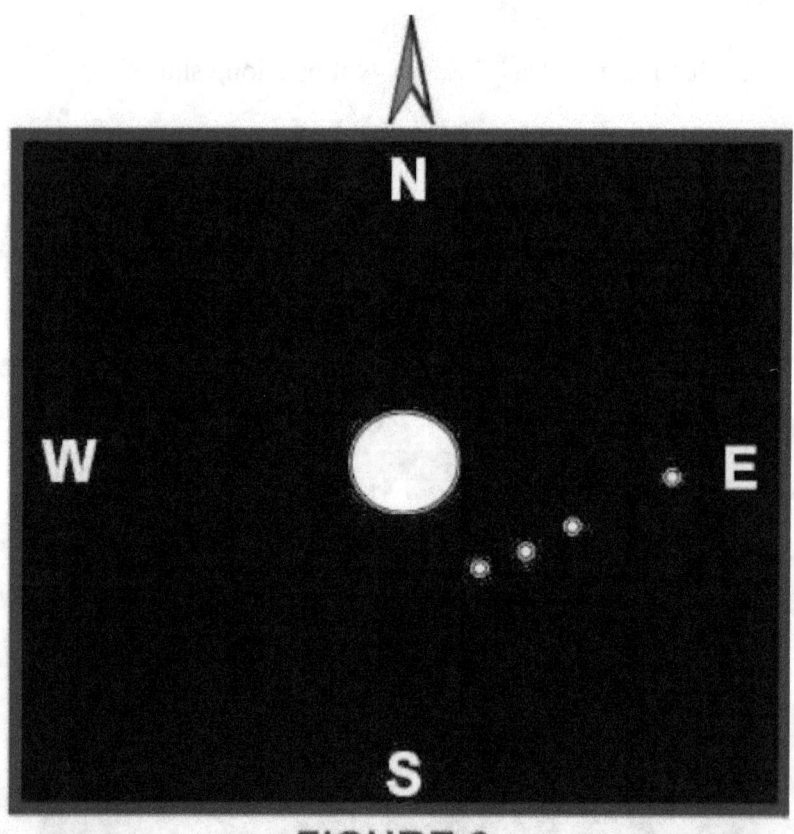

FIGURE 3
Sirius "B" seen in very high magnification.

Now....Figure 4 is the same field of view, but notice what I have done to eliminate all the glare...the diffraction rings...the light scatter: **SIRIUS HAS BEEN MOVED OUT** of the field of view on the edge OPPOSITE the direction of the tiny companion star! (and you're saying, "....gosh, why didn't I think of that!?").

It's a good and helpful idea, but it is very tricky and does not always work. There are two keys to using this "trick" to spot Sirius B:

1) the air must be absolutely perfectly steady (no star twinkling even on the horizons) and Sirius must be nearly on the Meridian (as much overhead as it's going to get);

2) you must use as MUCH magnification as your telescope can "hold;" I know this goes against everything I have preached on every other subject, but not on this one. It takes

POWER to see Sirius' elusive companion - and lots of it. Use Sirius as a guide on your magnification limits. Turn the star just slightly out-of-focus until the bright star becomes an extended disk with a few rings within it; if the overall image is moving ("wiggling") then forget it, you're not going to see anything but bright blue Sirius A. If, on the other hand, the unfocused disk is steady with no motion....well, you have a "shot."

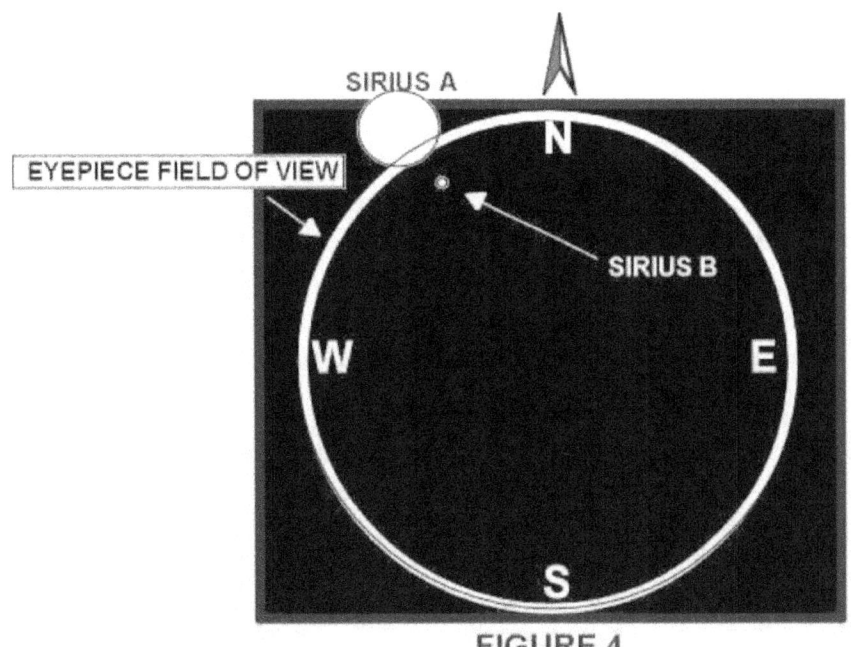

FIGURE 4
Move "Sirius A" out of view to see "Sirius B"

Using your highest practical magnification (where your star images are still faint points of light and not scattered - if they are "fuzzy", back down on magnification until they are points again - slowly move Sirius up and to the slight left (North-West) until it suddenly touches the NW edge of your eyepiece field.

Gently, either manually or with your electronic controller, tap the telescope until bright Sirius A is SUDDENLY just right over the edge and not visible (you will still see its bright light very vividly however); just to the lower right (SE) of where Sirius is "amost there", and very, very close to the edge of your field of view next to Sirius you should barely see Sirius B in the position marked on my accompanying chart (Figures 2 and 3).

You have "occulted" the bright star by the edge of the eyepiece, eliminating most of its glaring light that prevents Sirius B from being seen. You will probably have to repeat this technique time and time again within a short period of time until you FINALLY can actually spot the companion star.

But it CAN be done, and YOU can do it. You have the telescope, you know where to look, and now you know HOW to look for the star. If you don't see it tonight, just remember.....the conditions just keep getting more and more favorable for the next 25 years! By then, like me in 1966, you'll be a master and conquer this giant of all our celestial splendors!

Chapter 34

Observing the Variable "Demon Star" Algol

Using your telescope to for the "Minimum of Algol" to Start You on your Way to Observing Variable Stars

Scrolling and then "entering," and slewing through your computerized telescope keypad, you no doubt have come across the prompt under "Event" entitled "*Minimum of Algol*."

We all know that Algol is a bright variable star (*beta Persei*) in the constellation of Perseus. Perseus is "circumpolar", or tightly placed relative to the north celestial pole so that it circles nightly, monthly and even yearly in such a way that it never "rises or sets", but rather seems to be "tethered" in a neat little circular path about Polaris, the north star.

This allows its second-brightest star, ALGOL, or "*Demon Star*" to be visible throughout the course of a night and even throughout the year. Hence, it is an ideal subject to begin your adventure into the observations of VARIABLE STARS: stars that are not fixed in brightness, but rather change as they brighten and dim, usually somewhat predictably.

It is possible that ALL stars are somewhat variable, even our own sun if seen from some remote part of our galaxy, but some are more pronounced in their changes, sometimes varying as much as 10 to 12 magnitudes in a short period of time! Indeed, even a NOVA or SUPERNOVA is a variable star of sorts, rapidly surging in brightness and then steadily fall off to sometimes invisibility.

ALGOL gets its name from the Arabic *Al Ra's al Ghul*, meaning "the demon's head" in reference to early associations of the star to the serpentine-laced head of the Gorgon Medussa; to the ancient Hebrews, the star also had a rather ominous association, "*Satan's head*."

Although the star's dramatic brightness changes have been known since antiquity, it was not until only 1667 that careful attention (scientifically, not mythologically, at least!) was given to Algol; even then it took ANOTHER 120 years before its nearly-three day cycle was determined accurately!

ALGOL and your Telescope

You obviously do not need a computer or AutoStar to find Algol, nor to find out when it is a minimum (about 3.5 magnitude). Indeed, you can compute it yourself, because the star runs one complete cycle **EVERY 2 days 20 hours 49 minutes**, or you can find reference to it in *Sky and Telescope* magazine each month.

Also, you do not need a GO TO telescope to observe Algol's brightness changes....you do not need a telescope AT ALL! In fact, monitoring the brightness changes requires that you scan wide areas of the sky for bright stars with which to compare Algol's brightness at any given time (see ASO attached chart for Cassiopeia and Perseus comparison stars). You can keep up with the CYCLE of the star's brightness with AutoStar, but the actual observing is best done with the naked eye.

Bulletins of the Arkansas Sky Observatory
COPYRIGHT 1999, P. CLAY SHERROD

Observing the Variable Star ALGOL
(Beta Persei - Magnitude 2.2 to 3.5, Period: 2 days 20 hours 49 minutes)

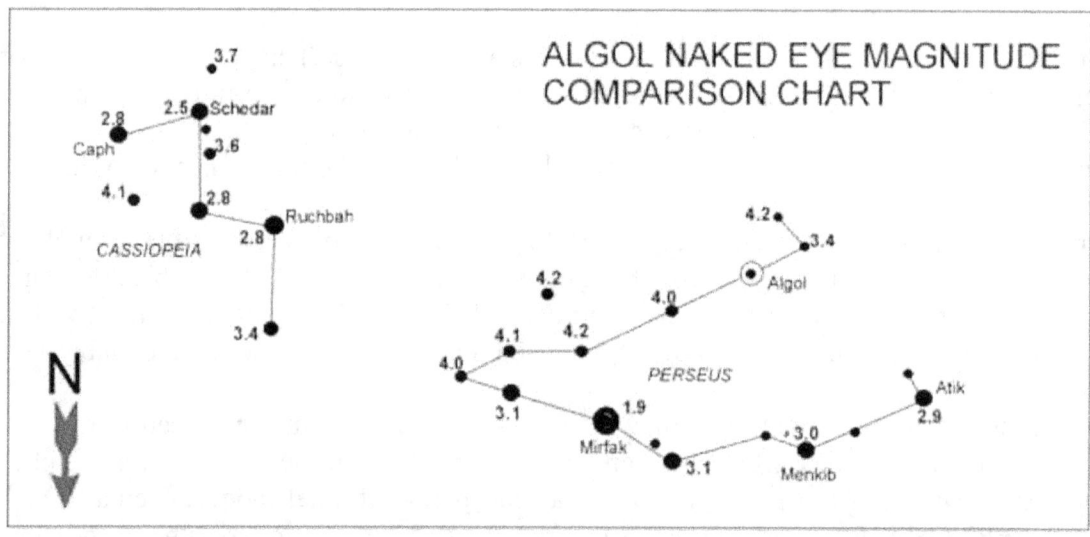

OBSERVING NOTE: use stars in BLUE to compare with the brightness of the star ALGOL. Binoculars will help in the comparison, by taking each star slightly out of focus to make a bright "disk" which can be more easily compared than can the "point" of a star.

When comparing magnitudes (start yourself a little chart and plot the star's magnitude yourself over a month-long period - makes a great family or club project), it is advisable to use BINOCULARS that are slightly out-of-focus on both Algol and the comparison star(s) to make small "disks" of light that are easier to compare than "points" of light. With such a rapid variation of only a couple of days, it is wise to monitor Algol as often as possible; write down the time (get from your AutoStar), date and your magnitude determination at every possible outing.

ABOUT THE STAR ALGOL

Algol is undoubtedly the best known, and one of the most rapidly-changing variable stars; it is an ideal naked eye variable that will allow you to acclimate yourself into the exciting study of variable stars (pull up www.AAVSO.org for the charts and information available free of charge for variable star observing through the American Association of Variable Star Observers).

The cause of Algol's rapid brightening and subsequent fading are well understood. Algol is TWO stars, not just one, as seen from Earth being aligned as an "eclipsing binary" star in which one star periodically occults, or covers up, the other in their tight orbit around one-another. Hence, with Algol, this system "eclipses" on a very predictable cycle; the MINIMUM BRIGHTNESS (about 3.5) is likely the brightness of the two stars seen aligned toward Earth, while the MAXIMUM BRIGHTNESS results when both stars are "in full view" (even though we cannot resolve the double nature as the stars are too close).

Algol is a relatively close star to our sun, at a distance of only 115 light years. The total luminosity of the double star system is on the order of 140 times that of the sun; when comparing the mass of both stars in the Algol system to the sun, we find that the smaller of the two stars is only .86 times the mass of the sun, while the larger star is 4.5 times more massive.

Observers who have never attempted variable star observing are missing a treat; it is one field in amateur astronomy that you can actually witness CHANGE taking place in an otherwise seemingly "fixed" deep space. Over time, you may want to move up to telescope variable stars that demonstrate wide ranges of variation, sometimes showing a bright star only - weeks later - the star has faded to near-invisibility. A good example and starter-star for your telescope might be **SS CYGNI**, in the constellation of Cygnus. Log onto the AAVSO website to download the charts for this variable.

REGARDING AAVSO CHARTS: for the catadioptic scopes, always download the "Reversed" charts which provide (thank goodness) star patterns as they appear in the mirror right angle eyepiece of the "CAT" scopes; this provides direct comparisons. Be sure and get the "A" charts (wide field finder charts) for each star you desire, and THEN get the "G" charts (higher magnification with fainter comparison stars) for the same star for telescopic viewing.
https://www.aavso.org/apps/vsp/ . Note for these charts, simply type in the NAME of the variable at top to generate your choice of chart.

Enjoy yet another wonderful exploration that provides you with exciting stellar changes and gives you some "purpose" to your telescope after AutoStar has given your grand tour and you have "....seen all there is to see!" Enjoy!

* * *

The Star Filled Globular Cluster Messier 80
As photographed by the Hubble Space Telescope

Chapter 35

Observing the Magnificent Spring Globular Clusters
Spectacular Objects for any Telescope

INTRODUCTION -

There are two events that occur every spring and continue through the hot days of summer that have some very much in common. Both I am sure you can relate to.

The first is that first "tinkle" of the little bell or the music that plays through a very bad bull-horn atop an old beat up panel van painted white...as it slowly progresses down your neighborhood streets, swarms of kids encircle it, follow it and surround it in a huge circular mass until it comes to a stop. The all-American Sno-Cone truck, attracting kids like bees around honey for decades.

Springtime also brings another similar-looking phenomenon, although not quite as "*Norman Rockwell*-ish" as the humble Sno-Cone truck. Indeed, this sight is awe-inspiring and the facts behind it even more.

It is the springtime and summer "appearance" of the hundreds of fantastic deep sky objects that we fittingly call '*GLOBULAR CLUSTERS*', strange globs of stars which encircle our galaxy (and others as well) just like kids encircling the Sno-Cone truck!

Historical   -

Believe it not the first globular cluster was "discovered" by *Edmund Halley*, of comet fame, in 1715 yet he was not able to distinguish the many thousands of stars which we know today comprise the cluster. It was not until about 50 years later that Messier added into his famous catalog, "M-13" denoting his thirteenth entry of about 100. Halley himself noted that Messier 13 was visible as a fuzzy star to the naked eye (it is 6th magnitude if that attests to early sky conditions!).

It was not resolved until the very late 18th century by Sir William Herschel with his great "40-foot reflector" into many scores of what appeared to be tiny stars....yet, after that, many people - T.W. Webb with a 4" refractor and Admiral Smyth with his famous 6" glass - were able to clearly see many of the brighter individual stars in this cluster.

Soon, it became recognized as one of many and closely associated with two other wonderful globulars, Messiers 3 and 5, both nearby.

Herschel suggested that Messier 13 perhaps had over 14,000 stars, and indeed this cluster contains well in excess of 35,000 stars to limiting magnitude 23, ALL of which are gravitationally bound together around a common orbital center.

ABOUT THE GLOBULAR CLUSTERS

There are two images that are permanently fixed in my mind from my over 30 years of observing: the first is my first look at Saturn through my 4" Unitron in 1966, which I had saved for for two years to get; the second and equally memorable is the sight of Messier 13 through a 40" *Boller & Chivens* Cassegrain from the dark cornfield in central Nebraska (darkest skies I have EVER seen) in the summer of 1973. With that telescope and those inky-black sky, I could actually make our the orange, yellow and red giant stars among the thousands that stood in what appeared to be a true three-dimensional presentation.

Your highly capable telescope is fully capable of searching out and giving you fantastic and memorable views of most of the brighter globular star clusters; in the medium aperture telescopes (say, 4-6 inch), many of the brighter stars become evident on very clear nights....the larger 8" and above can show hundreds of individual stars in the larger clusters and begin to break down the "fuzzy glow" into stellar components on many of the fainter and more distance clusters as well.

You observational appetite of these "gems" will most definitely be piqued by some knowledge of their locations, physical make-ups and distances.

We see <u>MOST</u> of the Milky Way galaxy's globular clusters (well over hundreds of them, only about a hundred clearly visible using common telescopes) during the spring and summer months, mostly in the constellations of Bootes, Hercules, Scorpious, Sagittarius, Ophiuchus and Ara. This is misleading, however, as the globulars ARE NOT concentrated in space ONLY in one location as they appear.

Our sun and its planets are about 30,000 light years from the center of the Milky Way, and thus our "view" toward the center is skewed somewhat; when we look at the sky in the summer/spring months we are looking INWARD toward the galactic center, into a nearby area where the clusters appear to be more concentrated; in winter months, our nighttime skies are peering AWAY from the center where the actual number of globulars decreases rapidly.

Knowing that, you will realize how "skewed" our vantage of the globular clusters really is when you note the apparent spherical and very UNIFORM distribution of globulars around the Milky Way. Just like those kids hovering around the Sno-Cone truck, the globulars - for whatever reason - hover like a swarm of bees around our galaxy in all directions! They are EVERYWHERE in our galaxy, but WE are located far to one side!

The distances of globular clusters from Earth, hence, vary greatly; we are afforded the BEST views only of those located very near the vantage point surrounding the center of our galaxy; since we are on "one side" of the galaxy, those opposite are VERY distant, very faint and very difficult to see in our telescopes. One of the most spectacular (and yes it CAN be seen by observers in the southern U.S., not to mention our observers in Australia, South America, Mexico and Africa) globulars is Omega Centauri, a clear 2nd magnitude naked eye "fuzz ball" about 17,000 light years away. It can be cleanly resolved into stars by a 90mm scope and even a 60mm during very good conditions and high magnification (see magnification guidelines below).

By contrast the closest is only 8,000 light years and the most distant (all the way to the "opposite" edge of the Milky Way from us) is an incredible 180,000 light years away.

A typical globular cluster contains some 120,000 stars, mostly olding or aging stars, and is a whopping 150 light years in diameter! Consider that size for ONE cluster: that is 75 times LARGER than the distance we are from our next closest star!

Without going into theoretical astrophysics here, you should know that little is known about the actual origin of the globulars and why they have this "hovering" relationship with our galaxy; it appears that perhaps all typical galaxies - even the elliptical ones - have globular clusters hanging around in similar fashion. Astronomers do no know how they got there, if they are from the common "stuff" of the galaxies, or even which got here first.....the galaxy or the clusters?!

OPENING THE DOOR: OBSERVING THE CLUSTERS

Following is a concise and quick guide to your spring and summer tour of globular clusters. Not all of these may be on your computer control, but most are. All major designations and coordinates are provided so that you can quickly access these curious objects.

Scope Size and the Clusters

as an "overall guide" the following pretty well sums up what you can expect to see with your particular telescope. Note that I have divided the globular clusters in to Class I and Class II, with "Class II" being the faintest and more difficult to see/resolve. Class I globulars are bright, large and fairly close typically; Class II are very tight and compact, usually very small and faint. This quick list summarizes for 1) dark sky conditions (no moon or lights), 2) the cluster being nearly overhead, and 3) a magnification of about 25x per inch of aperture which is PERFECT for maximum contrast, resolution and brightness (as well as image scale) for your telescope:

TELESCOPE	CLASS I	CLASS II	DESCRIPTION (overall)
60 / 70mm	Super	Select	Larger clusters appear mottled, very 3-D
90 mm	Super	Select	Larger clusters resolved
125 mm	Excellent	Super	More stars than -90; faint clusters mottled detail
200 mm+	Excellent	Super	Many stars in Class I, brighter stars in II

No matter what size scope you have, there is a true adventure with the globulars. They offer a rare opportunity to visualize a 3-dimensional aspect not seen in other deep sky objects and a chance to really put the optics of any size scope to the acid test, both in light gathering and resolution.

Selecting Your Magnification - I have always said that there are two ways of observing a globular: 1) the grandeur of the magnificent "fuzzy ball" suspended in a wonderful wide-field dark sky among splattered stars; and, 2) a remarkable close-up that begins to reveal stars like "star dust" as they seemingly flicker in and out of view. BOTH experiences are incredible and worth your distant trip to the globular clusters.

Obviously, for wide-field (1) observing any of our telescopes will do in a very dark sky location. You want a very low (but not overly so) power magnification. About 15 power (15x) per INCH of aperture is an IDEAL low power for globular observing. Hence, the 3-inch needs a wide field power of around 38x; 55x for a 4-inch will do nicely; 75x is ideal with a 6-inch scope; and 120x (or less) is suitable for an 8-inch.

These magnifications will show you a relatively wide field of view (to afford a panorama of sorts with other objects in the field) yet still provide enough image scale (size of the globular) to give satisfactory hints of the shape and size of the object.

Now, for (2) above....to really get "into" the cluster and attempt to make out individual stars....I have found that globulars are the exception to the "deep sky" rule of magnification. You use whatever it takes! typically, a magnification of around 50x per inch is the limit. What you are trying to do is: find an eyepiece for your scope that will provide: 1) maximum resolution (i.e. show or begin to show individual stars or mottling); 2) with still very dark background sky (contrast); and, 3) no loss of brightness (magnification fall-off). You likely have just such an eyepiece! Experimentation will tell you if you are using too much or too little power when observing a globular cluster.

Now, let's pick out the best of the best and I will tell you what to expect.

YOUR TELESCOPE GLOBULAR CLUSTER BONANZA (by constellation)

The following alphabetical list includes the most "observable" globular star clusters as found on your AutoStar "Star Cluster" Index. When possible all designations (Messier/Caldwell/NGC, etc.) are given. The brightness of the object, its size, and the concentration of stars within the cluster have led to this scale for telescope users.

THE RATING ("A" though "D") is provided for your telescope aperture; a rating of "A" denotes an excellent and easy globular cluster for ALL telescopes (all sizes and types); "B" on the other hand will be more difficult and best seen in at least a 3"; "C" puts the limit no smaller than the 6-inch, and "D" denotes a globular that is best seen with the 8-inch or perhaps disappointing in some scopes. A very brief description of the globular seen in some of these telescopes is also listed.

The rating is a simple way to eliminate frustration if you think you should be seeing "more" than you do in your telescope....for example, if the rating is "C" and you are observing with a 60mm, then don't expect too much! On the other hand, if the rating is "A" you should be very pleased with the view no matter WHAT size scope you have.

Each globular has an "R =..." rated within its description.

This list is broken into <u>NORTHERN HEMISPHERE</u> globulars (first list) and <u>SOUTHERN HEMISPHERE</u> GLOBULARS. (NOTE: there is some carry-over between, as some can be ideally or at least partially seen in both southern and northern hemisphere, depending on proximity to the Earth's equator. For example, Omega Centauri, even though it is far south at -47 degrees, can be seen as high as latitude +35 because of its large size and brightness! Likewise, Messier 13 at + 36 degrees is a good target for those at -35 degrees southern latitude with an unrestricted northern horizon!

<u>NORTHERN GLOBULARS IN GO TO programs</u> (all in Spring/Summer skies)

AQUARIUS –
R = C: ngc6981 / Messier 72 - RA 20 53; DEC -12 32 / Very faint and tight, no resolution.
R = A: ngc7089 / Messier 2 - RA 21 33; DEC +00 49 / Spectacular, bright, large, one of the best!
CANES VENATICI -
R = A: ngc5272 /Messier 3 - RA 13 42; DEC +28 23 / Fantastic, one of the best, very large.
CAPRICORN –
R = B: ngc7099 / Messier 30 - RA 21 40; DEC -23 11 / Very tight but fairly bright, no resolution.
COMA BERENICES -
R = A: ngc5024 / Messier 53 - RA 13 13; DEC +18 10 / Small but bright; partial resolv. in 6-8 inch
DELPHINIUS –
R = D: ngc6934 / Caldw. 47 - RA 20 32; DEC +07 24 / Very dim, compact, no resolution in any size telescope.
R = D: ngc7006 / Caldw. 42 - RA 21 01; DEC +16 11 / Even dimmer, very faint and difficult, tiny.
HERCULES –
R = A: ngc6205 / Messier 13 - RA 16 42; DEC +36 28 / The "standard", resolves partially in all scopes.
R = C: ngc6341 / Messier 92 - RA 17 17; DEC +43 08 / Very small, medium bright; part. res. in LX.
LYNX –
R = D: ngc2419 / Caldw. 25 - RA 07 38; +38 53 / 10th mag. very difficult, even with 8-inch tiny.
LYRA –
R = B: ngc 6779 / Messier 56 - RA 19 17; DEC +30 11 / Nice even tho small; fairly faint,no resolution.
OPHIUCHUS (are you ready for all these?) –

R = D: ngc6171 / Messier 107 - RA 16 32; DEC -13 03 / VERY small, med. bright, tough one!
R = B: ngc6218 / Messier 12 - RA 16 47; DEC -01 57 / Small but bright, some resolution in LX.
R = B: ngc6254 / Messier 10 - RA 16 57; DEC -04 06 / Almost a carbon copy of M-12 above!
R = A: ngc6286 / Messier 62 - RA 17 01; DEC -30 07 / Tiny, but nice even in small scope; no resolution.
R = B: ngc6273 / Messier 19 - RA 17 03; DEC -26 16 / Very nice, bright, some resolu. in 6- to 8-inch
R = A: ngc6333 / Messier 9 - RA 17 19; DEC -18 31 / Excellent in all, but dim; some resol in -90 +.
R = C: ngc6402 / Messier 14 - RA 17 37; DEC -03 15 / VERY small and difficult; no resolution.

PEGASUS –
R = A: ngc7078 / Messier 15 - RA 21 30; DEC +12 10 / Beautiful, resolves in 4-inch; a small one!

SAGITTARIUS –
R = A: ngc6626 / Messier 28 - RA 18 24; DEC -24 52 / Bright, medium size; resolves some in -125.
R = C: ngc6637 / Messier 69 - RA 18 31; DEC -32 21 / Very small, med. bright, tough even in LX.
R = A: ngc6656 / Messier 22 - RA 18 36; DEC -23 54 / One of the best, large, resolves in 8-inch and up.
R = D: ngc6681 / Messier 70 - RA 18 43; DEC -32 18 / Very hard, small, faint, no resolution
R = C: ngc6715 / Messier 54 - RA 18 55; DEC -30 29 / Very hard, small, interesting in 6-inch scope.
R = A: ngc6809 / Messier 55 - RA 19 40; DEC -30 58 / A very nice one! Large and easy; res in -125+.
R = D: ngc6864 / Messier 75 - RA 20 06; DEC -21 55 / Very faint and small, no resolution at all.

SAGITTA –
R = C: ngc6838 / Messier 71 - RA 19 54; DEC +18 47 / Small and very faint; some res. in the 8-inch and larger

SCORPIUS –
R = B: ngc6093 / Messier 80 - RA 16 17; DEC -22 59 / VERY small but interesting in all scopes.
R = A: ngc6121 / Messier 4 - RA 16 24; DEC -26 32 / One of the best, very large & bright; resolves 90+.

SCUTUM –
R = B: ngc6712 - RA 18 53; DEC -08 42 / Faint but very large; some mottling in 8-inch at high power.

SERPENS –
R = A: ngc5904 / Messier 5 - RA 15 19; DEC +02 05 / Maybe the best; res. in 3-inch, very large, bright!

SOUTHERN GLOBULARS IN MOST SKY PROGRAMS (all in Spring/Summer skies)

ARA –
R = C: ngc6352 / Caldw. 81 - RA 17 25; DEC -48 25 / A very faint and small one; resolves in 8-inch.
R = A: ngc6397 / Caldw. 86 - RA 17 41; DEC -43 40 / Very large and bright; one of the best for all scopes
APUS –
R = D: ngc6101 / Caldw.107 - RA 16 26; DEC -72 12 / Maybe the toughest on the list; "transparent."
CARINA –
R = B: ngc2808 - RA 09 12; DEC -64 52 / Tightly packed in center, bright but small.
CENTAURUS –
R = A: ngc5139 / Caldw. 80 - RA 13 27; DEC -47 29 / *Omega Centauri*....the BEST of them!
R = C: ngc5286 / Caldw. 84 - RA 13 46; DEC -51 22 / Fairly bright but small, resolves in LX.
COLUMBA
R = C: ngc1851 - RA 05 14; DEC -40 03 / 5-inch resolves stars around edge, looks like a glow.
CORONA AUSTRALIS –
R = B: ngc6541 / Caldw. 78 - RA 18 09; DEC -43 42 / Bright, but fairly small, good in ETX 90 +.
HYDRA –
R = D: ngc 4590 / Messier 68 - RA 12 39; DEC -26 45 / Faint and small, mottled in the 8" scopes.
R = D: ngc5694 / Caldw. 66 - RA 14 40; DEC -26 332 / VERY tiny and faint...a real test, even for the LX.
HOROLOGIUM –
R = D: ngc1261 - RA 03 12; DEC -55 13 / A very tough, compact one. Extremely small, no resolution.
LEPUS –
R = B: Messier 79 - RA 05 24; DEC -24 33 / Partially resolvable in 8-inch and above; very small.
LIBRA –
R = D: ngc5897 - RA 15 17; DEC -21 01 / Faint and very small, rather un-interesting in all scopes.
LUPUS –
R = B: ngc5986 - RA 15 46; DEC -37 47 / Fairly bright but small, partial resolve in 4-inch
MUSCA –
R = A: ngc4372 / Caldw.108 - RA 12 26; DEC -72 40 / Faint but very large, good in 4-inch and larger scopes

R = A: ngc4833 / Caldw.105 - RA 12 59; DEC -70 53 / Good in all scopes, large & med. bright.
PAVO –
R = A: ngc6752 / Caldw. 93 - RA 19 11; DEC -59 59 / A Beaut! Very large & bright, easy in all scopes.
SCULPTOR –
R = B: ngc0288 - RA 00 53; DEC -26 35 /. Very loose and nice sized, some stars in larger scopes.
TUCANAE –
R = A: ngc0104 / Caldw.106 - RA 00 24; DEC -72 05 / Magnificent! Very large and easy! ("47 Tucanae")
R = B: ngc0362 / Caldw.106 - RA 01 03; DEC -70 51 / Bright and large, resolves in 3" scope.
VELA –
R = B: ngc3201 / Caldw. 79 - RA 10 18; DEC -46 25 / Small but bright, some res. in 5-inch and larger

Now, armed with that complete PC library summary of GLOBULAR STAR CLUSTERS, you can go out there and get all of them in one night! Remember...there are MANY more such clusters (IC and NGC objects) than are listed here, but these are the primary listed ones in all modern popular planetarium programs. With their seemingly concentration in the spring and summer skies, why not make a "party" out of it....

.....and have "*GLOBS*" of fun!

The globular Messier 13 photographed in only 35 seconds at ASO

Chapter 36

Your Guide to Observing Variable Stars
Watching Evolution in the Skies

Introduction

We look out at the vast nighttime sky on a pristine, clear and dark night and sense a nature of "permanence." The constellations - stars construed in patterns hundreds, thousands of light years away - appearing to eyes and our emotions as the "firmament" of a never-changing creation. Such stability appears to be important to the comfort of the human psyche, but indeed....does NOT exist.

Of all things....the wonderful and seemingly static stars are NOT stable. In fact, they are ever-changing: all stars, every one that you can see including even our sun itself.

In many stars, this change can be quite dramatic and certainly visible in modest telescopes or to the naked eye. We have **CATACLYSMIC** changes such as the "novae" and "supernovae" in which a single event (we'll call it an "explosion" to keep from getting too deep into astrophysical jargon) transforms nearly the entire mass of a huge sun into a bubble of energy and light that spews outward at the speed of light from where a star used to be. Such novae occurred recently in July, 2000 with a sudden flare-up of Delta Scorpii from naked eye magnitude 2.3 to 1.9, and again in February 2001 with a telescopic nova in the constellation of Sagittarius which brightened from fainter than visibility in even the world's largest telescopes to a "star" nearly visible in binoculars, at magnitude 7.7!

The energy released from such catastrophic events is astronomical, literally. When any star changes and YOU are there to witness it - whether it be through the supernovae explosion, or the slow and methodical pulsating of a red giant as it gradually ages - YOU are watching first-hand the actual evolution of something that, in our lifetimes, appears to never change.

But these dramatic and sudden star events are NOT common and there are not too many to choose from. On the other hand, there ARE some wonderful stellar events that YOU can monitor over time periods of just a couple of days, to a year or even several years for the complete event to take place.

Such events are the **VARIABLE STARS**, stars which change noticeably in brightness. We have seen how the bright star ALGOL, in Perseus, changes its brightness every 2.3 days ( ), a result of a second and darker star eclipsing the first. And among those seemingly never-changing millions of other stars out there visible in your telescopes, the great majority of them also demonstrate such changes.

Such stars include: *ECLIPSING BINARY* STARS - like Algol, stars which change brightness as one star moves in front or behind the other; REGULAR *LONG-PERIOD* AND *SHORT PERIOD* variable stars, many of which are "red giant" stars that actually pulsate in size; the *CEPHEID VARIABLES*, also pulsating stars but "supergiant yellow" stars; and the *CATACLYSMIC* VARIABLE STARS, stars which - like the **novae** and **supernovae** but not to such an extreme -suddenly brighten without any warning nor predictability. The cataclysmic stars are monitored around-the-clock by astronomers as they cycle from very dim to brighter over a very erratic period of time.

GETTING STARTED IN VARIABLE STAR OBSERVING

You pretty much already have what you need to start a serious (but still "fun".....as opposed to at "work serious") adventure into observing variable stars, something that still remains primarily in the hands of amateur astronomers armed with everything from binoculars to modest telescopes.

One of the best things that you can do when getting started in Variable Star Observing is to study the Tutorial at the AAVSO posted at:
https://www.aavso.org/10-star-training

You have: 1) a telescope (probably); 2) yourself and a (hopefully) good set of eyes; 3) a yearning to do something besides just "look" through your capable telescope; 4) an accurate watch and 5) a computer, or you wouldn't be reading this right now. What you don't have - the CHARTS to find and compare these objects to stars of KNOWN and steady brightness - you ARE about to have access to.

So let's get started paving the way for YOU to do some "serious" scientific research, contribute to astronomy, and have an exciting time cataloging the life of some star that is 1,350 LIGHT YEARS away! What you are about to embark upon is a lesson taught easily to you by one star: R URSA MAJORIS, an easy-to-find long-period variable star in the constellation of Ursa Major, just north of the "big dipper." Figure 1 shows this position of R UMa and how easy it is to find.

The star varies in magnitude from a faint 13.9 visual (about the limiting magnitude under VERY DARK SKIES with the LX 90) to a very bright 6.6, easily visible in nearly all of the smallest telescopes. The entire "period" of R UMa is some 302 days with minor fluctuations along the way (that's why it is "fun" to keep up with!). The "period" of any variable star is the length of time it takes to cycle, i.e., from minimum brightness, to maximum and BACK to minimum again to start all over.

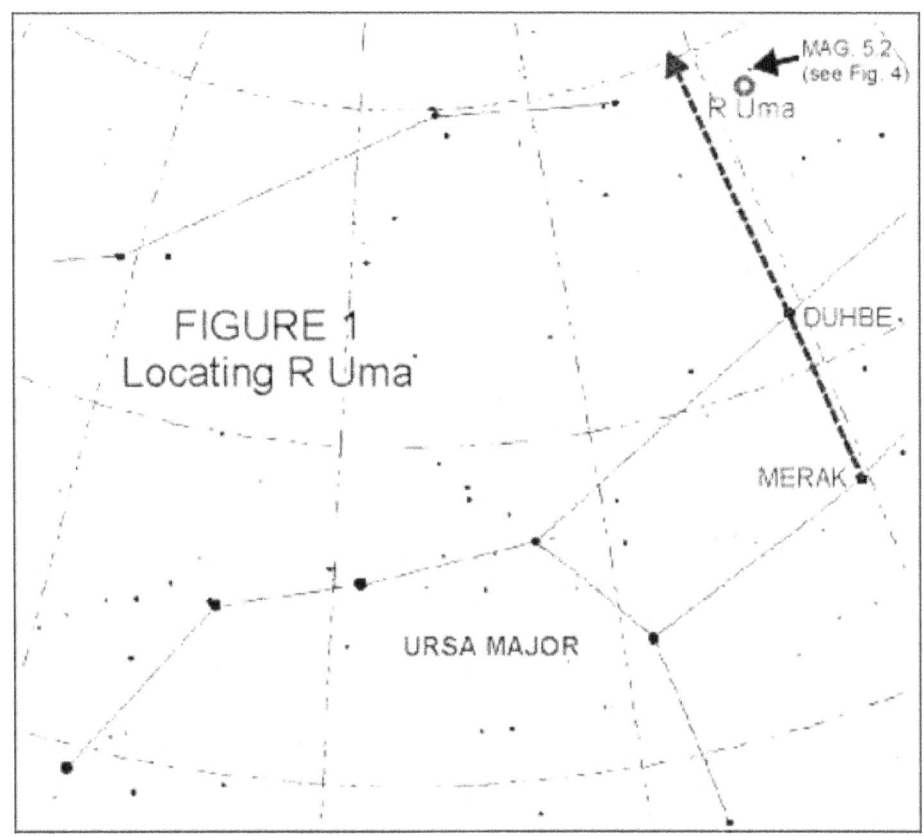

For setting circle use, or to program into your planetarium program, the coordinates (epoch 2000) of this variable star are:
Right Ascension: **10h 44m 34s**
Declination: **+68 degrees 46m.5**

*   *   *

R Ursa Majoris is easy to find, even without the GO TO or setting circles. Note in Figure 1, and again in Figure 3 the star marked "52" (the decimal point - 5.2 - is omitted to keep it from being confused with a very faint star!). This star is easy to find by using Duhbe and Merak as a guide (Figure 1); the "52" star is easy to locate with a small finderscope, but not distinct with the naked eye. NOTE that this magnitude 5.2 star will remain visible in the SAME field of view as the variable star, even with moderately high magnification (see Figures 3 & 4) and thus assist in your proper identification of the variable.

This is one of the REDDEST long period variables, and the red color can be clearly noted when at brightest. Likely a much older star than our sun, R UMa is about 250x brighter than our sun.

Note in Figure 2, the light curve (you can construct your own by simply plotting the observed magnitude - ON A LONG-PERIOD VARIABLE, OBSERVE ABOUT ONCE PER WEEK through the entire cycle - against the elapsed time) demonstrates that this star brightens TWICE as fast as it dims, taking only about 116 days to rise from minimum brightness to magnitude 6.6!

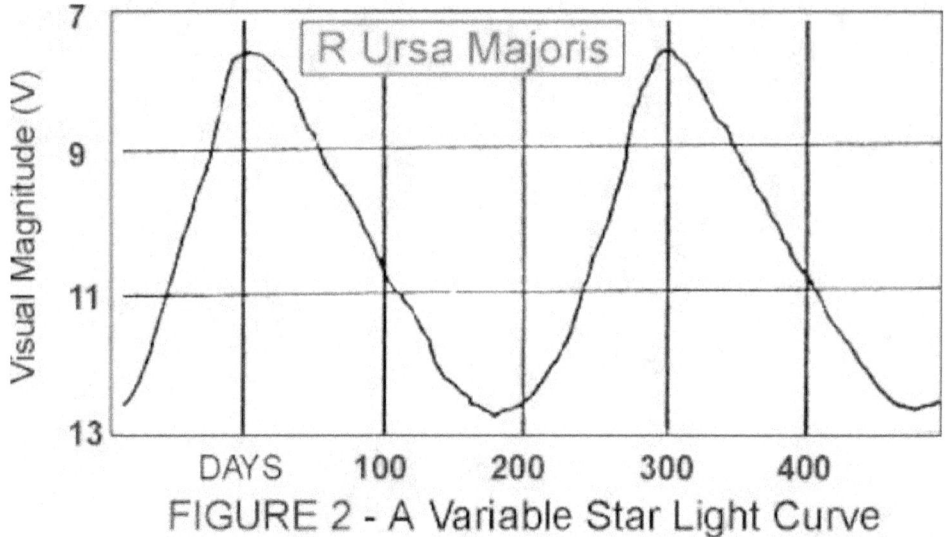

FIGURE 2 - A Variable Star Light Curve

To observe R Ursa Majoris - or any of the thousands of variable stars out there - a wonderful organization, the American Association of Variable Star Observers (A.A.V.S.O.) has compiled excellent and FREE star finder charts for you to: 1) locate your target star; and 2) find stars of KNOWN brightness with which to compare and evaluate YOUR estimate of the variable's brightness every time you observe it. Samples of these charts are given here, and more is said about obtaining these charts and the AAVSO throughout this observing Guide.

FINDING YOUR VARIABLE STAR

To find your target star is easy; once you have located it correctly each time thereafter the star can be spotted even quicker. For those not using a GO TO or setting circles to find the star (as might be the case with R UMa since it is so easily found as noted above), the AAVSO has several DIFFERENT charts available for nearly EACH variable, and each is indexed alphabetically. The **"a" charts** are FINDER CHARTS, showing a very wide field of sky as you would see it with the naked eye or with your finderscope; NOTE that the variable star will nearly ALWAYS be in the middle of the finder chart (unless more than one variable is denoted on that chart). From there, once you can locate the brighter stars around it, you can "zero-in" on your target by moving up to AAVSO **chart "b"** (shown in Figures 3 and 4) which is a narrower field and shows even fainter comparison stars (REMEMBER: the decimal point is left out of the magnitude numbers!). For most observing - except when the star reaches a VERY FAINT magnitude - you will be using **charts "c"**. When very faint, you might need to move up to **chart "d"** (Figure 8), which normally shows a telescopic field (about 1/2 to 1 degree across...they always have a scale at the BOTTOM) and a "limiting magnitude" of comparison stars of about 13th or 14th.

https://www.aavso.org/apps/vsp/ . Note for these charts, simply type in the NAME of the variable at top to generate your choice of chart.

The field and limiting magnitude of Chart "G" is ideal for your medium-to-high power magnitude estimates in your telescope, necessary when R UMa is reaching its dimmest point of the cycle.

Note carefully two stars that are marked with red ARROWS in BOTH Figures 4 and 8 (NOTE: both of these charts are "reversed charts", specially made for telescopes like yours, either Cassegrain or Maksutov to allow the field of view to be matched with that provided by catadioptic telescopes; these reversed charts are discussed in detail following). I have drawn in these arrows to allow you to have a "reference point" between two very different charts, both of which you will be using with your telescope. Figure 4 shows the two stars - magnitude "105" and "110" clearly marked; this chart is the 'b' chart which has a total field of view of about 3 degrees. Figure 8 shows the SAME two stars in relation to the variable star, but it has a much more narrow field of view - only 1 degree total - so that you can increase your magnification to see the fainter stars.

These and all other variable star charts may be downloaded from the AAVSO web site for FREE by logging onto the link above.

So, simply "find" the star's position (even if you can't SEE it) using the "a" chart and your finderscope (since you are using your naked eye or perhaps a straight-through finder to locate the star in a constellation and wide field, you do NOT use a "reversed chart" for an "a" chart); then tweak in your finderscope crosshairs even closer with the "b" (reversed) chart until you have found the field of view that MATCHES the stars seen on the chart (see "Using Reversed Charts", following). At this point, if the star is near maximum brightness, you may be able to use the wide field and brighter comparison stars of the "b" chart to make your estimate; if not, then proceed up to the "c" chart and make the estimate. With this link you can custom make your chart to fit YOUR telescope's parameters: catadioptic, straight through, reversed, mirror image, refractor with diagonal, etc.

The procedure for accurately estimating a star's brightness visually is very simple and is given following.

That's all there is to it. Using an accurate watch (to within a minute - log onto nist.time.gov for your time accurate to 0.7 second) merely write down the date and time (in Universal time) and the magnitude that you determine for the star.

You observation is VERY IMPORTANT. The AAVSO and professional astronomers world-wide are relying on amateur astronomers to make these observations! If you cannot observe regularly, that is fine; but every time you DO observe a variable, be sure to report it! The correct reporting format so that your observation can be included (you can even get an INSTANT light curve on each star that will update to include YOUR observation!) by logging on to: aavso.org/adata/curvegenerator.shtml

VARIABLE STAR CHARTS - *Using Reversed Star Charts*

Until recently, all available star charts including those for variable stars, were printed "astronomically-correct," whereas SOUTH is at the TOP, EAST is to the RIGHT and WEST is at the LEFT of the chart. This orientation matches: 1) Newtonian reflectors; 2) refractors with no right-angle prism; 3) most astronomical photographs taken at prime focus of a telescope. Such an AAVSO chart is the "standard" chart and is shown for R UMa in Figure 3.

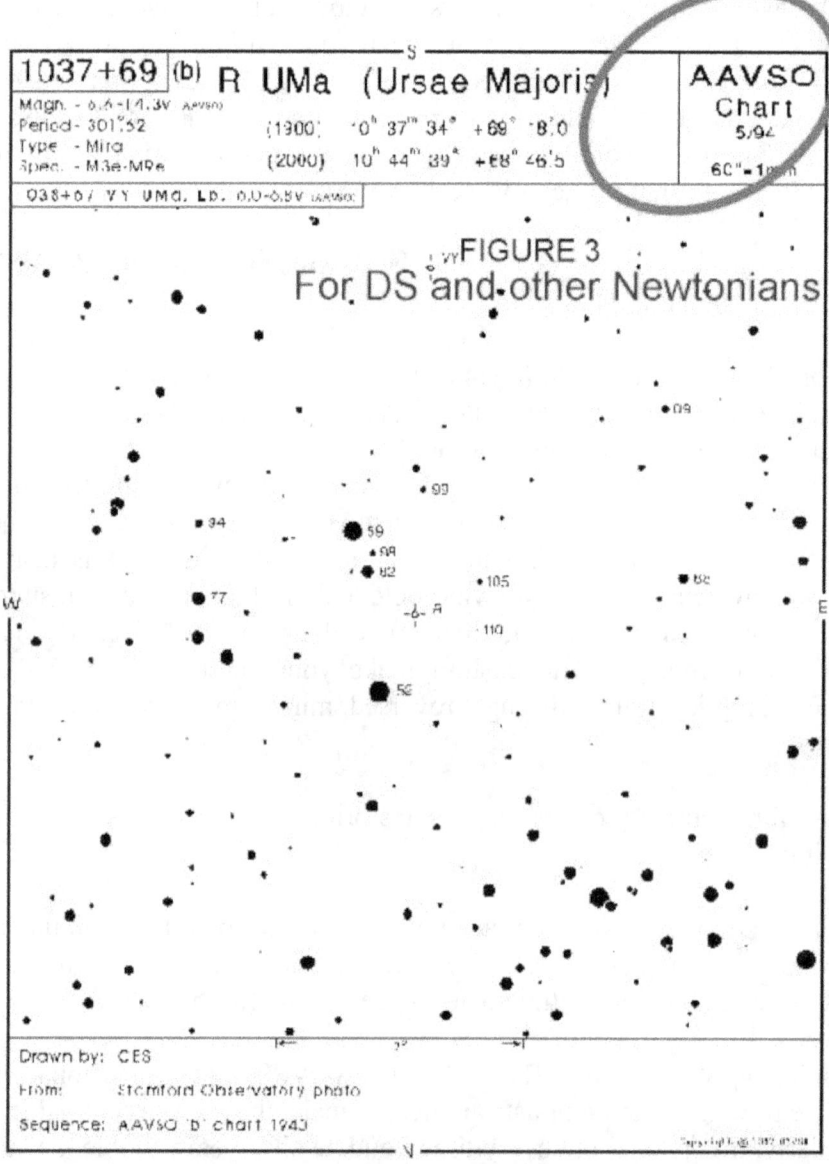

This was bad news for the popular Cassegrain and Catadioptic telescopes, the very design of all popular GO TO and portable telescopes on today's market! With the compound telescopes, a mirror or prism is used to divert the optical path at a right angle for viewing comfort and convenience; this provides a MIRROR IMAGE (or an "odd number of reflections" of the light path). Hence, attempting to use the older charts for serious variable star indentification was nearly impossible with a catadioptic scope.

Today, the AAVSO - realizing the growing number of these telescopes and the importance of the great many observers who would like to contribute estimates on a regular basis - provides "REVERSED" charts to give the exact orientation as one might see through the eyepiece of a catadioptic telescope. Compare Figure 3 with Figure 4: BOTH are "b" charts for R UMA, same scale and same field of view. Figure 4 is the newer "reversed" chart for catadioptic orientation. It will match the sky as you see it in your eyepiece....Figure 3 will NOT. So, when you download from the AAVSO Chart Index, ALWAYS go to the charts that are marked with an "R" at the end of the chart designation for the proper chart for YOUR telescope: https://www.aavso.org/apps/vsp/ . Note for these charts, simply type in the NAME of the variable at top to generate your choice of chart..

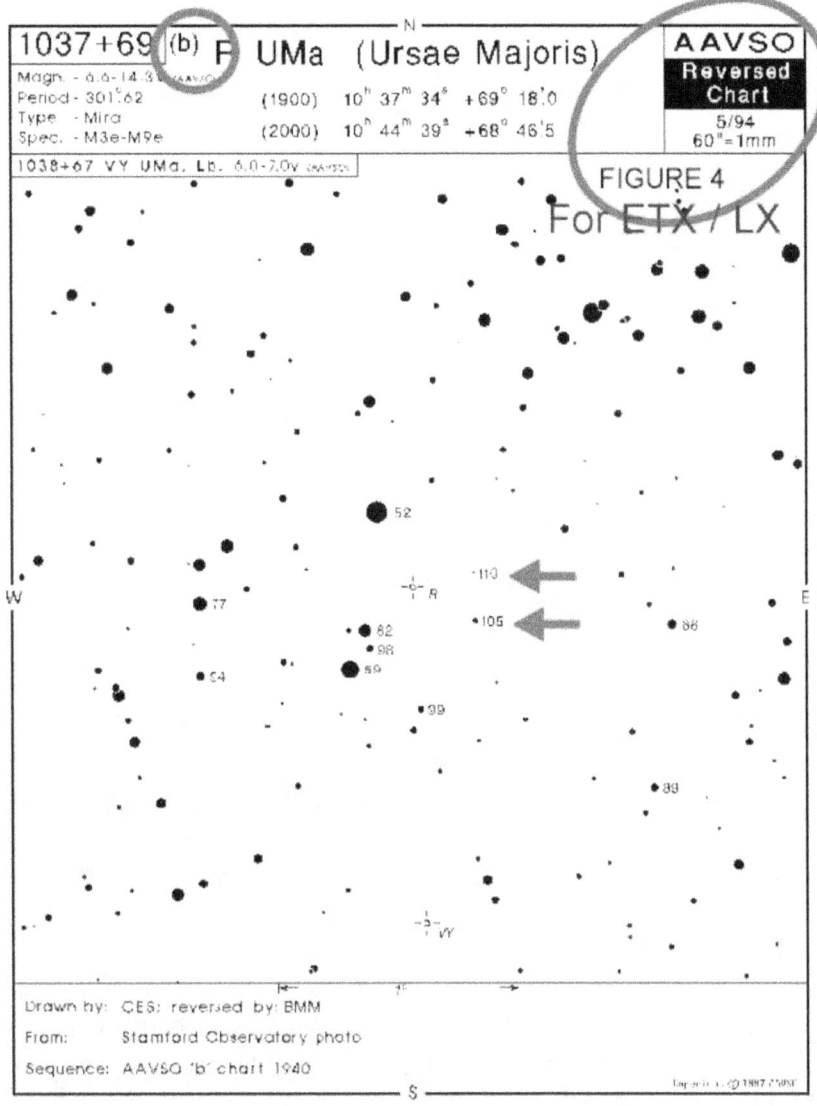

A close-up from BOTH the regular and "reversed" charts of the immediate area surrounding R UMa is shown in Figure 5 so that you may get an idea of just how different the two charts are. Using the older chart with the CAT scopes would be difficult, if not impossible to most people. In the new charts, the orientation is as follows to match catadioptic telescope:

NORTH - "TOP"
SOUTH - "BOTTOM'
EAST - "RIGHT"
WEST - "LEFT"

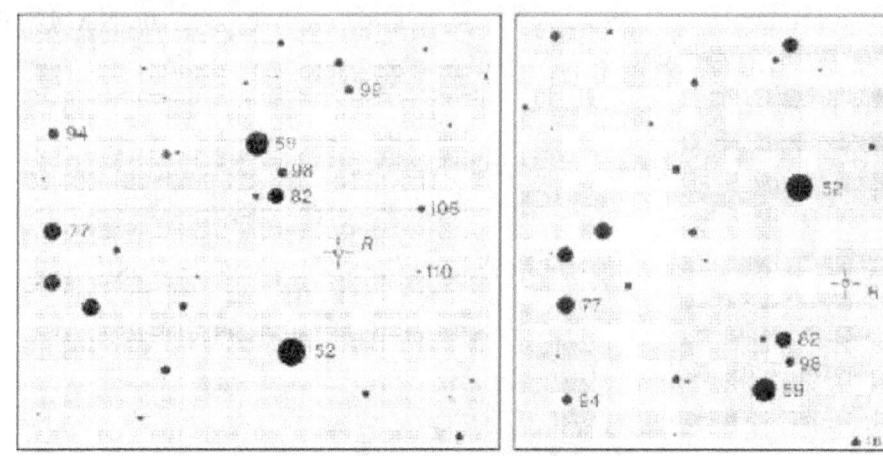

FIGURE 5
Comparing two close-up views of the standard (left) and "Reversed" (right) AAVSO charts for R Ursa Majoris. ETX and LX users will want to use the "Reversed" chart for correct eyepiece orientation

MAKING YOUR BRIGHTNESS ESTIMATE

Estimating the brightness of a variable star is simple and fun. It requires that you carefully locate a suitable "comparison star" for which the AAVSO has assigned a magnitude, and make sure that the star you are estimating is, indeed, the RIGHT star - it is easy to become confused! After several observations requiring you to locate the star, you will become familiar with both the field of stars surrounding it, and the proper identification of the star itself....like so many things, the first time is always the most frustrating and difficult!

Variable star observing should be "fun" as well as rewarding. Over a period of time, you will see the fruits of your labor as you compile your own light curves like the one in Figure 2. The following steps will get you started in variable star observing. A complete guide is available from the AAVSO at charts.aavso.org that will take you even further once you have mastered this high-in-the-sky and easy to monitor variable star, R Ursa Majoris:

1) It is important to not worry with stars you cannot see; that is why I have selected R Ursa Majoris for TWO reasons: a) it can be monitored both when bright and at its faintest light with our telescopes in just a little less than one year; and, b) it is in a "circumpolar constellation" meaning that - from ind-northern and higher latitudes, Ursa Major is up all night long, 365 nights a year! If you decide to observe variable other than R UMa, DO NOT select a star with a magnitude range mostly out of your limiting magnitude range when it is at faintest; do not bother - concentrate on stars you KNOW that at least most of the time you can see;

2) only use the star magnitudes given on the AAVSO star charts; these have been carefully checked and okayed for precision in visual work; do not use the GSC or other sources for estimates;

3) always use two stars and extrapolated between the magnitudes, unless you find a comparison star that is EXACTLY a match in brightness to the variable. Position both the variable and the comparison star so that they are both located at the same distance from the edges of your field of view (Figure 6);

FIGURE 6
Offsetting Variable and Comparison
Star Equally in Telescope Field of View

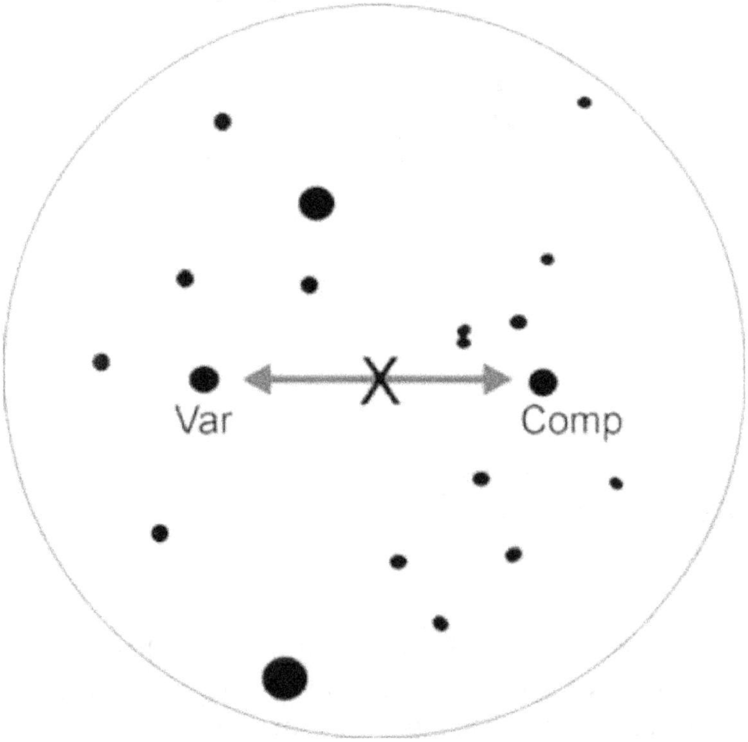

4) a lot of variable stars, by nature, are RED, just as is R UMa; there is an optical effect known as the "Perkinje Effect" that allows reddish objects to appear brighter and brighter to the human eye theylonger they are "looked at." Therefore, do not STARE at the variable, but rather keep your eye moving between comparison star and the variable at all times;

5) only use stars for comparison that are CLOSE IN MAGNITUDE to the variable....do not guess; if you cannot find a suitable star, then you should use a wider-field chart (i.e., if you are using a "c" chart and there is not a suitable comparison star handy, then you should try the "b", wider-field chart). Figure 7 shows correct comparison stars that I would select if R UMa was, say, magnitude 9.6. (the reason I DID NOT pick the magnitude "98" (decimal omitted!) star near the center and nearer to the variable is that it is too close to the magnitude "82" star and thereby difficult to get a clear brightness reading in my opinion);

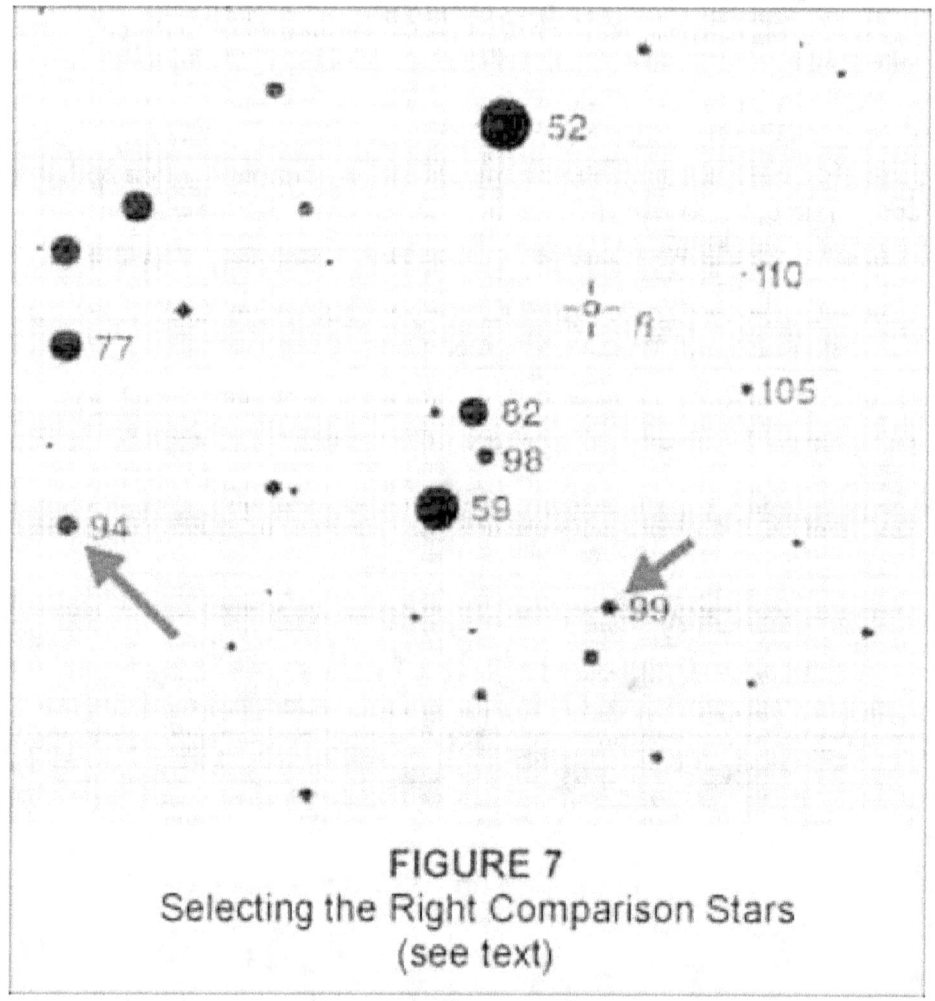

FIGURE 7
Selecting the Right Comparison Stars
(see text)

6) use the "OUT OF FOCUS" comparison method if you are having difficulty deciding; do not get frustrated if you can't quite make up your mind; **DEFOCUS** your telescope slightly and make a "disk" rather than a point out of your star image; diffusing the light helps you to discern very slight differences in brightness!

7) if you cannot see the star on a given night - even though you have been tracking it for some time - then the star has dropped BELOW YOUR LIMITING MAGNITUDE. If your limiting magnitude (see my Limiting Magnitude Test on this website) is, say 11.7 as it might be in a 3-inch, then enter your estimate for the evening as "<11.7, meaning it is

FAINTER than 11.7, which is the faintest star you can SEE ON THAT NIGHT. FIGURE 6 shows the center sections of the R UMA "d" chart (much smaller field of view and much fainter comparison stars) so that you can appreciate the detail at which the AAVSO has gone to supply you with accurate star magnitude estimates to very faint stars).

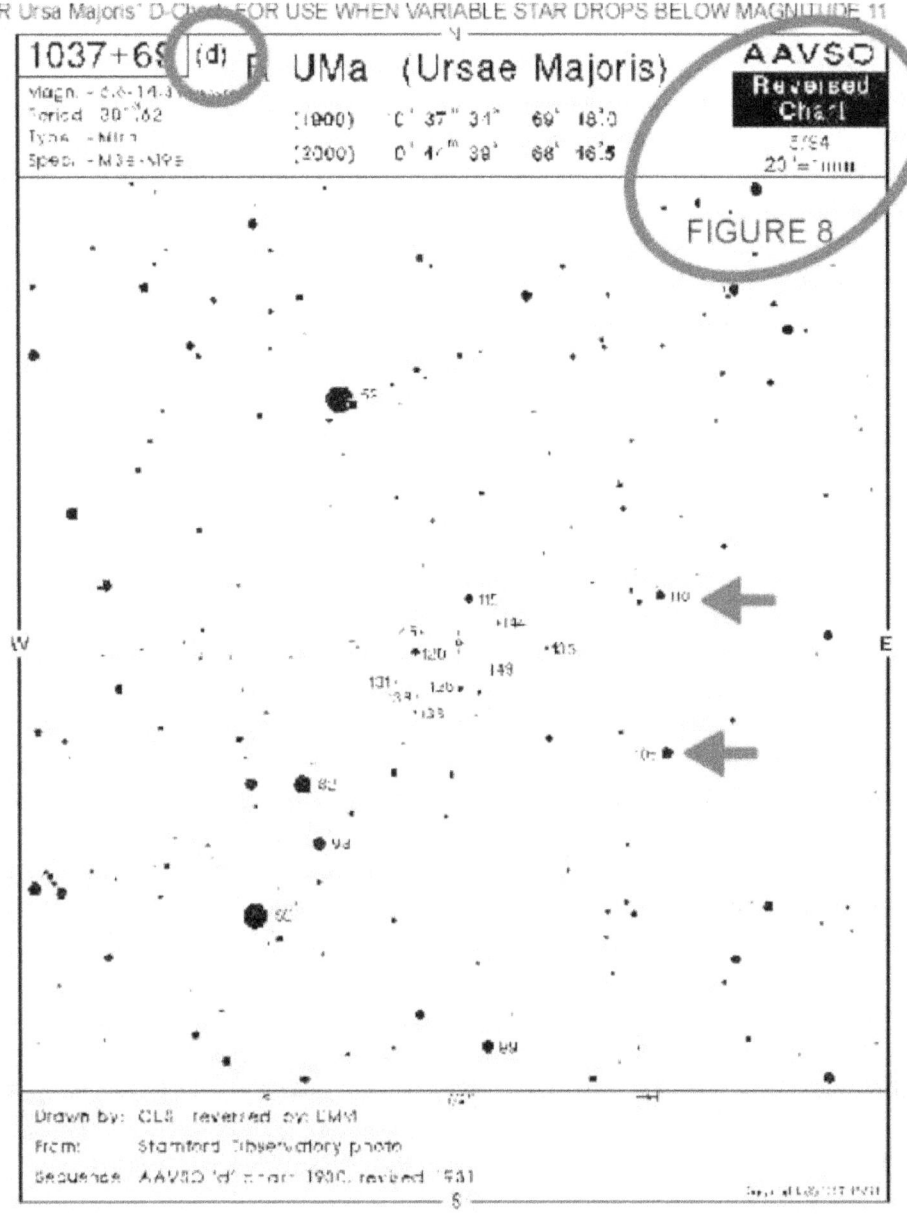

8) NEVER estimate the brightness of the variable when it is on the EDGE of the field of view; eyepieces and telescopes scopes have a "fall-off" from 100% illumination toward the edge of the field; therefore, when possible have both the variable and comparison star (Figure 7) as near to center when making our final estimate as possible.

9) always remember to check your watch or clock at the time you make your estimate; never leave anything to memory; a flashlight and clipboard at your side will encourage you to WRITE down your time and estimate at the time you make it!

This has been but a "crash course" introduction into variable star observing that I hope will whet your appetite for more. Log onto the AAVSO web site and learn more about this exciting and rewarding offshoot of your already-exciting hobby, and contribute a bit toward the future of science.

Too many times we all find ourselves in that "...well, I've had my look-around now....what's next? is THAT all there is to do?" and merely leave the telescope indoors when you COULD open up an entirely new realm of astronomy. Go ahead: Take the plunge! Find R Ursa Majoris tonight and make that FIRST estimate....you'll never forget the day that you did!

And just think...in only 302 days, you can compile a complete scientific light curve showing the actual evolution of this star in your own lifetime. And you just THOUGHT that things never changed in space!

* * *

"I love to revel in philosophical matters and especially astronomy. I study astronomy more than any other foolishness there is. I am a perfect slave to it. I am at it all the time. I have got more smoked glass than clothes. I am as familiar with the stars as the comets are. I know all the facts and figures and have all the knowledge there is concerning them. I yelp astronomy like a sun-dog, and paw the constellations like Ursa Major....."

— *Mark Twain, letter to the San Francisco* Alta California *newspaper, 1 August 1869.*

Chapter 37

The Thrill of Discovery: A Search for Novae and Supernovae

The thrill of discovery in astronomy is second to no other experience, and it is why the comet searcher religiously scans the skies in the early mornings of the year, spending hundreds or thousands of hours peering at the same regions of sky. All the time is well spent, but not forgotten, at the moment a new object is found. To help in understanding the physics involved in the evolution of stars, no event is as important as the explosion of a star - the **nova** or **supernova**.

These phenomena can occur instantaneously, and valuable records can be lost forever if the beginning of the nova is not monitored by the professional astronomer. Consequently, many amateur astronomers dedicate themselves to searching methodically through the sky for these new stars, either photographically or visually. If they are motivated by serious purpose, eventually they will find a previously undiscovered nova or supernova giving the professional a jump on a newly erupted star.

Lately there has been a renewed interest in the search for both novae within our own Milky Way galaxy - the "Galactic Novae" - and those that occur seemingly more frequently in other galaxies, or the "Extragalactic Supernovae." This overview is NOT intended to be all-inclusive, but to invite telescope users to examine the possibilities and the remote chance of the actual discovery of a "new star" in the heavens.

Novae and Supernovae searches can be conducted:

1) visually, using good star charts and the naked eye, binoculars or a telescope;
2) photographically, patrolling the same selected area(s) of sky at every opportunity and comparing images over time;
3) electronically, with CCD imagers which can provide not only rapid discovery information, but also serve as a photometer to accurately measure the brightness and color (hence an early indication of spectral type) of the new star.

Being Ready for the Nova Event –
It is likely that no amateur will be fortunate enough to be viewing, at just the right time, a starfield out of which one star will rapidly increase in brilliance by a magnitude of thousands. The rise to maximum light of the nova is quite rapid, requiring only hours to increase perhaps as much as 15 to 20 magnitudes. For discovery work, you should be concerned only about detecting a new nova as soon after the event takes place as possible. Others may jointly discover and report the new star, but it takes no worth away from your discovery.

This photograph of the constellation Cygnus (the outline of the 'Northern Cross" is indicated) shows the bright stars Deneb (top) and Albireo (bottom). Note that the nova indicated by the arrow can easily be compared in magnitude to the nearby brighter stars of the constellation. This was a 15-minute photo on Tri-X film taken through a Nikon 50mm f/1.2 lens guided atop the Observatory's 6" Refractor.

This very situation happened to me in the summer of 1975 when *Nova Cygni*....just a short distance from the "top" of the northern cross in bright Cygnus....popped into view. While I hesitated, opening up the observatory's roof, thinking that I might have been mistaken about that "new" third magnitude star being something I had not seen before.....others were reporting the event. At the time I was doing novae and comet confirmation for the Smithsonian Astrophysical Observatory and needed to call **Dr. Brian Marsden** that night anyway about Comet Bradfield 1975.....in our conversation, Dr. Marsden asked, ".....by the way have you seen the nova tonight?" Nova? What nova? Then it hit me and I asked him: *"By any chance, would this nova be in Cygnus?"*

The rest is history. *Sherrod's Nova 1975* was never history.

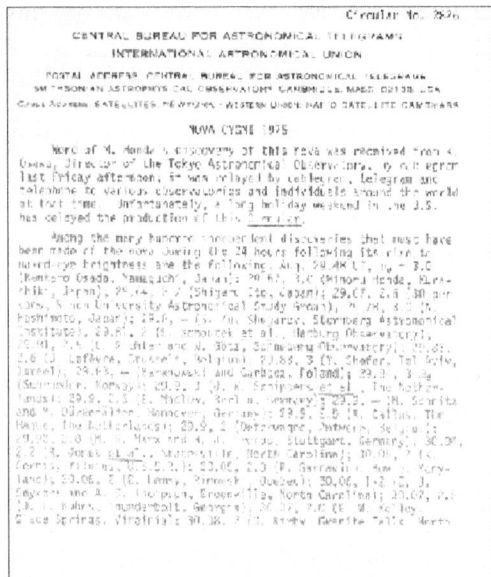

A Discovery that wasn't:
Nova Cygni 1975
These IAU Circulars are the only memory of a lost discovery

Here are two very faded Astronomical Telegrams from the International Bureau of Astronomical Telegrams from 1975. The one on the right announces the discovery of Nova Cygni 1975 (shown in photo above) and the one on right is the confirmation of the nova by the Smithsonian confirmation observers at that time. Note that the Author is one of the confirming stations for the event.....but he was not credited with its discovery!

It is quite important for amateur astronomers to continue to monitor any nova - whether YOU find it or not - as long as it is visible since maximum light of a nova remains for only a short time. If it is a very bright outburst, this maximum light is the signal that draws your attention to the area in which the new star is located. However, the outburst soon subsides, and a bright nova can be quickly lost in a field of third and fourth magnitude stars if you are not familiar with the star fields and patterns of the constellations. Such a bright nova can result in the simultaneous "discovery" of the star by perhaps hundreds of independent observers. The competition - and chance of being first to report such a discovery - is slim and it is not this type of discovery that we should concentrate on....Rather, we should think about the planned, scheduled and methodical searching of specific areas over intervals of time for novae that are at or near the naked-eye limit. Of the many novae that occur each year in our galaxy, only a small fraction are ever detected. The brightest ones are discovered usually by happenstance while those fainter than the naked-eye limit go largely unnoticed. Galactic novae, or those occurring in our galaxy, occur more frequently in the dense clouds of the Milky Way, such as seen in the early morning skies of spring or the evening skies of summer. However, the detection of the novae in these regions is VERY difficult because of the enormous numbers of stars in your field of view, no matter what instrument or equipment you might be using. Ordinary novae that occur in other galaxies are well beyond the range of amateur equipment. However, the most violent of all stellar events - the supernova - can be, and often is, detected in the smallest of telescopes even though the event might occur in a galaxy millions of light years distant.

A Visual Search for Galactic Novae - "New Stars" within our own Milky Way galaxy

If you have only a small telescope, or even only binoculars, you can still search for novae within our Milky Way galaxy with some degree of success. If you have a telescope, select and area of the sky and examine it visually at low power and with as wide a field of view as possible for many nights in a row. Although some type of star chart or "planetarium PC program" is necessary to identify both the stars that are "supposed to be there" and any suspected interloper, you might wish to concentrate on three or four general areas and dedicate your time to them. After several nights of NOT finding your nova, you will be very tempted to move on to another part of the sky....don't do it: the grass is NOT always greener over there!

Sketch or photograph the field of view as it appears in the telescope. This method has the distinct advantage that all stars seen in the telescope will be on the chart you have drawn. If later, you see any that do not appear on the chart then they are candidates for further scrutiny. Concentrate on the densely packed Milky Way regions of our galaxy; most discovered novae have been found more frequently there than in the sparsely populated areas of sky, such as the star-poor spring skies. If you examine four areas at least every other night, about 15 minutes for each area will be required once you become familiar with the star fields. Thus, plan to spend about one hour total in your search nightly.

It is remarkable how familiar you become with a star field once you have searched it a few times. If you have binoculars or use the naked eye, you can look for the brighter novae if you examine a dedicated swatch of sky each night as soon as twilight ends. For such naked eye searches, concentrate on regions of the Milky Way and learn the constellations and the stars within them to the naked-eye limit. (see my constellation GO TO star charts as listed under the Guides of the ASO web page) A great help is a good star atlas such as Hans Vehrenberg's *Handbook of the Constellations* (no longer in print but can be found on the used book market via the Internet), or *Atlas 2000* by Wil Tirion, both of which has each constellation indexed for quick reference and shows stars down to the naked eye limit.

NOTE: with all of the excellent computer "planetarium-type" programs with star chart capabilities, I highly recommend their use as it allows quick and easy determination of an EXACT Right Ascension (R.A.) and Declination (DEC) by merely placing the cursor at the exact point of a suspected new object. Also, if the object is bright enough, you can obtain an INSTANT magnitude from any nearby star by merely clicking on that star or moving your screen pointer to it.

A Photographic Program For Galactic Novae –

A more consistent and more rewarding search for galactic novae can be initiated through photographic (or digital camera/CCD) searching. The considerable benefits of such a program include the following:

- No memorization of star fields is necessary - the photograph will provide you with a permanent record and verification of any discovery, from which a magnitude and position may be derived from your star charts/PC planetarium program. Furthermore, examination of the field can be done at leisure, allowing you to take the image(s), store it or process it from film and look for suspect objects at a later time.

- More stars can be recorded on the photograph than can be discerned by the human eye and the limiting magnitude threshold will be greater than a visual examination of the field along.

- Hence, the chance of discovery increases because the number of stars on the photograph is greater than can be seen by the eye. Attempting to photograph a star field through an average telescope in search for novae is a perplexing and tedious task. A better method is to photograph as wide a field as possible on a single photograph and reach perhaps magnitude 10 or 11 with merely a good camera and standard lens attached (50mm up to 200mm f/2.8 to 3.5 if possible).

- If using a CCD system with a three-color capability (i.e., filter wheel) you can actually obtain spectral (color) data on the star, thereby giving a very early and valuable indication of the star's spectral type at or near brightest light.

- Also, with CCD, the telescope you are using can reach far fainter magnitudes that either visually or photographically, allowing even an 8" telescope to reach even magnitude 16 in very dark skies.....therefore the number of potential novae visible to you will dramatically increase.

The Equipment Necessary –

To systematically search for novae using the camera or CCD, only two things are required - a camera/CCD and some provision for tracking on the stars. A simple clock-driven equatorial mounting provides accurate tracking for standard 50mm camera lenses for up to 10 minutes if the polar axis has been accurately aligned to north and set at the proper latitude of your observing location.

However, telephoto lenses are advantageous and highly recommended for this study. Even so, you still must have some way to correct the tracking of the instrument. The piggyback method and a good camera lens is recommended with lenses of 100mm to 200mm focal length for the search, and exposure times on fast monochrome or color

CCDs should be around 1 minute. If your observing station is located in a bright suburban area, exposure times should not be greater than about 20 to 30 sec, although the limiting magnitude achieved will be considerably less than if the photograph was taken in dark skies and exposed for a longer period of time.

If using the CCD, it is to your advantage to do PATROL searches in black and white and - if something is found - then do three-color imaging to provide accurate color characteristics of the early nova. With CCD equipment, even the most modest, the imaging can be done effectively in suburban skies which offers a tremendous advantage over traditional photographic means. Exposures can be kept quite short and many areas can be imaged with INSTANT results via the computer link in an extremely short period of time.

The Method of Search –

In order to convincingly search the sky for nova there must be two successive photographs, taking them no more than about 10 days apart. Any new object in the 10 day (or preferably shorter) interval will appear on one photograph but not on the other, and thus give you initial findings that something new has appeared on the last photograph (or actually in some cases, it could appear on the first photo only to have faded significantly on the last one!) If any object IS suspected, further confirmation using a telescope visually must be done without hesitation.

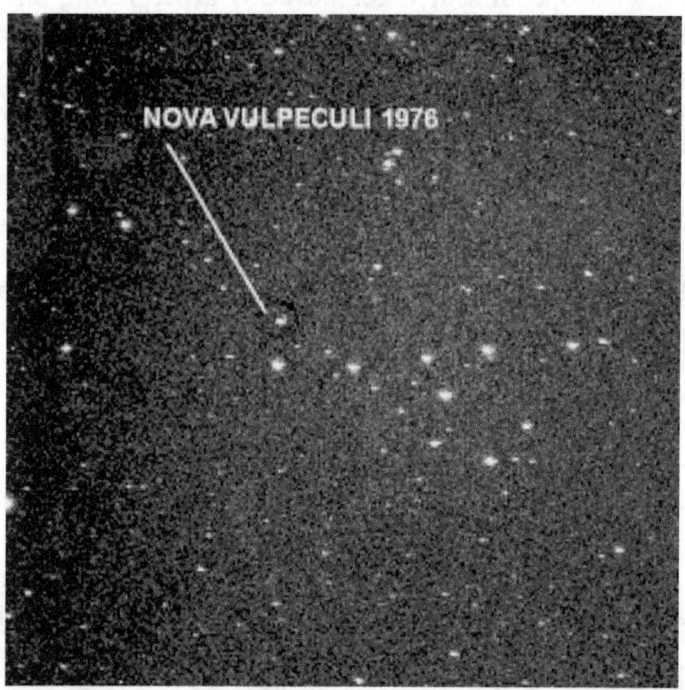

The photograph preceding shows a close-up of the region of *Nova Vulpeculae 1976* taken with the Arkansas Sky Observatory 6" f/5 Sky Patrol Refractor. This photograph was taken on the same night as the discovery after a phone call from the Smithsonian Observatory requested early verification through the Observatory You can see how important making sure that the star you have identified is the CORRECT star.....almost every star in this small asterism in Vulpecula just happens to be about the same magnitude as the nova!

This all sounds rather simple until you try to compare two photographs, each containing thousands of stars. Keep in mind that the faint stars are the ones under surveillance! No one can possible inspect each and every star, one by one. The professional astronomer uses either a computer which can digitally search a field for "visitors" OR a large desk-mounted instrument known as a blink comparator, a device in which two negatives are centered on the same star field and can be alternately compared through a binocular viewing eyepiece. If the two negatives are precisely lined up in the comparator, the effect is that the field is merely "flashing" on and off rapidly as the comparator shows first one negative and then the other. However, if a new object is apparent on one film and not the other, that new object will alternately flash on and off as the comparator shows first the negative with the object and then the negative without it.

This method is fast and reliable, but the instrument is quite expensive and beyond the budget of most amateur astronomers. Therefore, as in other cases of amateur research, some makeshift method must be devised by which YOU can accomplish similar results. Several small portable blink comparators aimed at the amateur astronomer have been developed by non-professional astronomers in past years. Most of these are adequate, although somewhat difficult to use.

A simpler method is to devise a light box using a translucent material on top of which you can position the two negatives (or color transparencies). Then precisely aim two viewing eye-pieces, one at one slide or negative and the other at the second film. Install within the box, under the translucent material, a timer (either mechanical or electronic) that switches the current back and forth between a light under both slides (or negatives). HOWEVER, today there are excellent computer programs (i.e., ***CCDsoft from Bisque***) which allow two or more individual digital photos to be aligned and blinked in an instant!

Another method that has had a great deal of success will allow you two use two SLIDE PROJECTORS, each of which is aimed so that it centers the projected image atop the other as they are projected onto the same screen. The method has the nice distinction of allowing you to sit back and enjoy the view as the slides flash onto and off of the simple projection screen As in the blink comparator, this method makes an object visible on at one moment, but not the other so that only one image appears to be blinking on and off. But in these more modern days, by all means, find a good computer program to do the job for you!

NOTE: Any method of comparing one image to the other requires exacting care to align the camera on the star field at the time the photograph is taken. The orientation and the center of each photograph should be the same, and of course,you must use the same optical system through-out. The formats of .fit, .tiff or in some cases raw or even .jpg is preferred for computer programs providing blinking programs in that the field will be dark when viewed by whatever comparator is used. Always blink white stars on black, rather than "negative, because a stark white background will quickly tires the eyes. I highly recommend Bisque's *CCDsoft* program for this: http://www.bisque.com/sc/media/p/27995.aspx

Studying External Galaxies For Supernovae –

Although only one supernova per century on an average is thought to occur in our Milky Way galaxy, there is a way in which a hundred times that many super-novae explosions might be found in the same timeframe: Look for them in a hundred galaxies! There are hundreds of galaxies within reach of modest amateur equipment that are large and close enough to show any supernova that might occur within them. When most supernovae occur in these external galaxies, they appear as VERY faint stars embedded in the seemingly nebulous outer portions of that galaxy. The brightest are on the order of magnitude 12 or 13, so telescopes of 8 inches (20 cm) or larger are required to see them visually. However, the "typical" supernova in an external galaxy might be more expected on the order of 15th or 16th magnitude, still within the photographic or CCD-grasp of modest amateur equipment.

A rare supernovae blazes in the skies of the distant galaxy Messier 51 (Whirlpool Galaxy). This 90-second set of images shows "before and after", with the June 7 supernova event clearly recorded in the image at right. Copyright Arkansas Sky Observatories, P. Clay Sherrod

To search for supernovae in external galaxies requires that you know the appearance of the galaxies as they should appear in your telescope. Photographs of galaxies are often published that have had the central regions burned in with an effort to expose the outer regions of the galaxies.

Any bright knots or star-like images that might normally be seen visually are thus obliterated. Consequently, it is necessary that you set up a schedule for recording the fields of the galaxies that you plan to observe before you start your actual observations.

Recording the Galaxy Field –

You should choose a number of galaxies large enough to allow you to examine 15 galaxies for supernovae each night. With the great number of galaxies visible in our spring skies, there is a predominance of excellent subjects during that season. At the advent of the project, examine each galaxy on your list, preferably in sequence by right ascension. Using a circle about 4 inches (10 cm) in diameter, sketch the field as you see it through the telescope eyepiece, including every star within the field of view. Because most galaxies appear small to us, draw the galaxy field with moderate magnification, say about 100x to 150x, and only on the very best night. It will take you much longer to make the initial field sketch than it will to search or the galaxies on every subsequent night. Draw the fields with the telescope that you will use for the searches. Once you have drawn the fields, it is a simple matter to examine as many galaxies in sequence as possible or practical for the night. Eventually, you will become so familiar with the fields that referring to the charts you have drawn will be necessary only when you find an unfamiliar object that might indeed be a "new star" within that galaxy.

Galaxies Recommended –

For supernovae surveillance the galaxies listed in the following table are recommended as part of your regular patrol for the possible supernova event. Be aware that many of them might have star-like points near the galaxy, most if not all of these distant stars located in our own Milky Way between the observer and the distant galaxy. By all means - don't take a chance! - record all star-like objects on your initial drawing. Monitor each galaxy on every available night because the detection of a supernova in its early stages is of the utmost importance to the professional astronomer. Notice that only spiral galaxies are listed in the table, since the occurrence of the supernova is considerably more common in the spiral than in the elliptical galaxies.

REPORTING and CONFIRMING ANY DISCOVERY

So let's be optimistic and assume that you are going to find that "new star" you are searching for....or in the process you accidentally just happen to find a new COMET or asteroid. What do you do?

    1. DON'T immediately assume that you have discovered something....this is the first mistake. Unfortunately there are TWO possible scenarios that are much more possible that you actually having discovered a nova or supernova: a) someone else has actually discovered it before you; b) it is an object that is "supposed to be there", but is not showing up for whatever reason on your star atlas, computer sky program or other source.

2. DO go back and check the same location after looking at your charts for verification or processing your photo/CCD image; if it is still there, then image it again or estimate its magnitude again visually.

3. WRITE DOWN EVERYTHING: including the time, date, telescope/equipment used, sky conditions and EXACT coordinates (R.A. and DEC) of the object.

4. If you are convinced that the object is "not supposed to be there" then you should proceed to notify someone. If there is a local public observatory handy, then check with them; if not, then check immediately via the Internet at: http://www.cbat.eps.harvard.edu/RecentCBETs.html which is the net's posting of ALL International Astronomical Union (IAU) telegrams announcing new discoveries, activities or events in the skies. Even if the "circular" is not active yet, it WILL be posted here and you can immediately see if there is a prior announcement of "Discovery of Nova in Cygnus" for example. That would be your first red flag that someone has beat you to the punch.

5. If you have been assured by your local observatory and/or the IAU telegrams posted that nothing exists where you "think" you have found your object, then by all means - with ALL data at hand - go ahead and report it as a "possible" discovery. Do NOT over-embellish what you have found....be very specific and realistic as to magnitude, color, position, etc. To report any astronomical discovery, go to the IAU page:
http://www.iau.org/public/themes/discoveries/ .

6. They will alert observatories around the world to confirm or deny the existence of what you might have found. Within a short period of time, usually less than one day, you will be notified if this checks out.

7. If it does, it's party time.

This is merely a very concise capsulated introduction into this exciting endeavor; cater YOUR observing plans to meet your equipment and needs; certainly CCD cameras can patrol in a mere snapshot many faint galaxies that cannot be accessed visually, and will record the faintest of stars possible in your equipment. But one does NOT necessarily need that sophistication....indeed, many bright novae have been discovered by a casual stargazer who happens out on a warm summer night and just happens to look skyward at a familiar constellation.....and there may be a bright star that causes him or her to freeze....is it or is it NOT an interloper in the heavens?

Only YOUR research might reveal the true answer!

* * *

APPENDIX ONE

An Introduction into Piggyback Photography

Many people are mesmerized by the fantastic photographs they see published of celestial objects....indeed, many people actually BUY a telescope *because* of these photographs. Likely, the very package that YOUR telescope came in - the catalog that advertised it.....the instruction manual that came with it....are ALL embellished with glorious full-color photographs of celestial splendors.

You look at these images, say of Saturn and its magnificent colors, and you ask yourself: *"When is it time for ME to take pictures like that?"*

We have a lot of inquiries concerning the suitability of the small GO TO telescopes for deep sky photography; the many published and beautiful photographs of the moon, sun and planets taken through lightweight compact telescopes suggests the fact that these telescopes CAN "do" astrophotography.

But - in the harsh, hard reality of the proverbial "nutshell" - these telescopes ARE not suited for long-exposure PRIME FOCUS astrophotography. Don't give up just yet....there is PLENTY of good photography that CAN be done in addition to lunar and planetary imaging.

PRIME FOCUS ASTROPHOTOGRAPHY

Photographing at the "**prime focus**" of a telescope is a very simple and straightforward procedure, yet all telescopes are not suited for LONG EXPOSURE prime focus work. Long exposures are required to capture the images of nebulae, galaxies and faint clusters unless a CCD imager taking multiple short "subframes" and subsequently stacking those via a computer program is used . In reality a large CCD or adapter Digital Single Lens Reflex (DSLR) camera is also very heavy and awkward to use on a compact and small telescope.

Imaging a faint celestial object is akin to soaking up raindrops with a sponge: a drop at a time and eventually the sponge becomes "wet," holding as much water as it possibly can. The film or electronic sensor in a CCD is the same way....over longer and longer periods of time (or with more and more exposures in the case of CCD), the light eventually "accumulates" and becomes brighter and more clear on the film. Remember that you will always be able to reveal more detail and fainter stars with photography via CCD sensors than your eye can ever see.

In **prime focus** photography, the telescope (say with a focal ratio of f/10) becomes the CAMERA LENS; with the standard camera lens removed with an 8-inch SCT f/10 telescope for example, and the camera attached at the rear of the telescope you have essentially a **2000 mm** telephoto lens!

This is great for relatively short exposures of the sun, moon and planets and certainly for nature photography, but the long focal length ALSO "magnifies" the length of time required for that "sponge" to soak up enough light, it adds to the vibration and motion as the picture is being taken, and it requires VERY CAREFUL guiding using a secondary guiding system or piggybacked telescope which is not normally an option with the compact computerized telescopes.

So, except for the wonderful experience of lunar and planetary photography - which I certainly urge you to undertake and experiment with - let us explore what other option is available to you to capture images of the wonders of DEEP SPACE!

MOUNTING A CAMERA "PIGGYBACK"

The small telescope may not be ideally suited for long exposure photography at the prime focus, but it does have some advantages when it comes to "*piggyback astrophotography*." The short, stubby tube on its sturdy mounting can be ideal for long exposures using your CAMERA and a regular or long telephoto lens and the telescope/mount as nothing more than a driving platform which will support and guide the camera during exposures.

In such photography, you may chose to "not guide" (for regular or relatively short-focus lenses up to 135mm and for shorter exposures - say less than 8 minutes) or "guide" (for telephoto lenses and any shot that you may take that exceeds 10 minutes, regardless of the lens used).

This following photo shows an Meade SCT telescope fully equipped with a piggyback camera mount holding a heavy SLR camera and its regular lens. Note how well suited for balancing and carrying the camera this design of telescope actually is. Not every telescope has a piggyback adapter commercially made and available for each model; some ingenious thinking can easily lead you to come up with some manner by which you can allow your camera to "ride" atop your telescope or equatorial mount.

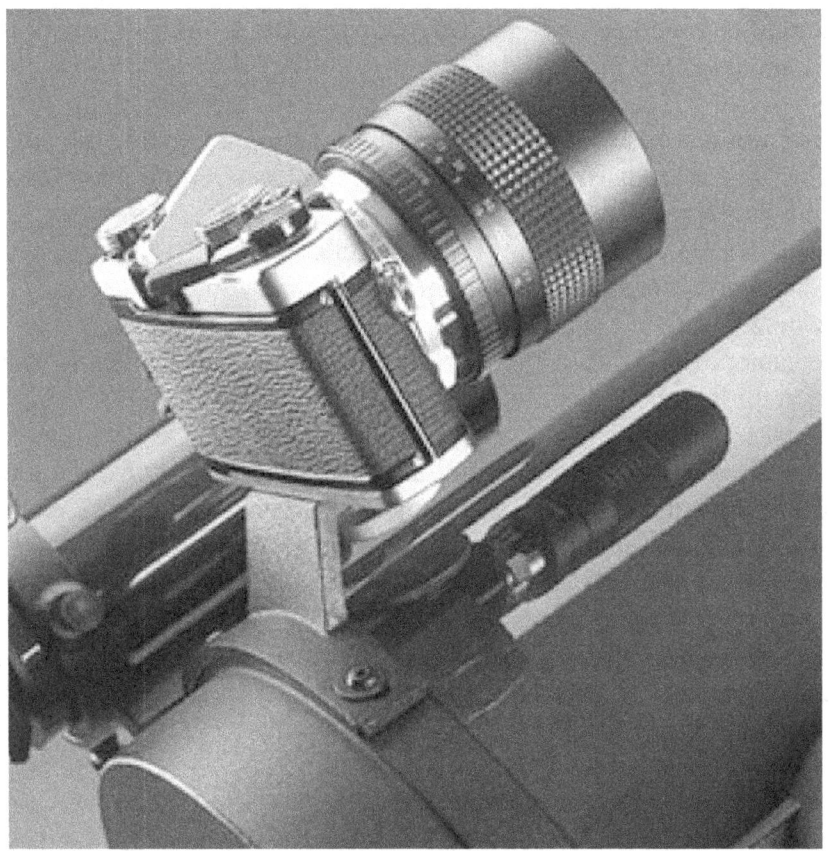

*A SLR camera riding Piggyback on a SCT telescope
Courtesy Meade Instruments*

Following are the items that are needed to begin your exploration into piggyback astrophotography and your first steps toward compiling and personal collection of deep sky astrophotographs you may never have considered possible with your telescope!

1) A motor-driven TELESCOPE operated in very dark sky conditions; for this, the portable small modern scopes are ideal with their internal (and auxiliary DC) power source. With the ease of celestial alignment and tracking that these telescopes offer, you can be photographing in no time....and with VERY little actual guiding required if your polar alignment is right on target!

2) A piggyback camera MOUNT; although many manufacturers to not make piggyback adapters for every telescope the brackets are available, either through third-part vendors or through you OWN creativity. The one shown in Figure 1 is easily constructed from scrap metal and an old tripod head . The mount merely clamps around the tube and the camera positioned on TOP of the tube as shown and balance by adding weights to the flat "tripod" plate already installed under your eyepiece section of the telescope.

YOU MUST BALANCE the telescope carefully if you piggyback a camera...NO exception. If you unclamp either axis and the telescope moves on its own, you are out of balance and need to add weights in various places until it stops moving....be clever: bags of pennies, an old hammerhead, scrap steel.....anything that you can harmlessly attach to the tube or the mount is fair play.

3) Obviously, you must have a CAMERA to do photography and some are more suited than others for these reasons (if you have a choice....if not, simply use what you have!):
   a) it should be as light as possible;
     b) it MUST be a Digital Single Lens Reflect (DSLR) camera so that you can actually SEE what you are photographing through the viewfinder;
     c) it would be best if the camera had removable lenses; and
     d) the camera MUST have a "bulb" or "time exposure" setting so that the shutter can be left open for long periods of time.

4) A CABLE RELEASE or remote shutter activator is required; this is a long extension cord or controller that allows you to trip the shutter open, lock a small clamp to keep it open for a desired length of time (usually 5 minutes or longer) and then release the shutter to close. Imagine the "western movies" you have seen with the wagon-drawn old traveling photographer who sets up his box camera, exploding flash and hood - he always has one hand over his head, and the other holding his "cabel release." All major modern cameras accept these, and some have to be special ordered; nonetheless, you cannot do photography without one.

5) A sensitive CCD sensor (or Fast FILM). By "fast" I mean a high ASA rating, at least 400 ASA but higher if possible; nearly all modern DSLR camera brands can produce good celestial images and true colors with rich reds and blues, ideal for deep sky subjects; however, at the time of this writing, the two commercial DSLR cameras of choice for astrophotography applications are the Canon and the Nikon.

6) GUIDING - for exposures greater than 8 minutes and all done with 150mm focal length and longer lenses, you must be able to either manually or electrically make corrections as the objects move; your motor drive will account for SIDEREAL MOTION, but your alignment may be off, so corrections will be required in BOTH right ascension (azimuth) and declination (altitude). The better you track (use a very high power eyepiece and merely find a bright star near the area you are photographing to center and keep centered) the better will your astrophograph will be....by FAR.

CHOOSING YOUR SUBJECT

The sky is your limit regarding piggyback photography, with everything from comets and meteors to the Orion Nebula and Andromeda Galaxy (Figure 1, below). Taken in 1970 with an old Mamiya camera and experimental film, modern results in full color will put this old photograph to shame.....in 1/10th the time!

FIGURE 1
Messier 31 - Andromeda Galaxy
200mm F/2.8, 40 minutes - SO 115 Kodak

Following are the possible subjects (certainly NOT all-inclusive), the best choice of camera lens and the APPROXIMATE length of time using a sensitive CCD/fast ASA for which your shutter must remain open:

1) PLANETARY / LUNAR CONJUNCTIONS - when objects "pair up" in the sky they usually are bright and many times not very close together. Therefore, use your standard 50mm lens or no more than say a 135mm lens and an exposure time of about five (5) minutes MAXIMUM. Guiding is NOT necessary through the telescope. For conjunction photographs, as well as bright naked eye comets, always attempt to place SOMETHING on your horizon in the photograph; this adds beauty and gives a wonderful sense of reference for your photograph!

2) BRIGHT COMETS - For really long comets, your standard 50mm lens will work with exposures of only 5 to 10 minutes; the comet will move slightly different than the stars so for any longer exposures, it will be necessary to TRACK ON THE COMET through your telescope! The fainter and smaller a comet, the longer both the focal length of your lens and the time of your exposure; the photograph of Halley's Comet - even with a huge 6" f/2.8 aerial lens required a very long time to capture the full extent of the comet's tail (Figure 6).

3) CONSTELLATIONS - Always use either a WIDE ANGLE (28-35mm) or your normal 50mm lens to photograph constellations; since most were originally outlined with the naked eye, most key stars are bright and thus require only about five (5) minutes of exposure to capture the outline.

4) METEOR SHOWERS - You really do not even need a piggyback arrangement for this, but only a standard camera lens and exposures of from 10 to 20 minutes with NO tracking. For piggyback, it is interesting to aim the scope and camera directly at the shower radiant for a "depth" effect as the meteors appear to emanate from that point.

5) DEEP SKY - The following photographs show just a small sample of what can be done with a regular camera and telephoto lens on your telescope. Of course, your object will appear larger and brighter if it is one of the more notable deep sky objects; the star clouds of the summer Milky Way are incredible and best captured via piggyback photography (see the shot of M-8 and the Great Sagittarius Star Cloud). Even a common camera can capture more stars than can be counted!

The accompanying photographs will provide some frame of reference both the image scale and to the length of exposure time on ASA 400 film. With today's modern CCD sensors, these subjects can be captures with far greater detail and depth in a fraction of the time that these images were taken with in the 1970's on fast film.

Take those "first steps" into astrophotography. You can literally spend the rest of your life enjoying piggyback photography without EVER exhausting the supply of celestial subjects!

As my dad always reminded me.....we may not have what all the other folks have. But use what we DO have, and always strive to use it a lot better than they do with what THEY'VE got!

*   *   *

FIGURE 2
Messier 8 - Lagoon Nebula
and the Sagittarius Star Cloud
150mm F/3.5, 30 minutes - Tri-X Kodak

FIGURE 3
North American Nebula / Pelican Nebula
near Deneb in Cygnus
200mm F/2.8, 40 minutes - SO 155 Kodak

FIGURE 4
Messier 45 - The Pleiades
200mm F/2.8, 20 minutes - SO 155 Kodak

FIGURE 5
Halley's Comet - May 5, 1986
6" Nikkor lens @ f/2.8
20 minutes - SO 115 Kodak

"Ralph Waldo Emerson once asked what we would do if the stars only came out once every thousand years. No one would sleep that night, of course. The world would create new religions overnight. We would be ecstatic, delirious, made rapturous by the glory of God. Instead, the stars come out every night and we watch television….."

— Paul Hawkens. Commencement address, University of Portland, 2009

www.ingramcontent.com/pod-product-compliance
Lightning Source LLC
Chambersburg PA
CBHW080904170526
45158CB00008B/1981